T0302367

Bayesian Nonparametrics for Causal Inference and Missing Data

Bayesian Nonparametrics for Causal Inference and Missing Data provides an overview of flexible Bayesian nonparametric (BNP) methods for modeling joint or conditional distributions and functional relationships, and their interplay with causal inference and missing data. This book emphasizes the importance of making untestable assumptions to identify estimands of interest, such as missing at random assumptions for missing data and unconfoundedness for causal inferences in observational studies. Unlike parametric methods, the BNP approach can account for possible violations of assumptions and minimize concerns about model misspecification. The overall strategy is to first specify BNP models for observed data and then to specify additional uncheckable assumptions to identify estimands of interest.

The book is divided into three parts. **Part I** develops the key concepts in causal inference and missing data and reviews relevant concepts in Bayesian inference. **Part II** introduces the fundamental BNP tools required to address causal inference and missing data problems. **Part III** shows how the BNP approach can be applied in a variety of case studies. The datasets in the case studies come from electronic health records data, survey data, cohort studies, and randomized clinical trials.

Features

- A thorough discussion of both BNP and its interplay with causal inference and missing data
- How to use BNP and g-computation for causal inference and non-ignorable missingness
- How to derive and calibrate sensitivity parameters to assess sensitivity to deviations from uncheckable causal and/or missingness assumptions
- Detailed case studies illustrating the application of BNP methods to causal inference and missing data
- R code and/or packages to implement BNP in causal inference and missing data problems

The book is primarily aimed at researchers and graduate students from statistics and biostatistics. It will also serve as a useful practical reference for mathematically sophisticated epidemiologists and medical researchers.

MONOGRAPHS ON STATISTICS AND APPLIED PROBABILITY

Editors: F. Bunea, R. Henderson, N. Keiding, L. Levina, N. Meinshausen, R. Smith,

Recently Published Titles

Multistate Models for the Analysis of Life History Data
Richard J. Cook and Jerald F. Lawless 158

Nonparametric Models for Longitudinal Data
with Implementation in R
Colin O. Wu and Xin Tian 159

Multivariate Kernel Smoothing and Its Applications
José E. Chacón and Tarn Duong 160

Sufficient Dimension Reduction
Methods and Applications with R
Bing Li 161

Large Covariance and Autocovariance Matrices
Arup Bose and Monika Bhattacharjee 162

The Statistical Analysis of Multivariate Failure Time Data: A Marginal Modeling Approach
Ross L. Prentice and Shanshan Zhao 163

Dynamic Treatment Regimes
Statistical Methods for Precision Medicine
Anastasios A. Tsiatis, Marie Davidian, Shannon T. Holloway, and Eric B. Laber 164

Sequential Change Detection and Hypothesis Testing
General Non-i.i.d. Stochastic Models and Asymptotically Optimal Rules
Alexander Tartakovsky 165

Introduction to Time Series Modeling
Genshiro Kitigawa 166

Replication and Evidence Factors in Observational Studies
Paul R. Rosenbaum 167

Introduction to High-Dimensional Statistics, Second Edition
Christophe Giraud 168

Object Oriented Data Analysis
J.S. Marron and Ian L. Dryden 169

Martingale Methods in Statistics
Yoichi Nishiyama 170

The Energy of Data and Distance Correlation
Gabor J. Szekely and Maria L. Rizzo

Sparse Graphical Modeling for High Dimensional Data: Sparse Graphical Modeling for High Dimensional Data
Faming Liang and Bochao Jia

Bayesian Nonparametrics for Causal Inference and Missing Data
Michael J. Daniels, Antonio Linero, and Jason Roy

For more information about this series please visit: https://www.crcpress.com/Chapman--Hall CRC-Monographs-on-Statistics--Applied-Probability/book-series/CHMONSTAAPP

Bayesian Nonparametrics for Causal Inference and Missing Data

Michael J. Daniels
Antonio Linero
Jason Roy

CRC Press is an imprint of the
Taylor & Francis Group, an **informa** business
A CHAPMAN & HALL BOOK

First edition published 2024
by CRC Press
6000 Broken Sound Parkway NW, Suite 300, Boca Raton, FL 33487-2742

and by CRC Press
4 Park Square, Milton Park, Abingdon, Oxon, OX14 4RN

CRC Press is an imprint of Taylor & Francis Group, LLC

© 2024 Michael J. Daniels, Antonio Linero, and Jason Roy

Reasonable efforts have been made to publish reliable data and information, but the author and publisher cannot assume responsibility for the validity of all materials or the consequences of their use. The authors and publishers have attempted to trace the copyright holders of all material reproduced in this publication and apologize to copyright holders if permission to publish in this form has not been obtained. If any copyright material has not been acknowledged please write and let us know so we may rectify in any future reprint.

Except as permitted under U.S. Copyright Law, no part of this book may be reprinted, reproduced, transmitted, or utilized in any form by any electronic, mechanical, or other means, now known or hereafter invented, including photocopying, microfilming, and recording, or in any information storage or retrieval system, without written permission from the publishers.

For permission to photocopy or use material electronically from this work, access www.copyright.com or contact the Copyright Clearance Center, Inc. (CCC), 222 Rosewood Drive, Danvers, MA 01923, 978-750-8400. For works that are not available on CCC please contact mpkbookspermissions@tandf.co.uk

Trademark notice: Product or corporate names may be trademarks or registered trademarks and are used only for identification and explanation without intent to infringe.

ISBN: 978-0-367-34100-8 (hbk)
ISBN: 978-1-032-45694-2 (pbk)
ISBN: 978-0-429-32422-2 (ebk)

DOI: 10.1201/9780429324222

Typeset in Nimbus font
by KnowledgeWorks Global Ltd.

Publisher's note: This book has been prepared from camera-ready copy provided by the authors.

MJD: To Marie, Mary, and Mia
AL: To Clara and Katie
JR: To Trevor and Mitchell

Contents

Preface

In this book, we explore Bayesian nonparametric (BNP) methods for tackling two classes of challenging problems: those involving causal inference and those with substantial amounts of missing data. Both types of problems require analysts to rely on assumptions that cannot be tested using the observed data. To address this issue, our overall modelling strategy is to first develop BNP models for the observed data-generating process, which can be assessed using the available data. We then supplement these observed data models with additional assumptions, such as the common *missing at random* (MAR) assumption for missing data and *unconfoundedness* for causal inference in observational studies. While untestable, these assumptions allow us to identify the estimands of interest. By adopting a Bayesian approach, we can naturally account for uncertainty about these uncheckable assumptions by specifying informative priors on a range of plausible assumptions. Unlike parametric approaches, our BNP approach can accommodate possible violations of assumptions while minimizing concerns about model misspecification.

In Part I, we provide a comprehensive review of the key concepts in causal inference and missing data, which will be essential for understanding the subsequent chapters. This section also covers relevant concepts in Bayesian inference that will be used when introducing Bayesian nonparametric methods and in our case studies. We conclude Part I with a detailed exploration of the identifiability of estimands, examining the different assumptions that can be used to identify the estimands of interest in causal inference and incomplete data problems. Of particular note is our emphasis on tools for performing a *sensitivity analysis*, which allows us to assess the robustness of inferences to violations of MAR, unconfoundedness, and other untestable assumptions. This chapter provides a critical foundation for the rest of the book, ensuring that readers have a strong grasp of the fundamental concepts and methods that underpin our BNP approach.

Part II of this book introduces the fundamental BNP tools required to address causal inference and missing data problems. We start by introducing *Bayesian additive regression trees* (BART), which provides a computationally efficient tool for nonparametric mean modelling and binary regression. We then move on to various *Dirichlet process mixture models* (DPMMs), which offer flexible nonparametric modelling of joint and conditional distributions. In the final chapter of Part II, we introduce *Gaussian process* (GP) priors, which can be used for problems similar to those addressed by BART. We then discuss the use of GPs with *dependent Dirichlet processes* (DDPs), which are a powerful nonparametric tool for modeling conditional distributions.

Part III of the book showcases the application of the BNP approach in a diverse range of case studies. These problems feature monotone and non-monotone missing data; causal inference with point exposures; implementation of marginal structural models; causal mediation analysis; and principal stratification in the setting of semi-competing risks. The datasets used in the case studies span a variety of sources, from observational datasets based on electronic health records (EHR) data to datasets obtained from randomized clinical trials.

Our focus will be on BNP approaches, as noted above. Because of this, we focus almost exclusively on estimands that can be defined in a purely *model-free* fashion. The effects we estimate will not correspond to, say, regression coefficients in a regression model, but instead correspond to contrasts in potential outcomes that can be defined for arbitrary distributions. While this strategy has an inherently nonparametric flavor, we note that the general strategy we propose could also be used with parametric Bayesian approaches as well; however, such approaches would not offer the same robustness to model misspecification as the BNP approach.

We will not go into detail on the many alternative approaches to BNP. Common alternative approaches for both missing data and causal inference problems involve either reweighting observations (via either the missing mechanism, propensity score, or both) [1, 2] or matching/stratifying observations (using the propensity score or missing data mechanism or, alternatively, the raw covariates). These approaches can be combined with regression approaches, such as outcome regression, to form *doubly robust* (DR) estimators [3, 4]. We direct interested readers to the substantial literature on these methods: see, for example, [5] for missing data, [6] for causal inference with longitudinal data, [7] for causal inference, [8, 9] for matching/stratifying approaches, or [4] for approaches based on machine learning and targeted learning. There have also been recent approaches that use Bayesian models for the propensity score and outcome regression in doubly robust estimators [10], and the use of flexible machine learning methods in the context of frequentist estimation has become very prevalent in recent years [11, 12, 13].

This book is well suited to a graduate course on Bayesian nonparametrics for missing data (Chapters 2-4, 5-7, 8, and 11-12) or causal inference (Chapters 1, 3-7, 8, 9, 10, 13, 14, and 15). Chapters in this book also provide a resource for review of causal inference (Chapter 1), review of missing data (Chapter 2), and a review of specific BNP methods (Chapters 5-7).

We would like to thank Rob Calver and his team at CRC Press for their support and patience through the long process for completing this book.

Code and `R` packages for the case studies in Part III are available at `https://github.com/theodds/CausalBNP`.

Part I

Overview of Bayesian inference in causal inference and missing data and identifiability

Chapter 1

Overview of causal inference

DOI: 10.1201/9780429324222-1

1.1 Introduction

Often a goal of researchers is to learn about the causal effect of exposures, treatments, or policies (referred to generically as "treatments") on outcomes of interest: for example, does smoking cause cancer? Or does exercising reduce depression? A key distinction between causal inference and standard statistical inference is that causal inference requires assumptions that cannot be checked from the data.

Example 1.1.1. Consider an observational study comparing two classes of medication for treating hypertension: angiotensin-converting enzyme inhibitor (ACEIs) and angiotensin II receptor blockers (ARBs). Suppose we are interested in systolic blood pressure (SBP) 3 months after the start of treatment (Y). From the data, using standard statistical methods, we could learn about the average value of Y for people who took ACEIs or for people who took ARBs. We could also learn about conditional averages, such as the average SBP among men, age 55, who took ACEIs. However, additional assumptions are needed to compare the average SBP in the hypothetical world where *everyone* in the population had been prescribed an ACEI to the average SBP in the world where everyone had been prescribed an ARB; this comparison requires assumptions that can only be assessed via subject matter considerations and that cannot be checked from data.

The fundamental problem in the above example is that we are attempting to draw conclusions about quantities that we did not observe: we cannot know what would have happened to someone who was prescribed an ACEI had they taken an ARB instead because physicians may systematically prescribe ACEIs to men with more severe symptoms or vice versa. One way to identify the effect of interest is to assume that we have measured and controlled for all *confounders*, i.e., variables that influence both the assigned treatment (medication here) and SBP, but it is important to recognize that this assumption cannot be tested strictly from the data itself.

In this chapter we review the *potential outcomes* framework [14] for causal inference, which will be used throughout. We will introduce the framework, use it to define commonly used causal estimands, and describe common estimation procedures. A more in-depth treatment of the potential outcomes framework itself can be found in [8].

1.1.1 Types of Causal Effects

Potential outcomes and point treatments

Suppose we are interested in the causal effect of a treatment A on an outcome Y. For simplicity, we begin with the simple case of a binary treatment where $A = 1$ if treated (prescribed ACEIs, say) and $A = 0$ if untreated (prescribed ARBs), and the outcome Y is the systolic blood pressure.

Define the *potential outcome* $Y(a)$ as the outcome that would be observed if the individual received treatment $a \in \{0,1\}$. We can then use these potential outcomes to define causal estimands of interest. For example, we might then be interested in the *average causal effect* (ACE) $E\{Y(1)\} - E\{Y(0)\}$. The ACE is a population-level parameter and can be interpreted as the answer to the question, "how much higher would the average systolic blood pressure be in the population if everyone had been treated with ACEIs compared to if everyone had been treated with ARBs?" This is a causal comparison because we are comparing outcomes on a *common population* where the only thing different is treatment received. Note the distinction between A (which is the random treatment assignment, i.e., whether an individual is prescribed ACEIs or ARBs, and is possibly correlated with $Y(\cdot)$) and a (which is a particular fixed value, either 0 or 1).

Other types of causal effects might be of interest, and we generally think of the step of defining the causal effect of interest as distinct from the step of modeling the observed data. That is, prior to performing any analysis, we (i) determine what scientific questions we would like to answer, (ii) define causal estimands so that they answer your questions, and (iii) determine if there is a reasonable way to design a study to identify and estimate these quantities. For example, rather than the ACE, we can define the *causal effect of treatment on the treated* as $E\{Y(1) \mid A=1\} - E\{Y(0) \mid A=1\}$. In words, this is a contrast between the average outcome under treatment and under no treatment within the subpopulation of treated individuals. This is a valid causal effect because it is still a contrast between potential outcomes on a common population (here, treated individuals). For treated individuals, we can think of $Y(0)$ as the *counterfactual* outcome – the outcome that would have been observed had the subject, contrary to fact, not been treated.

Causal effects do not have to be restricted to mean differences, and going beyond mean differences can expand the types of causal questions we can answer. For example, we can define the *quantile causal effect* $F_1^{-1}(p) - F_0^{-1}(p)$, which answers the question, "how much higher would the pth quantile of SBP be if everyone in the population was treated with ACEIs rather than ARBs?" Here, $F_a^{-1}(p)$ is the pth quantile of the cumulative distribution function $F_a(y) = \Pr\{Y(a) \leq y\}$. Quantile causal effects are particularly useful when the outcome of interest is skewed, or when the treatment only has a large effect for a small subset of the population (see the case study in Chapter 8).

For binary outcomes, one might instead be interested in the causal relative risk $E\{Y(1)\}/E\{Y(0)\}$, which answers questions of the form, "how many times more likely would high blood pressure be if everyone was treated with ACEIs rather than ARBs?" (see the case study in Chapter 9).

Lastly, one might be interested in asking questions about causal effects in *subpopulations* of the data; this might be important if a treatment is not very effective for most individuals but is highly effective for a small subset of individuals. If V denotes a set of covariates of interest, then the specific values of $V = v$ define subpopulations; we then define the average causal effect within levels of V as $E\{Y(1) - Y(0) \mid V = v\}$. This is known as the *conditional average causal effect* (CACE). One could similarly define conditional versions of the quantile causal effects and the causal relative risk.

Time-varying treatments

Above we considered only point treatments, where the treatment is given at one time and we observe the outcome at a later time. In many situations, however, the treatment an individual receives varies over time. Suppose instead that a treatment $A_t = 1$ or $A_t = 0$ is given at times $t = 1, \ldots, T$. We now have 2^T potential outcomes for every observational unit, each of the form $Y(a_1, a_2, \cdots, a_T)$. In time-varying settings, we can think of treatment *strategies* rather than treatments. A treatment strategy is a set of rules that assign treatments to individuals over time and can be either *static* or *dynamic* as described below. For a treatment strategy g, we denote the associated potential outcome as $Y(g)$.

Static treatment strategies are ones where the treatment plan does not depend on variables that are updated over time; that is, g can be defined in terms of a *fixed* collection of treatments $(a_1, \ldots, a_T)$. Thus, under a static treatment regime, we would know precisely what treatments would be received at each time for a given unit *prior to the start of the study*. For example, the static treatment strategy corresponding to "treat everyone at all times" is $Y(g) = Y(1, 1, \ldots, 1)$. An alternative static treatment strategy g' is to never treat at any time, in which case $Y(g') = Y(0, 0, \ldots, 0)$. One causal effect of interest in this case might be the average difference in the outcome if everyone was always treated versus if no one was ever treated: $E\{Y(g)\} - E\{Y(g')\}$.

An alternative type of treatment strategy is a *dynamic* treatment strategy. Dynamic treatment strategies are at least partially determined by one or more time-varying quantities (e.g., biomarkers such as blood pressure measurements collected at each time) that could be affected by earlier treatments.

Example 1.1.2. Consider a study where the interest was in determining when a treatment for type 2 diabetes mellitus should be intensified in response to changes in hemoglobin A1C value [15]. In this case, dynamic treatment strategies of interest were of the form: "patient should initiate treatment intensification at the first time their A1C level reaches or drifts above θ% and should remain on the intensified therapy thereafter." Causal contrasts of interest involving treatment strategies g_θ included $E\{Y(g_8)\} - E\{Y(g_7)\}$, where g_θ corresponds to setting $A_t = 0$ if A1C is less than θ% at all times up-to-and-including t and setting $A_t = 1$ otherwise.

1.1.2 *Identifiability and Causal Assumptions*

For most studies, we observe the treatment received (A_i) and the outcome (Y_i) for N total subjects ($i = 1, \ldots, N$). Notice that there are no potential outcomes here. To link observed data to potential outcomes, we need *causal assumptions*.

One assumption that we have already implicitly made is known as the Stable Unit Treatment Value Assumption (SUTVA). When we defined the potential outcome for subject i, $Y_i(a)$, it was defined only in terms of a treatment a potentially received by subject i. This would not be the case if there was *interference* between units, i.e., the treatment received by subject i' itself influences subject i. For example, the rate of earthquakes in one county might be dependent on whether a neighboring county is "treated" with hydraulic fracturing. As another example, this would also not be the case in vaccine trials with infectious diseases: the outcome for subject i, who is not vaccinated ("not treated"), is likely influenced by whether their neighbors are vaccinated (or not) [16].

Definition 1.1.1 (SUTVA). The *Stable Unit Treatment Value Assumption* (SUTVA) is said to hold when the potential outcome of any subject i does not depend on the treatment received by the other subjects. That is, if $\mathbf{a} = (a_1, \ldots, a_N)$ and $\mathbf{a}' = (a'_1, \ldots, a'_N)$ are any two possible assignments of subjects to treatments such that $a_i = a'_i$ and $Y_i(\mathbf{a})$ is the potential outcome of subject i under $\mathbf{a}$, we have $Y_i(\mathbf{a}) = Y_i(\mathbf{a}')$.

SUTVA justifies the notation $Y_i(a)$, as this notation relies on the potential outcome for subject i only depending on the treatment that subject i receives.

The next causal assumption that is usually made is known as *consistency*.

Definition 1.1.2 (Consistency). The consistency assumption holds if $Y_i = Y_i(a)$, if $A_i = a$. That is, if subject i is observed to have received treatment a then their observed outcome is just their potential outcome for treatment a.

Under the consistency assumption, we observe $Y_i = Y_i(A_i)$ for subject i, but $Y_i(1 - A_i)$ is unobserved. The consistency assumption is intended to capture the assumption that there are not "multiple versions of the treatment" or, in other words, that the formalization of the treatment through a captures all causally relevant features [17].

The fundamental problem of causal inference is that we only observe one potential outcome for each subject. However, with additional causal assumptions we can identify the causal estimands we have defined so far. For example, the reason that *randomized trials* are the gold standard for establishing causation is that randomizing the treatment assignment guarantees that $\{Y_i(0), Y_i(1)\}$ is independent of A; consequently,

$$E\{Y_i(a)\} = E\{Y_i(a) \mid A_i = a\} = E(Y_i \mid A_i = a)$$

so that $E\{Y_i(a)\}$ can be identified in terms of the observable quantities (Y_i, A_i).

In observational studies, in addition to Y and A, we also typically observe a set of covariates L that influence both Y and A; such covariates are referred to as *confounders*. The observed data is therefore $\{Y_i, A_i, L_i; i = 1, \cdots, N\}$. The selection of confounders L is, at least in part, based on subject matter knowledge [18]. These pretreatment variables should be chosen to completely capture the association between the treatment assignment and the outcome. This leads to the next causal assumption.

Definition 1.1.3 (Ignorability). The *ignorability* assumption holds if

$$\{Y(0), Y(1)\} \perp\!\!\!\perp A \mid L, \tag{1.1}$$

where the notation $Y(a) \perp\!\!\!\perp A \mid L$ means that $Y(a)$ is conditionally independent of A given L.

We can think of A as representing the random assignment of subjects to treatments. Ignorability requires that the treatment decision does not depend on the potential outcomes given the confounders L. For example, in a clinical study, sicker patients might be more likely to be treated, which could make the treatment appear less effective than it is; however, if we adequately capture measures of baseline health with L, then the treatment assignment might effectively be randomized *within levels of L*. This assumption would not hold if, for example, there were unmeasured variables U (i.e., not included in L) that are associated both with the outcome Y

and treatment A. Thus, this assumption is sometimes referred to as a *no unmeasured confounders* assumption, an *unconfoundedness* assumption, or as an *exchangeability* assumption.

The validity of ignorability cannot be assessed from the observed data, regardless of how large the sample size is. This is because we can never rule out the possibility that an apparent association between the treatment and outcome is due to some variable that we happen to not have measured. Because of this, we recommend that analysis under ignorability be taken as a starting point. A strategy to deal with untestable assumptions like ignorability is sensitivity analysis. This is discussed in more detail in Chapter 4.

At first glace, consistency and ignorability might appear to be enough to identify causal effects like the ACE. This is because, within the levels of L, treatment is randomized so that $E\{Y(a) \mid L = \ell\} = E\{Y(a) \mid A = a, L = \ell\} = E(Y \mid A = a, L = \ell)$. All of the random variables on the right are part of the observed data (not counterfactuals), and hence the right-hand side can ostensibly be estimated. However, if $\Pr(A = a \mid L = \ell) = 0$ for some (a, l), then we cannot estimate $E(Y \mid A = a, L = \ell)$ nonparametrically, as there would be no observed values of Y for this subpopulation. We therefore need an additional untestable assumption. The most common assumption for this is *positivity*.

Definition 1.1.4 (Positivity). The *positivity* assumption holds if

$$\Pr(A = a \mid L = \ell) > 0 \quad \text{for all } a \text{ and } \ell.$$

We have focused on identifiability in the point treatment setting; however, the same ideas hold in more complicated settings such as when the treatment is time varying. In that case, there are generalizations of the above causal assumptions, such as sequential ignorability (see Section 1.2.1).

Below we show how the combination of consistency, positivity, and ignorability can be used to identify the ACE with binary treatments.

Proof sketch of identifiability of the ACE. It suffices to show that $E\{Y(1)\}$ is identified, as the proofs for $E\{Y(0)\}$ and the ACE $E\{Y(1) - Y(0)\}$ are similar. For simplicity, assume that Y_i and L_i are discrete with finite support. Let (y_i, a_i, ℓ_i) denote the realized values of (Y_i, A_i, L_i); to show identification, it suffices to show that the ACE can be consistently estimated from the observed data.

Letting $f_a(y \mid \ell) = \Pr\{Y_i(a) = y \mid L_i = \ell\}$, $f(\ell) = \Pr(L_i = \ell)$, and $f(y \mid a, \ell) = \Pr(Y_i = y \mid A_i = a, L_i = \ell)$, we can write

$$E\{Y(1)\} = \sum_{y,\ell} y f(\ell) f_1(y \mid \ell) = \sum_{y,\ell} y f(\ell) f(y \mid 1, \ell),$$

where the second equality follows from *ignorability* and *consistency*:

$$f_1(y|\ell) = f_1(y|A = 1, \ell) = f(y|1, \ell).$$

By the law of large numbers, we can consistently estimate $f(\ell)$ with $\sum_i I(L_i = \ell)/N$, and, by *positivity*, we can estimate $f(y \mid 1, \ell)$ with $\sum_i I(Y_i = y, A_i = 1, L_i =$

$\ell)/\sum_i I(A_i = 1, L_i = \ell)$; positivity is required in order for $f(y \mid 1, \ell)$ to be well defined and for the denominator to be non-zero almost surely for large enough N. Since $E\{Y(1)\}$ can be consistently estimated from the observed data, it is identified.

□

1.2 The g-Formula

Under the causal assumptions of Section 1.1.2, we can do more than just identify particular causal parameters like the ACE $E\{Y(1) - Y(0)\}$. We can also identify the marginal *distributions* of the potential outcomes $Y(a)$. This is useful both for the purpose of deriving causal estimands that are not expressible as means and for generalizing the identification strategies for binary treatments to non-binary or time-varying settings. The *g-formula*, which we describe in this section, provides a systematic method for deriving the marginal distribution of the $Y(a)$'s in very general settings.

The g-formula [20] is an expression that allows us to obtain causal parameters for the potential outcomes under a particular treatment strategy from the observed data distribution. Consider an example where there are p binary confounders and we are interested in the ACE. We first obtain the distribution of $Y(a)$ for $a = 0, 1$ as

$$f\{Y(a) = y\} = \sum_{\ell_1=0}^{1} \cdots \sum_{\ell_p=0}^{1} f(Y = y \mid A = a, L_1 = \ell_1, \ldots, L_p = \ell_p) \times \quad (1.2)$$

$$\Pr(L_1 = \ell_1, \ldots, L_p = \ell_p). \quad (1.3)$$

This expression follows from the positivity and ignorability assumptions. Multiplying by and integrating over y, we obtain

$$E\{Y(a)\} = \sum_{\ell_1=0}^{1} \cdots \sum_{\ell_p=0}^{1} E(Y \mid A = a, L_\ell = \ell_1, \cdots, L_p = \ell_p) \Pr(L_1 = \ell_1, \cdots, L_p = \ell_p).$$

This expression standardizes the expectation of the potential outcome with respect to the population distribution of confounders. For continuous covariates, we would replace summations with integrals. In either case, the concept is the same: we condition the outcome on both the treatment and confounders, and then we average over the marginal distribution of the confounders.

The g-formula can be used to identify a much wider range of causal effects than the ACE in point treatment settings. Suppose, for example, that we are interested in the quantile causal effects, $F_1^{-1}(p) - F_0^{-1}(p)$. Using the g-formula, we can obtain $F_a^{-1}(p)$ as the solution to the equation

$$\int \Pr(Y \leq y \mid A = a, L = \ell) \, dF_L(\ell) = p,$$

where $F_L(\cdot)$ denotes the marginal distribution of the confounders. More generally, the g-formula (1.2) allows us to compute essentially any summary of $f\{Y(a) = y\}$ we might be interested in.

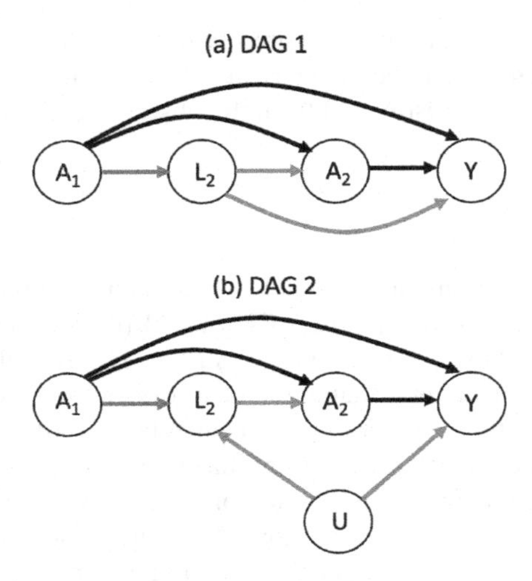

Figure 1.1 *Two hypothetical directed acyclic graphs in a longitudinal setting that illustrate time-dependent confounding. For DAG 1, treatment at time 1 affects the confounder at time 2 (blue arrow), while the confounder at time 2, L_2, affects the treatment decision at time 2, A_2, and the outcome Y (red arrows). In DAG 2, there is an unmeasured confounder U that affects L_2 and Y, while L_2 is affected by A_1.*

1.2.1 Time-Dependent Confounding

We now generalize the g-formula as described above to show how to control for time-varying treatments and confounders in a longitudinal data setting. We have treatments and confounders at times $t = 1, \ldots, T$ denoted $A_1, \cdots, A_T$ and $L_1, \cdots, L_T$, respectively. The major challenge here is that the confounders at time t, L_t, could have been affected by past treatments (e.g., A_{t-1}) and might affect future treatment decisions and outcomes [21, 22].

Example 1.2.1. Pre-exposure prophylaxis (PrEP) is a medicine designed to prevent HIV. However, there is a concern that PrEP users might have risk compensation, where they engage in riskier behaviors because they feel safer using PrEP. In that scenario, PrEP (the treatment) might affect sexual behaviors (e.g., having condomless sex), which itself might affect future decisions to use PrEP, and might affect an outcome such as sexually transmitted infections. If we record PrEP use, risk-taking behaviors, and STIs repeatedly over time, the risk-taking behaviors are time-dependent confounders.

Figure 1.1 illustrates two examples where time-dependent confounding is present when we have just two treatment times. The causal structure here is depicted with a *directed acyclic graph* (DAG), with causal relationships represented as arrows pointed from a cause to an effect. Suppose treatment at time 1, A_1 (e.g., PrEP), is randomized. After that, we observe a confounder at time 2, L_2 (e.g., a risk-taking behavior); treatment at time 2, A_2; and the outcome Y (e.g., STIs). In Figure 1.1(a),

L_2 is on the intermediate pathway from A_1 to Y ($A_1 \rightarrow L_2 \rightarrow Y$). However, we also see that L_2 confounds the effect of A_2 on Y, as it directly effect both of them. If we are interested in the effect of A_2 on Y, we need to control for L_2. Here, L_2 is a time-dependent confounder. Similarly, in Figure 1.1(b), L_2 is on the pathway from A_1 to Y. An unmeasured variable U affects A_2 (indirectly) and Y. Therefore, time-dependent confounding is also present in this DAG.

The challenge with time-dependent confounding is the following. Consider again the two time point examples in Figure 1.1(a), and suppose interest is in the joint effect of the two treatments. For the impact of A_1 on Y, we would not want to control for L_2, because L_2 is an outcome of treatment. However, for the effect of A_2 on Y, we need to control for L_2 since it confounds that relationship. Thus, we need to control for L_2 while not adjusting for the part of L_2 that was affected by A_1.

Robins' g-formula was developed specifically for this situation, where our interest is in causal effects that are comparisons of different treatment strategies g and g' (e.g., $E\{Y(g) - Y(g')\}$). Using "overbar" notation for history, define $\overline{A}_t = (A_1, \cdots, A_t)$ and $\overline{L}_t = (L_1, \cdots, L_t)$. To identify the causal effects, we need extensions of the assumptions from Section 1.1.2.

The consistency assumption is now that $Y_i = Y_i(A_1, \ldots, A_T)$. The positivity assumption extends to the longitudinal case as follows: $\Pr(A_t = a_t \mid \overline{L}_t = \overline{\ell}_t, \overline{A}_{t-1} = \overline{a}_{t-1})$ for all $(t, \overline{\ell}_t, \overline{a}_{t-1})$. for all t and for ℓ_t. Finally, the ignorability assumption is extended to what is known as sequential ignorability.

Definition 1.2.1 (Sequential Ignorability). The *sequential ignorability* assumption is

$$Y(a_1, \ldots, a_T) \perp\!\!\!\perp A_t \mid \overline{L}_t, \overline{A}_{t-1} \quad \text{for all } (a_1, \ldots, a_T) \text{ and } t = 1, \cdots, T. \tag{1.4}$$

That is, treatment assignment is random (independent of potential outcomes) given covariate history and treatment history at the time of every treatment decision.

Under the above assumptions, the average potential outcomes (and causal effects) are identifiable using the g-formula. The g-formula for static treatment strategies in the time-dependent confounding setting is

$$E\{Y(a_1, \cdots, a_T)\} = \int_{\ell_1} \cdots \int_{\ell_T} E(Y \mid \overline{A}_T = \overline{a}_T, \overline{L}_T = \overline{\ell}_T) \times \prod_{t=1}^{T} f(\ell_t \mid \overline{A}_{t-1} = \overline{a}_{t-1}, \overline{L}_{t-1} = \ell_{t-1}) d\ell_1 \cdots d\ell_T.$$

The key is that here we are not averaging over the marginal distribution of the confounders, but rather over the distribution of the confounders that would have been observed had treatment been set to $a_1, \cdots, a_T$. In this way, we adjust for the L's while not adjusting away the impact that earlier treatments had on the L's. Also note that the g-formula can be modified if interest is in dynamic treatment strategies. The modification involves conditioning on treatment values that are determined by the dynamic treatment rule, rather than to specific, pre-specified treatment values. For details, see [23].

1.2.2 Bayesian Nonparametrics and the g-Formula

The above versions of the g-formula involve conditional distributions and expectations for the outcome and marginal or conditional distributions for the covariates. Outside of very simple settings, one must specify models in order to estimate these distributions in a stable fashion. In the context of a point treatment, for example, we must specify the model $\{f(y,a \mid \ell,\theta) : \theta \in \Theta\}$ for the distribution of $[Y_i, A_i \mid L_i]$ and the model $\{f(\ell \mid \omega) : \omega \in \Omega\}$ for the distribution of L_i. For general choices of $f(y,a \mid \ell,\theta)$ and $f(\ell \mid \omega)$, the integrals in the g-formula are unlikely to be available in closed form, and so the g-formula is typically approximated using Monte Carlo integration.

Some degree of modeling is required to implement the g-formula in most settings, as nonparametrically estimating all of the required distributions is statistically challenging. Using models opens us up to the issue of model misspecification, which in turn can lead to poor (biased) inference about the causal effects. Therefore, a major focus of this book is on flexible models for distributions and functions. By flexibly modeling these distributions/functions using Bayesian nonparametrics, we will be able to implement the g-formula to compute causal effects of interest while mitigating any bias from model misspecification.

1.3 Propensity Scores

While conceptually straight forward, modeling the full conditional distribution $f(y \mid \ell)$ and using g-computation to obtain the estimates of the causal effects has drawbacks when p is large. First, it can be difficult to flexibly model this conditional distribution. Second, it can be computationally intensive.

Interestingly, rather than using the distribution $f(y \mid \ell)$, we can often obtain frequentist valid estimates of causal effects by considering the *propensity score* instead. In the case of a binary treatment, the propensity score is defined as the probability of treatment given confounders:

$$e(\ell) = \Pr(A = 1 \mid L = \ell).$$

The following identification result suggests an *inverse probability weighting* approach to estimating the average causal effect.

Proposition 1.3.1. *Suppose that the SUTVA, consistency, ignorability, and positivity assumptions hold and that the treatment A_i is binary. Then the average causal effect can be written*

$$\begin{aligned} \text{ACE} &= E\left\{\frac{AY}{e(L)} - \frac{(1-A)Y}{1-e(L)}\right\} \\ &= E\left\{Y \frac{A - e(L)}{e(L)\{1-e(L)\}}\right\} \end{aligned}$$

Proof. Note that, by positivity, the random variables $AY/e(L)$ and $(1-A)Y/\{1-e(L)\}$ are well defined. The second equality follows via simple algebraic manipulations. For the first equality, it suffices to show that $E\{AY/e(L)\} = E\{Y(1)\}$, with similar logic implying that $E[(1-A)Y/\{1-e(L)\}] = E\{Y(0)\}$. We show this using iterated expectation:

$$\begin{aligned}
E\left(\frac{AY}{e(L)}\right) &= E\left(\frac{AY(1)}{e(L)}\right) = E\left[E\left(\frac{AY(1)}{e(L)} \mid L, Y(1)\right)\right] \\
&= E\left[\frac{Y(1)}{e(L)}E\left(A \mid L, Y(1)\right)\right] = E\left[\frac{Y(1)}{e(L)}e(L)\right] \\
&= E\{Y(1)\}.
\end{aligned}$$

Above, the first equality follows from the fact that $AY = AY(1)$ due to consistency; the second equality follows from iterated expectation; the third equality follows from the fact that $Y(1)/e(L)$ is the constant conditional on $(Y(1), L)$; and the fourth equality follows from ignorability and the definition of the propensity score. □

Proposition 1.3.1 suggests a way to bypass the estimation of the outcome model $f(y \mid a, \ell)$: if we can estimate the propensity score sufficiently well, then the plug-in estimate

$$\widehat{\text{ACE}} = \frac{1}{n}\sum_i Y_i \frac{A_i - \widehat{e}(L_i)}{\widehat{e}(L_i)\{1-\widehat{e}(L_i)\}}$$

is consistent for the ACE.

The propensity score plays a prominent role in causal inference. It has the property of being a *balancing score*, which makes it a useful one-dimensional summary of L [24]. We discuss this feature in more detail next.

1.3.1 Covariate Balance

In a classic randomized trial, the propensity score is under the control of the experimenter and might be, for example, chosen so that $e(\ell) = 0.5$ (or some other fixed constant) for all ℓ. This means that the joint distribution of the covariates is the same in the treated and untreated groups, i.e., $f(\ell \mid A=1) = f(\ell \mid A=0) = f(\ell)$. This is important because any differences in the outcomes between a treatment and control group will not be due to *systematic* differences in the covariates L across the two populations. When combined with ignorability, covariate balance implies that $\{Y(0), Y(1), L\}$ is independent of A:

$$\begin{aligned}
f\{Y(0), Y(1), L, A\} &= f\{Y(0), Y(1) \mid A, L\} f(L \mid A) f(A) \\
&= f\{Y(0), Y(1) \mid L\} f(L) f(A) \\
&= f\{Y(0), Y(1), L\} f(A).
\end{aligned}$$

Hence, covariate balance allows us to use "naive" estimates of the treatment effect which do not control for the covariates:

$$\begin{aligned}
E\{Y(1)\} - E\{Y(0)\} &= E\{Y(1) \mid A=1\} - E\{Y(0) \mid A=0\} \\
&= E(Y \mid A=1) - E(Y \mid A=0).
\end{aligned}$$

By contrast, in observational studies $e(\ell)$ will virtually always depend on ℓ so that, in general, we expect that $f(\ell \mid A=1) \neq f(\ell \mid A=0)$. For example, in the PrEP study, we *expect* that risk-taking behavior is associated with the use of PrEP, so that risk-taking behavior is not balanced across the treatment groups. Suppose instead, however, that we can also condition on the realized propensity score $e(L)$. It turns out that, if we do this, then the covariates will be balanced among the treatment groups *within the subpopulations of subjects who have the same value of the propensity score*. This follows from a direct application of Bayes theorem.

Proposition 1.3.2. *Assume that positivity holds. If* $e(\ell^*) = \lambda$, *then* $f(L=\ell^* \mid e(L) = \lambda, A=1) = f(L=\ell^* \mid e(L)=\lambda, A=0) = f(L=\ell^* \mid e(L)=\lambda)$.

Proof. We again assume that all variables are discrete for simplicity. First note that

$$\begin{aligned} f(A=1 \mid e(L)=\lambda) &= \sum_{\ell} f(A=1, L=\ell \mid e(L)=\lambda) \\ &= \sum_{\ell: e(L)=\lambda} f(A=1, L=\ell \mid e(L)=\lambda) \\ &= \sum_{\ell: e(L)=\lambda} \underbrace{f(A=1 \mid L=\ell)}_{e(\ell)=\lambda} f(L=\ell \mid e(L)=\lambda). \end{aligned}$$

The second equality follows from the fact that $f(L=\ell \mid e(L)=\lambda)=0$ except when $e(\ell)=\lambda$. Then, because $f(L=\ell \mid e(L)=\lambda)$ is a mass function,

$$\sum_{\ell: e(L)=\lambda} \lambda\, f(L=\ell \mid e(L)=\lambda) = \lambda \sum_{\ell: e(L)=\lambda} f(L=\ell \mid e(L)=\lambda) = \lambda.$$

Hence $f(A=1 \mid e(L)=\lambda)=\lambda$. Because $e(\ell^*)=\lambda$, we also have $f(A=1 \mid e(L) = \lambda, L=\ell^*) = f(A=1 \mid L=\ell^*) = \lambda$. Using these facts, we then have

$$\begin{aligned} f(L=\ell^* \mid e(L)=\lambda, A=1) &= \frac{f(A=1 \mid e(L)=\lambda, L=\ell^*) f(L=\ell^* \mid e(L)=\lambda)}{f(A=1 \mid e(L)=\lambda)} \\ &= f(L=\ell^* \mid e(L)=\lambda). \end{aligned}$$

Finally,

$$\begin{aligned} f(L=\ell^* \mid e(L)=\lambda) &= f(L=\ell^* \mid e(L)=\lambda, A=1)\, f(A=1 \mid e(L)=\lambda) \\ &\quad + f(L=\ell^* \mid e(L)=\lambda, A=0)\, f(A=0 \mid e(L)=\lambda) \\ &= \lambda\, f(L=\ell^* \mid e(L)=\lambda) + (1-\lambda) f(L=\ell^* \mid e(L)=\lambda, A=0). \end{aligned}$$

Noting that $1-\lambda \neq 0$ by positivity, solving for $f(L=l^* \mid e(L)=\lambda, A=0)$ gives $f(L=\ell^* \mid e(L)=\lambda)$ as well. □

The above proposition states that, within levels of the propensity score, we have covariate balance between treated and untreated subjects. Because the covariates are balanced conditional on $e(L)$ (i.e., L is independent of A given $e(L)$), we refer to $e(L)$ as a *balancing score*. Balancing scores are useful because, as we will see, we can recover causal effects from the distribution of the outcome conditional on any balancing score.

While $e(L)$ is not the *only* balancing score, it is *minimal* in the following sense: $e(L)$ is necessarily a function of any other balancing score. Because $e(L)$ is just a scalar summary of the p covariates, it will typically be far easier to estimate the distribution $f(Y \mid e(L),A)$ than $f(Y \mid L,A)$, particularly when p is large. Causal inference procedures have been proposed that involve either stratifying, matching, or re-weighting based on the propensity score [25]. Although these approaches are usually not fully Bayesian, we consider inverse probability of treatment weighting as a comparison method in the case study in Chapter 9. Another approach is to directly condition on $e(L)$ in the outcome model [26, 27], which we demonstrate in the case study in Chapter 8.

1.3.2 Conditioning on the Propensity Score

Methods based on the g-formula require a model for $f(y \mid a, \ell)$. By contrast, methods based on conditioning on a propensity score instead model the outcome conditional on the propensity score: $f\{Y = y \mid A = a, e(L) = e(\ell)\}$. This reduces the dimension of the predictors in the conditional distribution from $p+1$ to 2. If p is large, then we have greatly reduced the challenge of modeling the response distribution.

Because this book focuses on Bayesian methods, we consider propensity score-based inference methods from the likelihood perspective. To do this, we specify models $\{f(y \mid a, e, \theta) : \theta \in \Theta\}$ and $\{e(\ell, \alpha) : \alpha \in A\}$ for the outcome distribution and propensity score, respectively. However, unless $[Y \perp\!\!\!\perp L \mid e(L)]$ (which is unlikely to hold), it is not true that $f(y, a \mid \ell, \theta, \alpha) = f(y \mid a, e(\ell, \alpha), \theta)\, f(a \mid \ell, \alpha)$. In other words, while $f(y \mid a, e(\ell, \alpha), \theta)$ and $f(a \mid \ell, \alpha)$ are, separately, valid probability distributions, their product is not equal to the likelihood of the data. Further, performing maximum likelihood on $\prod_i f(Y_i \mid A_i, e(L_i, \alpha), \theta)\, f(A_i \mid L_i, \alpha)$ results in *model feedback*, where poorer fitting propensity score models might lead to better fitting outcome models.

Typically, to avoid these problems, an approximate Bayesian approach is used. The propensity score and outcome model are either fitted separately (in two stages) or the models are specified in such a way to block feedback. This is discussed in a Bayesian context in [28, 29]. We demonstrate the usefulness of the propensity score as a covariate in Bayesian models in the case studies in Chapters 8 and 14.

1.3.3 Positivity and Overlap

As exploratory analyses, investigators often assess how similar the covariate distribution is between the treated and untreated groups. One way to quantify differences in the covariate distribution is with standardized differences (differences in the sample means in a covariate divided by the standard deviation). Large standardized difference on a particular covariate would suggest that the average value of that covariate is substantially different between the groups. Often, a rule-of-thumb is that a (absolute) standardized difference of greater than 0.1 is noteworthy. In some cases, standardized differences are reported before and after covariate adjustment (such as matching or weighting), to assess how much adjustment helped with covariate overlap.

An additional way the propensity score is used in practice is to assess the positivity assumption [30]. While we cannot formally test positivity, we can get an idea of whether it is nearly violated (i.e., whether the probability of treatment is very small for some observed values of L). Typically, the propensity score model will be estimated, and then the predicted probability of treatment will be plotted separately for each treatment group. Ideally, most of these values fall in such a way that there is someone in the opposite group with a similar propensity score [31, 32].

An example of using standardized differences and the propensity score to assess overlap is given in Chapter 10.

1.4 Marginal Structural Models

The g-formula, which identifies the causal effects implicitly through the specifications of $f(y \mid a, \ell)$ and $f(\ell)$, works well for situations where the set of causal contrasts of interest is small relative to the sample size. For example, in the point treatment setting, if there are just two possible treatments ($a = 0$ or 1) and interest is in the marginal causal effects, then the g-formula only needs to be carried out twice (once for each value of a). We do not need to directly parameterize the model in terms of the effects of interest in this case.

There are situations, however, where directly modeling the causal effects of interest might be practically necessary or lead to better statistical performance. Suppose, for example, that the treatment effect is continuous. In that case, it might make sense to model $E\{Y(a)\}$ directly as, a priori, we would expect this quantity to be smooth in a. Similarly, when there is a binary treatment but multiple treatment times (T), the number of unique potential outcomes is 2^T. For the sake of parsimony in the multiple treatment setting, we might wish to build some structural parametric assumptions that are supported by subject matter knowledge; for example, we might assume that $E\{Y(\overline{a}_T)\}$ only depends on the total exposure $\sum_{t=1}^{T} a_t$ or on only the two most recent values of the treatment (a_{T-1}, a_T).

We now give an example of a *marginal structural model* in the context of time-varying treatments. Let V be a vector of baseline covariates; V could be a subset of L but does not necessarily need to be. We then define a marginal structural model (MSM; [20]) as

$$E\{Y(\overline{a}_T) \mid V = v\} = h(\overline{a}_T, v; \psi),$$

where h is a known function and ψ consists of unknown parameters. Thus, on the right-hand side we have a model for the mean of the potential outcomes. This is a *marginal* model in the sense that the L's that are not part of V are marginalized out. In practice, one might choose V to be a set of covariates for which interactions with treatment (known as *treatment effect heterogeneity*) are of interest. As a very simple example of an MSM, we could assume that the impact of treatment is linear and strictly through the cumulative amount of treatment received, i.e., $E\{Y(\overline{a})\} = \psi_0 + \psi_1 \sum_{j=1}^{T} a_j$. Here, ψ_1 would represent how much the mean of the potential outcome changes with each additional dose of treatment over time. The same causal identifying assumptions introduced in Section 1.2.1 can be used with this MSM.

Frequentist methods for estimating parameters from MSMs include inverse probability of treatment weighting (IPTW) [20, 33] and augmented IPTW [34]. Fully Bayesian approaches for MSMs are not straightforward due to MSMs being a model for potential outcomes rather than observed variables. However, Bayesian solutions have been proposed for both the point treatment setting [35] and time-dependent confounding settings [36]. An example of a Bayesian approach to MSMs is described in the case study in Chapter 10.

1.5 Principal Stratification

In some situations, there is a post-treatment variable S that has an important role in defining the causal effects of A on Y. Consider, for example, a study on the effect of exercise on smoking abstinence, where A denotes being *assigned* to exercise; in this case, it might be desirable to take into account whether or not a subject *actually* exercised ($S = 0$ if an individual does not exercise and $S = 1$ if they do) in assessing the efficacy of exercise on smoking. However, conditioning on S in an outcome model can lead to bias, in part because S is itself an outcome of the treatment assignment. Another common instance of this problem occurs in clinical trials studying deadly diseases, where our interest is the effect of a treatment (A) on an outcome of interest (Y) that is only observed when a subject has survived a certain amount of time (S). However, S itself might be affected by A. Principal stratification has been proposed as a method to account for these types of post-treatment variables [37].

Example 1.5.1 (*Non-compliance*)**.** To model the setting with both an assigned treatment a and a received treatment s, we expand the potential outcomes notation and let $Y(a,s)$ denote the potential outcome that would have been observed under an assigned treatment a and realized treatment s. Let $S(a)$ be the indicator that the subject would receive the treatment if the treatment assignment was set to a. The intention-to-treat (ITT) effect in this setting is $E\{Y(1,S(1)) - Y(0,S(0))\}$. This estimand compares populations based on their assigned treatment, i.e., the mean outcome if everyone had been assigned $a = 1$ compared with the mean outcome if everyone had been assigned $a = 0$. When there is non-compliance, the ITT is the causal effect of being randomized to treatment (assuming treatment was randomized). While still potentially useful, the ITT is not the causal effect of *receiving* the treatment. Principal stratification is a way to define a causal effect of treatment in this setting.

Under consistency, the observed data is the assigned treatment indicator A, the observed treatment received $S = S(A)$, and the observed outcome $Y = Y(A,S) = Y\{A,S(A)\}$. The principal stratification approach begins by defining strata (called the *principal strata*) based on $S(1)$ and $S(0)$ and defining causal effects with these strata (called the *principal effects*). In the non-compliance setting, we typically define four principal strata: "compliers" ($S(0) = 0$ and $S(1) = 1$), "never-takers" ($S(0) = 0$ and $S(1) = 0$), "defiers" ($S(0) = 1$ and $S(1) = 0$), and "always-takers" ($S(0) = 1$ and $S(1) = 1$). A causal effect of interest in this problem is the *complier average causal effect* (CACE)

$$\begin{aligned}\text{CACE} &= E\{Y(1,S(1)) - Y(0,S(0)) \mid S(0)=0, S(1)=1\} \\ &= E\{Y(1,1) - Y(0,0) \mid S(0)=0, S(1)=1\}.\end{aligned}$$

This is a valid causal effect because it is a contrast of mean potential outcomes *on a common population* (compliers – subjects who would take treatment if assigned treatment, and would not take treatment if assigned to control arm). For this subpopulation, the causal effect of treatment assignment (intention-to-treat) is the causal effect of treatment received.

There is an identifiability problem caused by the fact that we are defining strata based on pairs of variables that are never jointly observed. For example, a subject who is observed to have $\{A=1, S=1\}$ could be a complier or an always-taker. To resolve this issue, the following two assumptions can be made. First, we assume that $Y(a,s) = Y(a',s)$, which is known as the *exclusion restriction*. This just means that what matters in terms of the potential outcomes is treatment received; for example, in the smoking cessation study, what matters for smoking cessation is only whether an individual *actually* exercises, and not whether they were *assigned* to exercise. This allows us to write potential outcomes with a single index $Y(s)$ (the outcome that would be observed if treatment received was s). The next assumption commonly made is *monotonicity*, $S(1) \geq S(0)$. This assumption is also known as the *no defiers* assumption. The no defiers assumption rules out the possibility that there are people who would always do the opposite of their treatment assignment. In the context of the smoking cessation study, the no defiers assumption states that there is nobody who would exercise if-and-only-if they were told not to. This assumption would automatically be met if subjects in the control group did not have access to treatment. By ruling out defiers, we can then identify subjects with $\{A=1, S=0\}$ as "never-takers." These assumptions are sufficient for identification of the CACE. To see this, first note that we can write the ITT effect as follows:

$$\begin{aligned}&E\{Y(S(1)) - Y(S(0))\} \\ &= E\{Y(S(1)) - Y(S(0)) \mid S(0)=S(1)=1\}\Pr\{S(0)=S(1)=1\} \\ &+ E\{Y(S(1)) - Y(S(0)) \mid S(0)=S(1)=0\}\Pr\{S(0)=S(1)=0\} \\ &+ E\{Y(S(1)) - Y(S(0)) \mid S(0)=0, S(1)=1\}\Pr\{S(0)=0, S(1)=1\}.\end{aligned}$$

In the above, $E\{Y(S(1)) - Y(S(0)) \mid S(0)=S(1)=1\} = E\{Y(S(1)) - Y(S(0)) \mid S(0)=S(1)=0\} = 0$ by the exclusion restriction. Next, note that by randomization we can write $E\{Y(S(a))\} = E(Y \mid A=a)$. Combining these facts, we have

$$E\{Y(S(1)) - Y(S(0)) \mid S(0)=0, S(1)=1\} = \frac{E(Y \mid A=1) - E(Y \mid A=0)}{Pr(S=1 \mid A=1) - Pr(S=1 \mid A=0)}.$$

That is, the CACE (left hand side) is the intention to treat effect divided by the proportion of compliers. The intuition here is that treatment assignment only affects treatment received for one of the subpopulations defined by principal strata – the compliers. Treatment assignment has no impact on treatment received for always takers and never-takers. By the exclusion restriction, treatment assignment therefore has no impact on the outcome for any group other than compliers. Thus, any effect of treatment assignment as measured by ITT must come from compliers. We explore sensitivity to the monotonicity assumption in Chapter 4.

Example 1.5.2 (*Censoring by, or truncation due to, death)*. Suppose we are interested in the causal effect of surgery (A) on quality of life (Y) one year later. However, some subjects die ($S = 0$) within the first year. If $S = 0$, then Y would be missing (and arguably not even well defined). However, treatment might affect the risk of death; hence, conditioning on survivors only would lead to an attenuation of the true treatment effect. For example, a treatment such as surgery could increase the risk of death, but lead to better outcomes for individuals who survive.

Principal stratification deals with the "censoring by death" issue as follows. We again form strata based on the pairs $\{S(0), S(1)\}$: "always survive" ($S(0) = 1, S(1) = 1$), "survive only if untreated" ($S(0) = 1, S(1) = 0$), "survive only if treated" ($S(0) = 0, S(1) = 1$), and "never survive" ($S(0) = 0, S(1) = 0$). Of interest is the *survivor average causal effect* (SACE):

$$\begin{aligned}\text{SACE} &= E\{Y(1,S(1)) - Y(0,S(0)) \mid S(0) = 1, S(1) = 1\} \\ &= E\{Y(1,1) - Y(0,0) \mid S(0) = 1, S(1) = 1\}.\end{aligned}$$

The subpopulation of interest here is the group of people who would survive under either treatment. Identification of the SACE can proceed along similar lines as the identification of CACE for non-compliance. For this subpopulation, the outcome Y is always observed and well defined. For any other principal strata, Y would not be entirely observed and possibly not well defined for any of the treatment groups. We use principal stratification in the setting of truncation due to death in the context of semi competing risks in Chapter 15.

Example 1.5.3 (Mediation). Lastly, we consider the situation where there is a binary intermediate variable; think of this variable as one that occurs after treatment is assigned (A) and before the outcome (Y) and might be affected by treatment and also causally affect the outcome. Researchers are sometimes interested in understanding these pathways from treatment to outcome; for example, consider a behavioral intervention to facilitate smoking cessation. If part of the intervention includes messaging and strategies to encourage healthier behaviors, then an intermediate variable of interest could be the change in exercise. Researchers might want to know about the *direct* versus *indirect* impact of the intervention on smoking cessation rates. Or, as another example, researchers might be interested in knowing how much of the causal effect of smoking on medical expenditures is indirect through the impact of smoking on overall health (see Example 1.5.3).

In this setting, principal strata can be formed based on whether or not the treatment affects this intermediate variable, which is typically referred to as a *mediator* [38]. The first causal effect of interest is known as the "disassociative effect": $E\{Y(1) - Y(0) \mid S(0) = S(1) = s\}$, $s = 0, 1$. Within these principal strata, there is no effect of treatment on the mediator. The treatment effect on the outcome must therefore be directly from treatment itself. This causal estimand is also sometimes referred to as the *principal strata direct effect*. We can also define causal effects for subpopulations of subjects for whom treatment does have impact on the intermediate variable. Associative effects are defined as $E\{Y(1) - Y(0) \mid S(0) = s_0, S(1) = s_1\}$, $s_0 \neq s_1$. Associative effects capture both direct effects of the treatment on the outcome and indirect effects through the effect on the mediator. Unlike other causal mediation methods (described next in Section 1.6), the principal stratification approach to mediation does not result in a decomposition of total effects into the sum of direct and indirect effects.

1.6 Causal Mediation

We now consider causal mediation more generally, not just as a special case of principal stratification. As mentioned in Example 1.5.3, often interest is not simply in the causal relationship between a treatment A and an outcome Y, but also in causal pathways between A and Y through intermediate variable(s) M, which are known *mediators*.

A mediator M is a variable that occurs in the time between the treatment and outcome. It might be affected by treatment and also can affect the outcome. We define $M(a)$ as the value the mediator would take if the treatment were set to a and $Y(a,m)$ to be the value the outcome would take if the treatment were set to a and the mediator were set to m. The potential outcomes can now be defined in terms of both the treatment and the mediator. Here, we will focus on what is known as *natural* direct and indirect effects [39], which are defined as contrasts in $\{Y\{a, M(a')\} : a, a' \in \{0,1\}\}$; for example, $Y\{0, M(0)\}$ is the outcome that would occur if treatment is set to 0 and the mediator takes the value that it would take under no treatment. These potential outcomes are "natural" in the sense that we consider only values of the mediator that could actually have been realized under one of the treatments (c.f. the *controlled* effects defined at the end of this section).

Causal Estimands We can now define the *natural direct effect* (NDE) and *natural indirect effect* (NIE). The NDE of a treatment, given covariates V (possibly a subset of the confounders but not necessarily), is

$$NDE(v) = E[Y\{1, M(0)\} - Y\{0, M(0)\} \mid V = v].$$

Note that the mediator is set to its "natural" value, i.e., the value that it would take under no intervention. Because the only difference between the arguments in the two potential outcomes is the treatment received by the outcome, the NDE isolates the effect of the treatment (not through its impact on the mediator). In other words, the NDE is the effect of treatment if we were able to fix the mediator at its natural value.

The NIE is defined similarly as

$$NIE(v) = E[Y\{1, M(1)\} - Y\{1, M(0)\} \mid V = v].$$

This is the effect of the intervention (with the treatment set to $a = 1$) through the mediator. Note that, if M is binary, then for some subjects we will have $M(1) = M(0)$ while for other subjects $M(1) \neq M(0)$. The key is that what is being contrasted are values of the mediator in response to different treatments. Recall that the goal is to capture the (indirect) effect that treatment is having through treatment's impact on the mediator.

The total effect of treatment on the outcome is

$$TE(v) = E[Y\{1, M(1)\} - Y\{0, M(0)\} \mid V = v].$$

This is just the usual (conditional) average causal effect and is identifiable through standard causal assumptions described earlier. It is easy to see that we can decompose the total effect as the sum of the natural direct and indirect effects:

$$TE(v) = NIE(v) + NDE(v).$$

Note all of the above could also be defined marginally (not conditional on V) by integrating over the distribution of V.

Identifiability The potential outcomes $Y\{1, M(1)\}$ and $Y\{0, M(0)\}$ are observable under the consistency assumption. However, $Y\{1, M(0)\}$ is never observable for a continuous mediator and is either not observable (if $M(0) \neq M(1)$), or we would not know we were observing it (if $M(0) = M(1)$ and $A = 1$) for a binary mediator; such unobservable counterfactuals have been called *a priori* counterfactuals [40]. The primary challenge that is *unique* to mediation is therefore identifying $E[Y\{1, M(0)\} \mid V = v]$.

The most common approach to identification is to make the following sequential ignorability assumptions [41]:

$$\{Y(a', m), M(a)\} \perp\!\!\!\perp A \mid L = \ell, \tag{1.5}$$

$$Y(a', m) \perp\!\!\!\perp M(a) \mid A = a, L = \ell \tag{1.6}$$

for $a, a' \in \{0, 1\}$. Assumption 1.5 is just the standard ignorability assumption (treatment assignment and potential outcomes independent of given confounders). Assumption 1.6 is similar in the sense that it assumes no unmeasured confounding between the mediator and the outcome. The confounders needed in (1.6) could include covariates not needed for (1.5), and all the confounders for both assumptions need to be measured at baseline (as opposed to post-treatment confounders). Under these assumptions, and positivity, we can identify the mean potential outcome using a "mediational" g-formula:

$$E[Y\{1, M(0)\} \mid L = \ell] = \int E\{Y \mid M(1) = m, A = 1, L = \ell\}\, dF_{M|A=0, L=\ell}(m).$$

Notice that on the right-hand side the conditional mean of the outcome has the treatment equal to 1, but we average the mediator over its distribution when the treatment is set equal to 0.

The sequential ignorability assumption is violated when there are confounders for the mediator-outcome relationship that are affected by treatment [42] (post-treatment confounders as noted above). The effects can still be identified in special cases with some additional assumptions [43]. Sensitivity analyses related to this assumption have also been proposed ([44, 45]). See Section 4.2.2 for details.

An alternative causal identifying assumption, called mediator induction equivalence, replaces (1.6) by [45],

$$f[Y\{1,M(0)\} \mid M(0)=m, M(1), X=x] = f[Y\{1,M(0)\} \mid M(0), M(1)=m, X=x].$$

Here X can include confounders as well as other covariates that are not confounders. Both sides of the equation involve distributions of potential outcomes where treatment is set to a and the mediator has value of m, but on the left-hand side the value of m comes from $M(0)$ and on the right-hand side the value of m comes from $M(1)$. So it does not matter whether the mediator value m was induced under treatment or control. Connections between sequential ignorability and mediator-induced equivalence assumptions are made in [40], and connections between the assumptions necessary for principal stratification mediation and full causal mediation can be found in [46, 40].

An alternative approach to mediation analysis uses *controlled* effects, where we consider contrasts in $Y(a,m)$ for differing values of (a,m). These effects take the mediator fixed at particular values of interest rather than fixing the mediator at its natural values [47]. To understand the distinction, consider a subject i for whom $M_i(0) = M_i(1) = 1$. From the natural effects perspective, we would only consider a mediator value of 1 for subject i. For controlled effects, we can imagine setting the mediator to either 1 or 0, even though we would not actually see a value of 0 regardless of what treatment they received. See [48] for recent work on this problem.

1.7 Summary

In this chapter we reviewed causal estimands and identifying assumptions for a variety of common study designs and research scenarios. As can be seen from the identification formulas, causal inference inevitably will require models for observed data distributions and/or their functionals (e.g., conditional means). In Chapters 5-7, we will introduce various Bayesian nonparametric models and show how they can be used to model the observed data distributions or their functionals directly. This allows for posterior inference on the causal estimands of interest without relying on strong parametric assumptions. Further, because causal assumptions are untestable, it is important to understand the degree to which inferential conclusions rely on these assumptions being exactly correct. We will show how causal identifying assumptions can be parameterized in ways that allow for sensitivity analyses to be carried out that capture uncertainty about the assumptions in Chapter 4.

Chapter 2

Overview of missing data

DOI: 10.1201/9780429324222-2

2.1 Introduction

A common occurrence is that the data one intended to collect is not the same as the data that was actually collected. Examples include surveys in which respondents fail to answer all questions, longitudinal clinical trials in which subjects are no longer in the study prior to the collection of all measurements, and observational studies of medical data where certain diagnostic tests are not run on all patients in the study. We discuss some specific examples next.

Example 2.1.1. A longitudinal clinical trial is being conducted to determine if a new drug is effective at reducing the symptoms of acute schizophrenia, as measured using the Positive and Negative Syndrome Scale (PANSS) [49]. Measurements of the PANSS score are scheduled to be collected at fixed times during the study, with the final end point being the change in PANSS score at the end of the study from its baseline measurement. Over the course of the trial, many participants are removed from the study; reasons include (i) adverse reaction to the drug, (ii) progression of symptoms, and (iii) patient withdrawal of consent or self-removal from the study. In addition to these events – which cause the subject to be removed completely – some measurements are missing for reasons that do not result in subjects being removed from the trial (say, a scheduling issue causes a patient to not be available on a particular day), with missingness in these cases not thought to be associated with the unobserved PANSS score.

Example 2.1.1 is typical of longitudinal clinical trials, where the majority of the missingness is due to *dropout*: events occur which remove the subject from the study entirely. Additionally, this study also features *intermittent* missingness, which is any missingness not due to dropout.

In Example 2.1.1, the response is missing while the covariates (which are recorded at baseline) are observed. While somewhat atypical, the opposite can also occur in practice, where there is missingness in covariates but the response is always observed.

Example 2.1.2. Consider a clinical trial in which one collects baseline measurements $X_i = (X_i^e, X_i^h)$ where the components of X_i^e are easy to measure (such as the sex of a subject) while the components of X_i^h are difficult to measure (such as the whole genomic profile of the subject). In this case, we might measure X_i^e for all subjects and measure X_i^h only for a small number of subjects. The response Y_i (such as the occurrence of some disease) is also always observed. In this case, there is missingness in a subset of the covariates, but not in the response.

In addition to being ubiquitous in practice, the missing data problem is in some sense a "dual" problem to the causal inference problem.

Example 2.1.3. In observational studies with a binary treatment (see Chapter 1), we have a binary treatment indicator A and potential outcomes $(Y(0), Y(1))$. If $A = 1$ then we observe the potential outcome $Y(1)$ while, in some sense, $Y(0)$ can be viewed as missing. The main difference between this problem and most other "missing data" problems is that we do not intend, going into the study, to measure both $Y(1)$ and $Y(0)$, as this would be impossible (except in certain designs under strong assumptions).

Because of this duality, many of the fundamental identifiability issues – such as determining when we can consistently estimate treatment effects – arise in both missing data and causal inference problems.

2.2 Overview of Missing Data

We now outline the main concepts that we will use to frame the missing data problem, including carefully discussing what missing data is and the processes that lead to missing data in practice.

2.2.1 What is "Missing Data?"

In essence, missing data is any data that we intended to collect, but in fact did not. In many situations, it is easy to unambiguously define the *full data* we intended to collect.

Example 2.2.1. Consider a list of job applicants for an academic position who all fill out a survey upon applying. Each is asked to identify their race, but some decline to answer the question. In this case, the full data consists of each applicant's race. The observed data will be the races of the applicants who answered the question, and the missing data will be the races of the applicants who did not answer the question.

Another example occurs in the schizophrenia study (Example 2.1.1).

Example 2.2.2. Consider an individual who drops out of the study due to withdrawal of consent. We define the full data as PANSS scores at each planned measurement time. The missing data here is the PANSS scores that we would have measured if the subject had not dropped out.

In this example, we are implicitly defining the missing data as the measurement that would have been obtained after dropout if we had actually measured the patient, regardless of whether they stopped taking the treatment or had switched to a different treatment; this is the typical definition of the missing data and leads to the estimation of the so-called *intention-to-treat* (ITT) effects. However, we might be more interested in the measurements that would have been obtained after dropout, *assuming the patient remained compliant* with the treatment (cf. Example 1.5.1). Such a response would be a counterfactual as defined in Chapter 1. It is important to be explicit about how the missing data is defined, as in the above example both the effect of the treatment and its interpretation depend on this; an analysis can be valid for one interpretation of the missing data but not the other. This is tied to defining the estimand of interest, as described in Section 2.3.

Another complication occurs when missing responses are not well defined, as in the following example.

Example 2.2.3. In cancer treatments, there is often a trade-off between how long a patient will live and their quality of life (QOL) while alive. Consider a study of cancer patients that aims to determine the effect of a treatment on both the survival and QOL of patients. Over the course of the trial, QOL is measured on a regular basis up to the time the individual dies. QOL after death is not well defined (or undefined).

The distinction between "missing" and "undefined" is important because, using the techniques we will discuss, it would be possible to treat the unmeasured QOL scores after death as "missing" in an analysis. This is not generally recommended, and (as discussed in Example 1.5.2 in Chapter 1) other techniques developed specifically for censoring by death problems should be used [50, 51].

2.2.2 *Full vs. Observed Data*

Following [52], we differentiate between the *full data*, the *observed data*, and the *full data response*. The full data refers to all measurements that we intended to collect, including the response of interest, missing data indicators, covariates of direct interest, and *auxiliary covariates* that are used to assist in controlling for missingness. The observed data is the component of the full data that was in fact measured, while the full data response refers to the response portion of the full data.

Example 2.2.4. Consider the schizophrenia clinical trial in Example 2.1.1. In this study the full data consists of

- The PANSS score for each individual at each time in the study;
- A binary variable indicating whether the PANSS score was recorded at each time;
- A variable indicating which treatment the individual was assigned (treatment, placebo, or active control);
- Baseline covariates, such as age and sex of the subject.

The full data response consists only of the PANSS score at each time. The observed data consists of all the quantities described above that were actually recorded (i.e., the measurements that were not missing).

2.2.3 *Notation and Data Structures*

We adopt the following notation. Let n denote the number of units on which we take measurements. For simplicity, we will assume the number of measurements intended to be collected per individual, J, is constant. The full data consists of J-dimensional response vectors Y_i, p-dimensional covariates of interest X_i (including any treatments), and missing data indicators $R_i \in \{0,1\}^J$ such that $R_{ij} = 1$ or 0 according to whether Y_{ij} is observed or not ($i = 1, \ldots, n$). Additionally, we may consider D-dimensional covariates V_i that are themselves subject to missingness; these covariates might be of interest or might be included so as to make certain assumptions about the missingness more plausible. For simplicity, we will subsume V_i into Y_i. The full data distribution is assumed to have the likelihood

$$L(\omega) = \prod_{i=1}^{n} f(y_i, r_i \mid x_i, \omega),$$

The parameter ω indexes the full data distribution and is potentially infinite dimensional. The density $f(y_i, r_i \mid x_i, \omega)$ is the density of a single replication of (Y_i, R_i) conditional on X_i and ω. We let ω_y denote the components of ω that parameterize the marginal distribution of Y_i given X_i, which we write $f(y_i \mid x_i, \omega_y)$. We also let ψ denote the part of ω that parameterizes $f(r_i \mid y_i, x_i, \omega)$ and let θ denote the part of ω that parameterizes $f(y_i \mid r_i, x_i, \omega)$.

To write down the observed data distribution, it is helpful to introduce a notation that picks out the missing and observed components of Y_i. Given any binary vector r, we $Y_{ir} = (Y_{ij} : r_j = 1)$ and $Y_{i-r} = (Y_{ij} : r_j = 0)$. So, for example, Y_{iR_i} and Y_{i-R_i} denote the observed and missing components of Y_i, respectively

For longitudinal settings ($J > 1$), it will be useful to introduce additional notation and terminology. Missingness is said to be *monotone* if $R_{ij} = 0$ implies $R_{i,j+1} = 0$ (that is, missingness is due to *dropout*). In this case, we can summarize the missingness with $S_i = \sum_{j=1}^{J} R_{ij} = \max\{j : R_{ij} = 1\}$, which we refer to as the *follow-up time*.

This notation does not quite capture missingness in complete generality. For example, when the times at which observations are measured vary across subjects, we might instead wish to model Y_i as a continuous-time stochastic process $Y_i(t)$, which is observed at times $t_1, \ldots, t_J$. In this case, even defining what one means by the data being "missing" is more difficult, as clinical studies typically do not intend to measure the entire stochastic process $Y_i(t)$.

2.2.4 *Processes Leading to Missing Data*

Missing data can be caused in a variety of different ways.

- Missingness might be by design; for example, there may be limited resources for following up on subjects, with missingness "generated" via a known mechanism (e.g., randomly choosing subjects to follow up on).
- Patients might withdraw consent or be removed from a study due to a lack of efficacy of treatment.
- Individuals might decline to answer certain survey questions, such as their gender or race, for a variety of reasons (possibly due to their gender or race).
- Among users of online rating websites such as `Yelp!`, users might only rate services that they feel strongly about; in this case, the rating a user would have provided had they rated a given service is often regarded as missing [53]. In this case, most ratings are conceptualized as missing because most users do not rate most services.

Missingness can also occur for different reasons within the same setting, in which case we may wish to incorporate the reason for missingness into the analysis. For example, one subject in a clinical trial might be missing due to lack of efficacy, while another may have died for a reason unrelated to the study; we might wish to treat these two reasons separately because the former is likely related to the response while the latter is not.

For many (but not all) longitudinal settings, missingness is monotone. For example, when a patient removes consent for the study, we cease measuring the response from that point on. Dropout is much easier to handle than intermittent missingness because we only need to model the follow-up time S_i (which takes J unique values) rather than R_i (which takes 2^J unique values). Monotone missingness can also simplify the identification of estimands; for example, the missing at random (MAR) assumption, which we discuss later, becomes very simple when missingness is monotone.

Because of its simplicity, monotone missingness has seen much more progress in the literature than non-monotone missingness. In recent years, however, substantial progress has also been made in the non-monotone settings [54, 55, 56]. In this book, we develop methods for both monotone and non-monotone missingness, as well as forms of missingness that are in some sense intermediate between the two (monotone with intermittent non-monotone missingness).

2.3 Defining Estimands with Missing Data

In Chapter 1, careful attention was given to defining the causal effects (estimands) of interest. Importantly, causal effects like the average treatment effect (ATE; under perfect compliance in randomized trials, this is the CACE) are defined in a model-free manner, which allows for the same estimand to be considered even if we change the modeling strategy or parametric assumptions used. The missing data problems we consider warrant a similar level of attention. For example, it does not matter whether we model $E(Y_{ij} \mid X_i = x, \omega)$ using a linear regression, a generalized additive model, or using Bayesian nonparametrics; the estimand $E(Y_{iJ} - Y_{i1} \mid \omega)$ is well defined regardless. By contrast, we will not focus on estimands whose definition is contingent upon a correctly specified model (such as a regression coefficient or MSMs for causal inference (cf. Section 1.4)).

The benefits of explicitly defining the estimands of interest are emphasized in reports from the National Research Council and the International Conference on Harmonization [57, 58]. The International Conference on Harmonization identifies the following key steps to defining a target estimand:

1. Define the population to be used to answer the scientific questions of interest.
2. Define the response variable (endpoint) of interest that would ideally be collected, including how these measurements are defined when the response is missing (this was discussed in Section 2.2.1).
3. Define a population-level quantity to address the scientific question, which will be estimated based on the response.

We mainly focus here on Steps 2 and 3. For example, suppose that $X_i = (A_i, W_i)$ where A_i is a treatment of interest in a randomized clinical trial and W_i is a vector of baseline covariates. A natural approach to assess the treatment is to estimate the treatment effect

$$\Delta = E(Y_{iJ} - Y_{i1} \mid A_i = 1, \omega) - E(Y_{iJ} - Y_{i1} \mid A_i = 0, \omega). \tag{2.1}$$

Even with this definition, the interpretation of Δ is contingent on how we define the missing data. We discuss several interpretations of Δ in a typical missing data situation next.

Example 2.3.1. Recall the schizophrenia clinical trial described in Example 2.1.1, where the PANSS score Y_{ij} is intended to be measured at times $t_1, \dots, t_J$. Suppose that, due to a lack of efficacy, subject i is removed from the study and an alternative treatment is applied between time t_{j-1} and time t_j. We typically assume Y_{iJ} is the PANSS score the patient has at the conclusion of the study, time t_J. As such, the definition of Δ takes into account the possibility that a patient might be removed from the study and assigned a new treatment. This corresponds to the standard "intention to treat" (ITT) effect, which measures the causal effect of the randomization procedure rather than the effect of compliance with the treatment. As discussed before, if Y_{ij} is taken to instead be the hypothetical (counterfactual) PANSS score we *would have* recorded had the patient remained on study and remained compliant with the treatment, Δ would instead represent a causal effect (in which case, we would instead prefer the causal notation from Chapter 1). When Δ represents the ITT effect, it would be possible to estimate Δ by continuing to measure the subject after they have been removed from the study; on the other hand, if Δ represents a causal effect, this is not possible and we would require additional causal assumptions introduced in Chapter 1.

Examples of possible types of estimands are given below.

- **Treatment policy effects** are defined with respect to the treatment regime assigned. For example, if a treatment policy was compared to a control arm, and no individuals on the treatment arm took the treatment, we might observe no treatment effect (even though the treatment could be very effective); such effects are clearly tied to the compliance pattern of the population under study. Additionally, if a subject is reassigned to a new treatment due to a lack of efficacy, that effect would be included as an effect of the whole policy. ITT effects are treatment policy effects: treatment effects are defined based on the group to which an individual was randomized, regardless of compliance or dropout.
- **Hypothetical effects** are causal effects that are defined with respect to the response that would have been observed had the subject complied with the assigned treatment despite being missing. For example, if a subject is assigned an alternative treatment due to a lack of efficacy, a hypothetical effect would be defined in terms of the response that would have been observed had the individual remained on the assigned treatment and not been given the alternative treatment. We discussed assumptions needed to estimate such effects in terms of principal stratum estimands in Chapter 1, Section 1.5.

As discussed above, different (numbers of) assumptions are needed to identify different estimands. The choice is often a trade-off between the quantities of interest and the assumptions required to identify them. Treatment policy effects can be estimated directly, provided that the response is measured even after the subject is removed from the study, but only measure the effect of being randomized to the treatment arm rather than the effect of the treatment itself. And if the response is not measured post-dropout (which is often the case), assumptions need to be made about the reasons for missingness and whether the missingness depends on the value of the unobserved response (see the next section). Hypothetical effects are arguably more clinically relevant, but require additional untestable causal assumptions to estimate (see Chapter 1).

2.4 Classification of Missing Data Mechanisms

In seminal work of Rubin [59], it was recognized that (loosely speaking) the *reason* for the missingness determines whether an analysis strategy is appropriate or not. Rubin classified missing data into three categories: missing completely at random (MCAR), missing at random (MAR), and missing not at random (MNAR, sometimes written NMAR). These categories are defined in terms of the missing data mechanism, defined next.

Definition 2.4.1 (Missing Data Mechanism). The *missing data mechanism* is the distribution of R_i given (X_i, Y_i); we will also refer to the model $f(r_i \mid y_i, x_i, \psi)$ as the missing data mechanism. In words, the missing data mechanism denotes the probability of observing $[R_i = r]$ given the (potentially unobserved) values of Y_i and X_i.

The importance of the MCAR/MAR/MNAR taxonomy is that it informs what analysis strategies yield valid inferences, as well as whether one can identify the desired effects.

2.4.1 *Missing Completely at Random (MCAR)*

The simplest form of missingness is *missing completely at random* (MCAR). Heuristically, the missing data is MCAR when the reason the data is missing is entirely unrelated to the observable data.

Definition 2.4.2 (Missing Completely at Random). We say that the missing data is MCAR at (r_i, y_i, x_i) if, for all values of ψ, $f(R_i = r_i \mid y_i, x_i, \psi)$ does not depend on (y_i, x_i); that is, for any values of (y_i, x_i) and $(\widetilde{y}_i, \widetilde{x}_i)$, we have

$$f(R_i = r_i \mid y_i, x_i, \psi) = f(R_i = r_i \mid \widetilde{y}_i, \widetilde{x}_i, \psi).$$

In words, the missing data is MCAR if the missingness pattern cannot be predicted from the response. MCAR typically only holds in cases where the missingness is by design (see the next example).

Example 2.4.1. A longitudinal study records the effect of a new diet on entering college freshmen. Participants report their sex X_i and weight Y_{i1} at the beginning of their freshman year. Due to budgetary constraints, only a subset of freshmen have their weight Y_{i2} recorded at the end of the year, and this subset is randomly sampled from the original group of students. Because missingness of Y_{i2} is independent of both Y_i and X_i, the missing values of Y_{i2} are MCAR.

Remark 1. Occasionally the term MCAR is used for the assumption that $f(R_i = r \mid y_i, x_i, \psi)$ does not depend on y_i (but might depend on x_i). We refer to this assumption as MCAR *conditional on covariates*.

Remark 2. An interesting feature of MCAR missingness is that a *complete case* analysis produces valid inferences – that is, we are free to restrict attention to the Y_i's such that $R_{ij} = 1$ for all j. For instance, the above example implies that

$$E(Y_{i2} \mid R_{i2} = 1) = E(Y_{i2}).$$

Hence, any consistent estimator of $E(Y_{i2} \mid R_{i2} = 1)$ is also a consistent estimator of $E(Y_{i2})$. The same argument holds for any functional of the distribution of (Y_i, X_i). However, even though we can obtain consistent estimators, there will typically be a loss of efficiency if we exclude the incomplete cases.

Remark 3. MCAR is quite strong in practice and is unrealistic in the majority of settings we consider. Typically, MCAR is only known to hold when the researchers have control over which responses will be missing (i.e., when missingness is by design as noted above), which is rarely the case. Our main interest in MCAR is twofold: (i) it serves as motivation for MAR and MNAR missingness and (ii) to make the point that it is essentially the *only* practically occurring situation in which a complete case analysis remains valid for estimating $E(Y_i)$.

2.4.2 *Missing at Random (MAR)*

An attractive feature of MCAR is that, because a complete-case analysis is valid, it does not require us to estimate the missing data mechanism to obtain valid inferences. As MCAR is a very strong assumption, one might seek other conditions that are (i) more realistic in practice, but (ii) still do not require the analyst to model the missing data mechanism. A first step in this direction is the MAR assumption. As we show in Section 2.5, the MAR assumption, along with a certain prior independence assumption, leads to *ignorability*, which allows likelihood-based inference to proceed without explicitly specifying the missing data mechanism.

Definition 2.4.3 (Missing at Random)**.** We say that the missing data is *missing at random* given observed data (r_i, y_{ir_i}, x_i) if, for all values of ψ, $f(R_i = r_i \mid y_i, x_i, \psi)$ does not depend on y_{i-r_i}; that is, for any values y_i and $\widetilde{y}_i$ of the response such that $y_{ir_i} = \widetilde{y}_{ir_i}$, we have

$$f(R_i = r_i \mid y_i, x_i, \psi) = f(R_i = r_i \mid \widetilde{y}_i, x_i, \psi).$$

In words, the probability of some missing data pattern $[R_i = r_i]$ depends only on the data which would be observed under that pattern.

One heuristic way of understanding MAR is that the probability of a response being missing does not depend on the value of that response, but is allowed to depend on the other (observed!) responses. When the response $Y_i = (Y_{i1}, \ldots, Y_{iJ})$ is a longitudinally observed vector subject to monotone missingness, MAR has a highly appealing interpretation: the probability of dropout at time j depends on Y_i only through the past outcomes $(Y_{i1}, \ldots, Y_{i,j-1})$ and is independent of the present (Y_{ij}) and future $(Y_{i,j+1}, \ldots, Y_{iJ})$ outcomes.

Example 2.4.2 (A Missing at Random Missing Data Mechanism)**.** Consider a longitudinal study in which subjects have the value of a biomarker Y_{ij} scheduled to be collected at times $t_1, \ldots, t_J$. Depending on the value of the biomarker at time t_{j-1}, the doctor may decide to remove the patient from the study due to a lack of efficacy of their treatment, resulting in $Y_{ij}, \ldots, Y_{iJ}$ being missing. Because the missingness at time t_j depends only on the recorded value $Y_{i1}, \ldots, Y_{i,j-1}$ (which form the entire basis for the doctor's decision), the probability of dropout (here removal by doctor from the study) at time t_j depends only on the observed data; hence, the missing data is MAR.

Inference based on MAR missingness is almost always more plausible in practice than MCAR missingness (and more efficient). For better or worse, MAR is very frequently assumed in practice both to simplify the modeling and to identify the effects one wishes to estimate (see Section 2.5). Nevertheless, we feel that in the absence of strong subject-matter expertise, it should be used with caution. Like MCAR, MAR should be viewed as the exception rather than the rule. In Section 2.6.4 we show how MAR can be used as a starting point for less-restrictive analyses.

2.4.3 *Missing Not at Random (MNAR)*

The last category for missing data is *missing not at random*, which encompasses all settings in which the missing data is not MAR.

Definition 2.4.4 (Missing Not at Random)**.** We say that the missing data is *missing not at random* given the observed data (x_i, y_{ir_i}, r_i) if the missing data is not MAR; that is, there exists a ψ and some response values y_i and $\widetilde{y}_i$ with $y_{ir_i} = \widetilde{y}_{ir_i}$ such that

$$f(R_i = r_i \mid y_i, x_i, \psi) \neq f(R_i = r_i \mid \widetilde{y}_i, x_i, \psi).$$

Our view is that, lacking strong subject-matter knowledge, MNAR should be viewed as the default assumption. This is not to diminish the value of analyses which assume MAR, but rather to make the point that an MAR analysis should be viewed as a starting point rather than a final product. Below we describe a common setting in which MNAR missingness can be expected to hold.

Example 2.4.3. In Example 2.4.2, suppose that instead of the doctor making a decision based on $Y_{i,j-1}$ to remove the subject from the study, the subject decides drop out at time t_j based on a decline in their current (unobserved) health. As their unobserved health status Y_{ij} is likely correlated with R_{ij}, even after controlling for their observed medical history $Y_{i1}, \ldots, Y_{i,j-1}$, the missing data is MNAR.

Remark 4. *Auxiliary covariates* are sometimes included in the missing data mechanism to either make MAR more realistic or to move MNAR closer to MAR. For inference on the model of interest, $f(y|x;\omega)$, the auxiliary covariates need to be integrated out [52, 60].

2.4.4 *Everywhere MAR and MCAR*

The definitions of MCAR, MAR, and MNAR given above coincide with the original definitions given in [59]. Several alternative definitions have been proposed in the literature (see [61] which provides a history of the confusion surrounding MAR). Definitions 2.4.2, 2.4.3, and 2.4.4 impose conditions *only at the realized values* (y_i, r_i, x_i)

and *not* on the random variables (Y_i, R_i, X_i). For example, we note that under Definition 2.4.2, if Y_i is completely observed for every subject then MAR is guaranteed to hold, irrespective of what the missing data mechanism is.

Many authors use definitions which impose conditions *for all values* of (y_i, r_i, x_i) [61]. This leads, for example, to the following non-equivalent definition of MAR and MCAR.

Definition 2.4.5. We say that the missing data mechanism is *everywhere missing at random* if, for all values of (r_i, y_i, x_i), we have

$$f(R_i = r_i \mid y_i, x_i, \psi) = f(R_i = r_i \mid \widetilde{y}_i, x_i, \psi)$$

for all $\widetilde{y}_i$ such that $y_{ir_i} = \widetilde{y}_{ir_i}$. The missing data mechanism is *everywhere missing completely at random* if this holds for all $(r_i, y_i, x_i, \widetilde{y}_i)$, even if $y_{ir_i} \neq \widetilde{y}_{ir_i}$; equivalently, missingness is MCAR everywhere if (Y_i, X_i) is independent of R_i.

Everywhere MAR and everywhere MCAR are sometimes easier to conceptualize than MAR and MCAR; in particular, whether the data is MAR or MCAR can depend on the particular values of (r_i, x_i, y_{ir_i}) which were observed. That is, everywhere MAR is a property of the missing data mechanism alone, whereas traditional MAR depends on both the missing data mechanism *and* the observed data.

2.4.5 *Identifiability of Estimands under MCAR, MAR, and MNAR*

Beyond the implications for modeling the missing data mechanism, whether the missing data is MCAR/MAR/MNAR has important implications for whether the estimands of interest can be consistently estimated at all. For simplicity, we assume that MCAR and MAR correspond to their "everywhere" variants. To illustrate, suppose that we are interested in the mean effect $\eta = E(Y_{iJ} \mid \omega)$. Under MCAR, we can consistently estimate this parameter using

$$\widehat{\eta} = \frac{\sum_i R_{iJ} Y_{iJ}}{\sum_i R_{iJ}}.$$

As $n \to \infty$, the law of large numbers implies $\widehat{\eta} \to E(Y_{iJ} \mid R_{iJ} = 1, \omega)$, which is equal to η under MCAR. It turns out that η can also be consistently estimated under MAR using an *inverse probability weighting* (IPW) modification of this estimator.

Example 2.4.4. Suppose that Y_i is a scalar and that missingness is everywhere MAR, which implies that R_i is conditionally independent of Y_i given X_i. Then, by iterated expectation

$$\begin{aligned} E\left(\frac{R_i Y_i}{f(R_i = 1 \mid X_i, \omega)} \mid \omega\right) &= E\left(E\left[\frac{R_i Y_i}{f(R_i = 1 \mid X_i, \omega)} \mid Y_i, X_i, \omega\right] \mid \omega\right) \\ &= E\left(\frac{f(R_i = 1 \mid Y_i, X_i, \omega) Y_i}{f(R_i = 1 \mid X_i, \omega)} \mid \omega\right). \end{aligned}$$

Because the missing data is everywhere MAR, we have $f(R_i = 1 \mid Y_i, X_i, \omega) = f(R_i = 1 \mid X_i, \omega)$ so that

$$E\left(\frac{R_i Y_i}{f(R_i = 1 \mid X_i, \omega)} \mid \omega\right) = E(Y_i \mid \omega).$$

Hence $\eta = E(Y_i \mid \omega)$ is identified. If $\widehat{f}(R_i = 1 \mid X_i = x)$ is a consistent estimator of $f(R_i = 1 \mid X_i = x, \omega)$, then it is reasonable to expect that

$$\widehat{\eta} = \frac{1}{N} \sum_i \frac{R_i Y_i}{\widehat{f}(R_i = 1 \mid X_i)} \tag{2.2}$$

is a consistent estimator of $E(Y_i \mid \omega)$. As the argument above can be repeated with Y_i replaced with an arbitrary transformation $g(Y_i)$ with finite expectation, it follows that the entire distribution of $[Y_i \mid \omega]$ is identified.

This example shows heuristically that, under minor (but non-trivial) assumptions, the entire distribution of Y_i can be estimated consistently under MAR. The argument in the above example can be made rigorous, but requires some additional assumptions; an important assumption made for theoretical purposes is that $f(R_i = 1 \mid X_i = x, \omega)$ is bounded away from 0 on the support of X_i (a type of positivity assumption).

The resulting estimator (2.2) is called the *Horvitz-Thompson (HT) estimator* [62]; a common issue with HT estimators is that they may be unstable when the positivity assumption is nearly violated, as this will result in small denominators in (2.2).

The situation under MNAR missingness is quite different, as under MNAR there do not exist consistent estimators of $E(Y_i \mid \omega)$ in general.

Example 2.4.5. Suppose $Y_i = (Y_{i1}, Y_{i2})$, $R_{i1} \equiv 1$, and $R_{i2} \sim \text{Bernoulli}(p)$, with no covariates. Suppose further that

$$Y_i \sim N(\mu_r, \Sigma_r) \qquad \text{given } R_i = r,$$

where

$$\mu_r = (\mu_{r,1}, \mu_{r,2})^{\mathrm{T}} \qquad \text{and} \qquad \Sigma_r = \begin{pmatrix} \Sigma_{r,11} & \Sigma_{r,12} \\ \Sigma_{r,21} & \Sigma_{r,22} \end{pmatrix}.$$

The parameters μ_1, Σ_1 can be estimated from the observations with $R_{i2} = 1$ while $\mu_{0,1}, \Sigma_{0,11}$ can be estimated from the observations with $R_{i2} = 0$ using the observations of Y_{i1}. The distribution of the observed data Y_{iR_i}, however, does not depend on $(\mu_{0,2}, \Sigma_{0,12}, \Sigma_{0,22})$, and therefore any statistic depending on the observed data will also not depend on these parameters. To see this, note that the distribution of Y_{iR_i} has density

$$f(Y_{ir_i} = y_{ir_i}, R_i = r_i \mid X_i, \omega) = \begin{cases} (1-p) N(y_{i1} \mid \mu_{0,1}, \Sigma_{0,11}) & \text{if } r_{i2} = 0, \\ p N(y_i \mid \mu_1, \Sigma_1) & \text{if } r_{i2} = 1. \end{cases}$$

Consequently, any estimator based on the observed data $(Y_{1R_1}, \dots, Y_{nR_n})$ will not depend on the parameters $(\mu_{0,2}, \Sigma_{0,12}, \Sigma_{0,22})$, implying that consistent estimation of these parameters is impossible.

The above example implies that consistent estimation of the distribution of Y_i is impossible under MNAR missingness, even in cases where we make rather strong distributional assumptions about Y_i (in this case, we assumed Y_i is normally distributed given R_i). Identifying the distribution of Y_i requires either very strong distributional assumptions (see Section 2.6.1 where identification is obtained in a class of "selection models") or further restrictions on the type of MNAR missingness we allow. This is in stark contrast to the situation for MAR missingness, where the distribution of Y_i can be estimated using essentially nonparametric models.

2.4.6 *Deciding between MCAR, MAR, and MNAR*

As our ability to construct consistent estimators of the targeted estimands depends on whether the missing data is MCAR, MAR, or MNAR, one might hope to use the data itself to decide which assumption to make. Specifically, one might hope to test one of the following two hypotheses:

- H_0 : the missing data is MAR versus H_1 : the missing data is MNAR.
- H_0 : the missing data is MCAR versus H_1 : the missing data is not MCAR.

An important limitation on our ability to test these hypotheses is that any MAR model admits many different MNAR models with the same distribution for the observed data [63]. In general, the full data distribution factors as

$$f(y_i, r_i \mid x_i, \omega) = f(Y_{ir_i} = y_{ir_i}, R_i = r_i \mid x_i, \omega) \tag{2.3}$$
$$\times f(Y_{i-r_i} = y_{i-r_i} \mid Y_{ir_i} = y_{ir_i}, R_i = r_i, x_i, \omega), \tag{2.4}$$

where (2.3) is the density of the observed data (Y_{iR_i}, R_i) and (2.4) is the density of the missing data Y_{i-R_i} given the observed data; (2.4) is sometimes called the *extrapolation distribution* [52]. Hence, given an MAR model, we can generate an infinite number of MNAR models with the same observed data distribution by varying (2.4) and leaving (2.3) fixed. This implies that the two hypotheses above are untestable in the following sense: let $\mathscr{M}$ denote either the family of MCAR or MAR complete data generating mechanisms. Then, for every $f \in \mathscr{M}$, there exists a $f' \in \mathscr{M}^c$ such that the power of any test at f' is equal to its Type I error rate at f.

Readers experienced with missing data may note that the statement above seems to contradict the existence of tests that purport to check the MCAR assumption, such as Little's MCAR test [64]. There is no contradiction, however, as it *is* possible to test the hypothesis H_0 : the missing data is MCAR against the alternative H_1 : the missing data is MAR; unlike testing against MNAR, there are differences between MAR and MCAR missingness that can be assessed from the data (for example, we could check whether the fully observed covariates X_i or observed outcomes are predictive of missingness).

2.5 Ignorable versus Non-Ignorable Missingness

Beyond identifying the estimands of interest, MAR is also important because it is required in order for the missing data mechanism to be *ignorable*. Under ignorability, likelihood-based inference for the full data response model does not require explicit specification of the missing data mechanism.

Definition 2.5.1 (Ignorability). Let $\omega = (\omega_y, \psi)$ where ω_y parametrizes $f(y \mid x, \omega)$ and ψ parametrizes $f(r \mid y, x, \omega)$. The missing data mechanism is *ignorable* if

1. The missing data is MAR
2. The parameters ψ and ω_y are "independent."

Remark 5. For Bayesian inference, the condition that ψ and ω_y are "independent" means that ψ and ω_y are independent with respect to their joint prior distribution Π; see Chapter 3. For non-Bayesian inference, independence is interpreted differently: ψ and ω_y must be *variationally independent*, meaning that the possible values of ψ do not depend on the particular choice of ω_y, and vice versa (more formally, the parameter space of (ψ, ω_y) must be a product of the respective parameter spaces of ψ and ω_y).

The name "ignorability" comes from the fact that the likelihood for (ψ, ω_y) factors into a term for ψ and a term for ω_y. Hence, for the purposes of likelihood-based inference about ω_y, the term for ψ is a constant and can be "ignored."

Definition 2.5.2. Given a full data model $f(y, r \mid x, \omega)$ with observed data (Y_{iR_i}, R_i) for $i = 1, \dots, n$, the *observed data likelihood function* is given by

$$L_{\text{obs}}(\omega) = \prod_{i=1}^{n} \int f(y_i, r_i \mid x_i, \omega)\, dy_{i-r_i}.$$

Here, the notation $\int g(y)\, dy_{-r}$ is understood to integrate over the coordinates of y for which $r_j = 0$ (with the coordinates for which $r_j = 1$ being fixed values).

Theorem 2.5.1 (Ignorability and Likelihood-based Inference). *Let*

$$f(y, r \mid x, \omega) = f(y \mid x, \omega_y)\, f(r \mid y, x, \psi)$$

be a full data model with observed data (Y_{iR_i}, R_i) for $i = 1, \dots, n$. Suppose that the missing data is MAR. Then the observed data likelihood $L_{obs}(\omega_y, \psi)$ factors as

$$L_{obs}(\omega_y, \psi) = L_y(\omega_y)\, L_{r|y}(\psi)$$

where

$$L_y(\omega_y) = \prod_{i=1}^{n} f(Y_{ir_i} = y_{ir_i} \mid x_i, \omega_y) \quad \textit{and}$$

$$L_{r|y}(\psi) = \prod_{i=1}^{n} f(R_i = r_i \mid Y_{ir_i} = y_{ir_i}, x_i, \psi).$$

Theorem 2.5.1 implies that we can compute the maximum likelihood estimator and Fisher information of ω_y using only $L_y(\omega_y)$. This requires only specification of the full data response model; the same answer will be obtained *regardless of the exact form of missing data mechanism.* This is the sense in which the missing data mechanism is "ignored." Ignorability has the consequence that ω_y and ψ are independent in any posterior distribution, implying that the choice of missing data mechanism is irrelevant for Bayesian inference in this case as well.

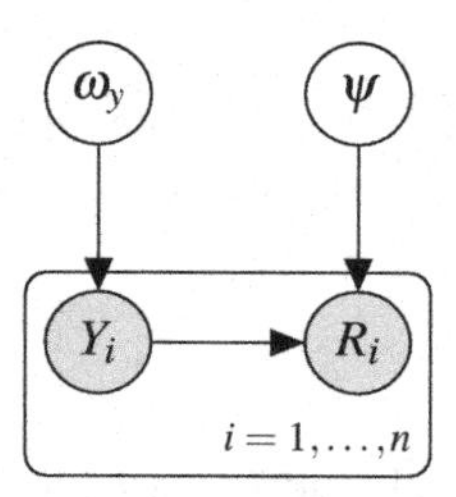

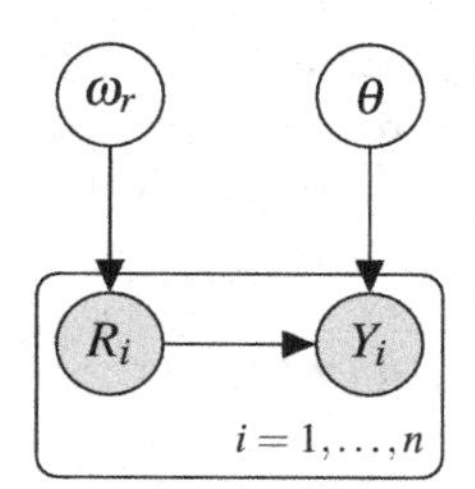

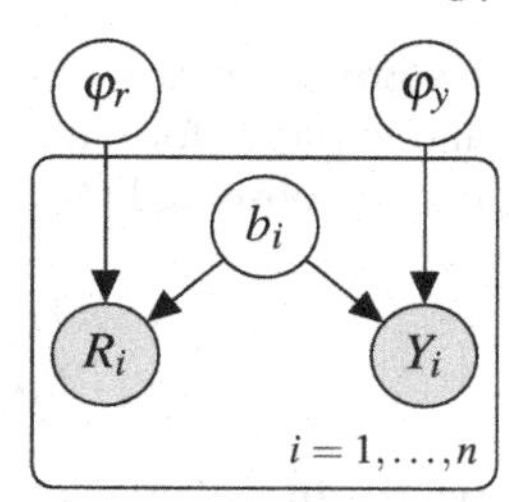

Figure 2.1 *Graphs depicting the different approaches to specifying nonignorable models. Left: the selection model factorization. Middle: the pattern mixture factorization. Right: the shared parameter model with independence given a latent variable b_i.*

Remark 6. While MAR and ignorability are clearly related, the assumption that ψ and ω_y must be independent is crucial. We will consider non-ignorable MAR models as part of a sensitivity analysis in Chapter 4. Further, as pointed out by [65], likelihood-based inference under ignorability seems to be at odds with the fundamental importance of propensity score-based methods, which are commonly used in both causal inference and missing data problems: if the missing data mechanism is ignorable, then likelihood-based methods *cannot* depend on the propensity score. Estimators like the one considered in Example 2.4.4 violate the likelihood principle if the missing data mechanism is ignorable and therefore would be precluded from a fully Bayesian analysis (at least if ignorability is assumed).

2.6 Types of Non-Ignorable Models

When missingness is non-ignorable, the analyst must specify a joint model $f(y,r \mid x,\omega)$ for Y_i and R_i. A first step for many approaches is to decide on a factorization of $f(y,r \mid x,\omega)$ to work with; for example, we can factor the joint model as either $f(y \mid x,\omega_y)\,f(r \mid y,x,\psi)$ or as $f(y \mid r,x,\theta)\,f(r \mid x,\omega_r)$. We refer to the former as the *selection model factorization* and the latter as the *pattern mixture factorization*. It is of course true that any joint model can be expressed in terms of both the selection model and pattern mixture model factorizations; however, adopting (say) the selection model factorization as a starting point for model specification will bring different advantages/disadvantages than starting from the pattern mixture model factorization.

We will consider three different types of models: selection models, pattern mixture models, and shared parameter models. A schematic illustrating the differences between these approaches is given in Figure 2.1. The graphs in Figure 2.1 depict how the complete data distribution is factored in the model; for example, the arrow $Y_i \to R_i$ indicates that we model $f(y \mid x,\omega_y)$ and $f(r \mid y,x,\psi)$.

2.6.1 *Selection Models*

Selection models are based on specifying separate models for the full data response $f(y \mid x,\omega_y)$ and the missing data mechanism $f(r \mid y,x,\psi)$.

Example 2.6.1. Consider the schizophrenia clinical trial described in Example 2.1.1, and suppose for simplicity that there are no covariates and that Y_i is univariate and may be observed ($R_i = 1$) or not ($R_i = 0$). A selection model in this setting is

$$
\begin{aligned}
Y_i &\sim N(\mu, \sigma^2), \\
R_i &\sim \text{Bernoulli}\{\Phi(\gamma + \lambda y)\} \qquad \text{given } Y_i = y,
\end{aligned}
$$

where $\Phi(x)$ denotes the distribution function of a standard normal random variable. This is known as the Heckman selection model [66]. In this case, $\omega = (\omega_y, \psi)$ where $\omega_y = (\mu, \sigma^2)$ and $\psi = (\gamma, \lambda)$.

Selection models have the benefit of directly modeling $f(Y_i \mid X_i = x, \omega_y)$, which is usually of primary interest. Unfortunately, parametric selection models have a tendency to identify all of the parameters in the model. Because of this, selection models are sometimes used to assess the MAR assumption. In Example 2.6.1, MAR corresponds to $\lambda = 0$, and it can be shown that λ is identified in this model. Hence one could form a test of MAR by testing $\lambda = 0$ against the alternative $\lambda \neq 0$. While at first glance this might be seen as a virtue of selection models, it is important to remember that equations (2.3) and (2.4) imply that it is not possible in general to test for MAR. To achieve identification, parametric selection models impose parametric assumptions which are themselves untestable. In Example 2.6.1, the distribution of Y_i given $R_i = 1$ is given by

$$
f(y \mid R_i = 1, \omega) \propto N(y \mid \mu, \sigma^2)\,\Phi(\gamma + \lambda y),
$$

This is a fairly simple parametric family in four parameters, and given samples from such a distribution, we can estimate them. The structure of this model is similar to a skew-normal distribution [67] with λ playing the role of a skewness parameter. Intuitively, one can learn λ from the skewness of the distribution $[Y_i \mid R_i = 1]$ and $\lambda = 0$ would correspond to a normal model. The distribution of $[Y_i \mid R_i = 0]$ (i.e., the extrapolation distribution) is a function of the same parameters:

$$
f(y \mid R_i = 0, \omega) \propto N(y \mid \mu, \sigma^2)\,\{1 - \Phi(\gamma + \lambda y)\},
$$

so it is identified even though we never observe Y when $R = 0$.

The perspective we take is that it is better to assess directly the sensitivity of our results to uncheckable assumptions. Parametric selection models are not well suited to this, as the assumptions about the missing data are confounded with uncheckable parametric assumptions about the full data response.

2.6.2 *Pattern Mixture Models*

Pattern mixture models are based on the pattern mixture factorization. The strategy used is the opposite of the one used for selection models: models are specified by positing separate models for the *marginal* distribution of the missing data $f(r \mid x, \omega_r)$ and the *conditional* distribution of the full data response $f(y \mid r, x, \theta)$.

Example 2.6.2. Consider the schizophrenia clinical trial described in Example 2.1.1, and suppose for simplicity that Y_i is univariate and may be observed ($R_i = 1$) or not ($R_i = 0$); we observe complete covariates X_i. A pattern mixture model in this setting might take

$$R_i \sim \text{Bernoulli}\left(\frac{\exp(\gamma + \delta^{\mathrm{T}} X_i)}{1 + \exp(\gamma + \delta^{\mathrm{T}} X_i)}\right),$$
$$Y_i \sim N(\alpha_r + \beta_r^{\mathrm{T}} X_i, \sigma_r^2) \qquad \text{given } R_i = r.$$

The pattern mixture model parameters are $\omega_r = (\gamma, \delta)$ and $\theta = (\alpha_0, \beta_0, \sigma_0, \alpha_1, \beta_1, \sigma_1)$.

Considering how one would estimate the pattern mixture model in Example 2.6.2 gives insight into the strengths and weaknesses of pattern mixture models. It is straight-forward to estimate $f(r \mid x, \omega_r)$ using a logistic regression because R_i is always observed. It is also straight forward to estimate $(\alpha_1, \beta_1, \sigma_1)$ by regressing the observed Y_i's on their respective X_i's.

The problem of estimating $(\alpha_0, \beta_0, \sigma_0)$ highlights a difference between pattern mixture models and (parametric) selection models. The observed data likelihood in Example 2.6.2 is given by

$$L_{\text{obs}}(\omega) = \prod_{i=1}^{n} f(r_i \mid x_i, \omega_r) \times \prod_{i:r_i=1} f(y_i \mid R_i = 1, x_i, \alpha_1, \beta_1, \sigma_1).$$

Note that $(\alpha_0, \beta_0, \sigma_0)$, the parameters in $f(y_i \mid R_i = 0)$, do not appear in $L_{\text{obs}}(\omega)$, i.e., these parameters are unidentified.

We regard the fact that pattern mixture models can potentially leave parameters unidentified an *advantage*, albeit one that carries potential risks if one is not careful. Evidently, information outside of the data is required to make use of these models. For example, we could resolve the issue by imposing restrictions such as $(\alpha_0, \beta_0, \sigma_0) = (\alpha_1, \beta_1, \sigma_1)$; this implies MAR missingness. Alternatively, we might assume $(\beta_0, \sigma_0) = (\beta_1, \sigma_1)$ but $\alpha_0 = \alpha_1 + \xi$ for some constant value of ξ. We refer to the parameter ξ as a *sensitivity parameter*. Sensitivity of results to the MAR assumption can then be assessed as ξ varies from its MAR value $\xi = 0$; this *sensitivity analysis* strategy is expanded upon in Chapter 4.

A drawback of pattern mixture models is that they do not give direct access to the full data parameter ω_y, which is typically a complicated function of both θ and ω_r; by comparison, selection models estimate ω_y directly. In Example 2.6.2, ω_y is a function of $(\omega_r, \{\alpha_j, \beta_j, \sigma_j : j = 0, 1\})$,

$$f(y \mid x, \omega_y) = \sum_{r=0}^{1} f(r \mid x, \omega_r) f(y \mid R = r, x, \alpha_r, \beta_r, \sigma_r).$$

Additionally, when missingness is non-monotone, it can be challenging to specify a parsimonious model for all possible patterns, and many patterns may have no observations associated with them. Nevertheless, we often find that the benefits of the pattern mixture approach can easily overcome these drawbacks. In the case study in Chapter 12 we show how these types of issues can be resolved using Bayesian nonparametrics.

2.6.3 Shared Parameter Models

Shared parameter models capture the dependence between Y_i and R_i by introducing a latent variable b_i (i.e., a random effect) that is shared between Y_i and R_i. A particularly common shared parameter model assumes conditional independence given the latent variable, with the joint density of (Y_i, R_i) being

$$f(y,r \mid \omega) = \int f(y \mid b, \varphi_y) f(r \mid b, \varphi_r) g(b \mid \theta_b) \, db. \tag{2.5}$$

The latent variable b is referred to as a *shared parameter*. This model captures the intuition that there is a (typically low-dimensional) summary b that captures all of the dependence between Y_i and R_i.

Example 2.6.3 (Latent Class Models)**.** A strategy that we employ frequently is to use *latent class models*. This is a special case of a shared parameter model in which b is discrete, taking values in $\{1,\ldots,K\}$. Under this model, the marginal distribution of $f(y,r \mid \omega)$ is the mixture model

$$f(y,r \mid \omega) = \sum_{k=1}^{K} w_k f(y \mid \varphi_{yk}) f(r \mid \varphi_{rk}). \tag{2.6}$$

An interpretation of this model is that the latent classes correspond to unknown/unmeasured groupings of the observations. The model (2.6) can be expressed in terms of latent variables as

$$\begin{aligned} Y_i &\sim f(y \mid \varphi_{yk}) && \text{given } b_i = k, \\ R_i &\sim f(r \mid \varphi_{rk}) && \text{given } b_i = k, \\ P(b_i = k) &= w_k. \end{aligned}$$

Latent class models are highly flexible and can approximate essentially any distribution given a sufficiently large number of latent classes.

Model (2.5) is very powerful in that it allows for highly complex dependencies between Y_i and R_i using a small number of parameters. Nonparametric versions of (2.5) are flexible enough to capture any dependency structure in (Y_i, R_i). We introduce Bayesian nonparametric versions of these models in Chapter 6. Shared parameter models, like pattern mixture models, do not give direct access to ω_y. This makes it potentially difficult to compute estimands. Shared parameter models also share the drawback of selection models in that it is unclear whether or not the parameters of the model are identified.

2.6.4 Observed Data Modeling Strategies

In this monograph we mainly take an approach we refer to as "observed data modeling" which does not uniquely fit into the pattern mixture, selection, or shared parameter modeling approaches, though they are most connected to pattern mixture models. The idea is to model the observed data distribution $f(y_{ir_i}, r_i \mid \omega)$ while leaving the *extrapolation distribution* $f(y_{i-r_i} \mid y_{ir_i}, r_i, \omega)$ unspecified. When missingness is monotone, this approach is fairly straight forward to implement in a sequential fashion.

Example 2.6.4. Consider Example 2.1.1 and suppose that Y_i is subject to monotone missingness. A possible model for the observed model is

$$
\begin{aligned}
R_{i1} &\sim \text{Bernoulli}(p_1), \\
[Y_{i1} \mid R_{i1} = 1] &\sim N(\alpha_1, \sigma_1^2), \\
[R_{ij} \mid R_{i,j-1} = 1, Y_{i1}, \ldots, Y_{i,j-1}] &\sim \text{Bernoulli}\left(\frac{\exp(\gamma_j + \sum_{k=1}^{j-1} \zeta_{jk} Y_{ik})}{1 + \exp(\gamma_j + \sum_{k=1}^{j-1} \zeta_{jk} Y_{ik})} \right) \\
[Y_{ij} \mid R_{ij} = 1, Y_{i1}, \ldots, Y_{i,j-1}] &\sim N\left(\alpha_j + \sum_{k=1}^{j-1} \beta_{jk} Y_{ik}, \sigma_j^2 \right).
\end{aligned}
$$

As the observed data modeling approach explicitly avoids modeling the extrapolation distribution, the model above is not fully specified. To address the extrapolation distribution, one can consider a collection of different *identifying restrictions* that pin down $f(y_{i-r_i} \mid y_{ir_i}, r_i, \omega)$ from the observed data distribution $f(y_{ir_i}, r_i \mid \omega)$. The most commonly used identifying restriction is given by the MAR assumption, but there are many other assumptions one can make. We refer the reader to Chapter 4 for details on identifying restrictions.

2.7 Summary and a Look Forward

The purpose of this chapter has been to introduce the basic concepts which have been used to analyze incomplete data under both MAR and MNAR missingness. There are two fundamental challenges the analyst must confront:

1. The *identifiability problem*, which requires us to make untestable assumptions to identify the estimands of interest and to assess the robustness of our conclusions to these assumptions. This issue has been discussed here extensively and will continue to be a theme moving forward.
2. The *curse of dimensionality*, which makes it difficult to flexibly estimate the effects of interest even after they have been identified. For example, fully nonparametric estimation of $f(y \mid r, x, \omega)$ is typically infeasible if the dimension of (Y_i, R_i, X_i) is of even moderate size; this issue is not particular to missing data and is simply a reality of nonparametric modeling.

Bayesian nonparametrics is a compelling tool to address both of these issues. To address the curse of dimensionality, we will specify flexible models that shrink toward parsimonious sub-models in a data-adaptive fashion. To address the identifiability problem, we will use the observed data modeling strategy described in Section 2.6.4. This allows us to specify complex models for the observed data distribution while placing no restrictions on the extrapolation distribution. Different identifying restrictions and approaches for introducing sensitivity parameters can then be used to assess the robustness of conclusions to untestable assumptions (see Chapter 4).

Chapter 3

Overview of Bayesian inference for missing data and causal inference

DOI: 10.1201/9780429324222-3

In this chapter, we provide a brief overview of Bayesian inference, primarily in the parametric setting that is most relevant to the causal inference, missing data, and Bayesian nonparametric developments in the rest of the book. In particular, we review posteriors and priors, Bayesian computations (including data augmentation and g-computation), and model selection and checking.

3.1 The Posterior Distribution

The posterior distribution is the basis for Bayesian inference. Given a parametric model $\{f(y \mid \theta) : \theta \in \Theta\}$ and a *prior* distribution $\pi(\theta)$ on Θ, the posterior $\pi(\theta \mid y)$ is defined as

$$\pi(\theta \mid y) = \frac{L(\theta \mid y)\,\pi(\theta)}{\int L(\theta \mid y)\,\pi(\theta)\,d\theta}. \tag{3.1}$$

It is composed of two pieces: the *likelihood* $L(\theta \mid y)$ and the prior $\pi(\theta)$. Assuming iid data with $y = (y_1, \ldots, y_n)$, the likelihood is proportional (as a function of θ) to $\prod_{i=1}^{n} f(y_i \mid \theta)$. The likelihood encodes the information contained in the data about the parameter; fully model-based frequentist inference often uses the likelihood, but typically does not make use of a prior. The prior distribution, by contrast, incorporates external, *a priori*, information about the parameters. This information could be based on previous studies or scientific knowledge, but at times little is known a priori.

The principle of Bayesian inference is to update the information about the parameters by combining the prior and the data (through the likelihood). Note that the posterior is proportional to $L(\theta \mid y)\,\pi(\theta)$, the likelihood times the prior. Computationally, performing Bayesian inference amounts to computing the integrals involved in the posterior distribution; for example, if θ is a scalar, then the posterior mean is given by

$$E(\theta \mid y) = \int \theta\,\pi(\theta \mid y)\,d\theta = \frac{\int \theta\,L(\theta \mid y)\,\pi(\theta)\,d\theta}{\int L(\theta \mid y)\,\pi(\theta)\,d\theta}. \tag{3.2}$$

In all but the simplest settings, neither of the integrals above will be available in closed form. Fortunately, there are computational algorithms available that allow us to obtain samples from the posterior; we briefly describe some of these in Section 3.3. We provide two simple examples next (one with a closed form solution and one without).

Example 3.1.1 (Likelihood for a normal model). Assume $f(y_i \mid \theta)$ is a normal density with mean θ and variance 1. Suppose we have prior information that the mean, θ, is roughly θ_0 with some degree of uncertainty. We might convey this information with a normal prior on θ, with mean θ_0 and variance σ_0^2. With this prior, the numerator of (3.1) is

$$\begin{aligned} L(\theta \mid y)\,\pi(\theta) &= \left[\prod_{i=1}^{n} (2\pi)^{-1/2} \exp\left\{-\frac{(y_i-\theta)^2}{2}\right\}\right] (2\pi\sigma_0^2)^{-1/2} \exp\left\{-\frac{(\theta_0-\theta)^2}{2\sigma_0^2}\right\} \\ &\propto \exp\left\{-\frac{1}{2}\sum_{i=1}^{n}(y_i-\theta)^2\right\} \exp\left\{-\frac{1}{2\sigma_0^2}(\theta-\theta_0)^2\right\}. \end{aligned}$$

By completing the square and ignoring proportionality constants, one can show that the posterior is proportional to

$$L(\theta \mid y)\,\pi(\theta) \propto \exp\left\{-\frac{(\theta-\theta_0^\star)^2}{2\sigma_0^{\star 2}}\right\},$$

where $\theta_0^\star = \frac{n\bar{y}+\theta_0/\sigma_0^2}{n+1/\sigma_0^2}$ and $\sigma_0^{\star 2} = \frac{1}{n+1/\sigma_0^2}$. Recognizing this as the kernel of a normal distribution, it follows that the posterior of θ is normal with mean $\theta_0^\star$ and variance $\sigma_0^{\star 2}$. Hence, both (3.1) and (3.2) are available in closed form for this example.

In the above example, we started with a normal prior for θ and ended with another normal distribution for the posterior. This is an example of *conjugacy*, and the normal distribution is referred to as a *conjugate* prior for the likelihood. In practice, when the posterior is available in closed form, it is usually because the prior is conjugate to the likelihood. As in the following example, when the prior is not conjugate, it is rare that the posterior is available in closed form.

Example 3.1.2 (Likelihood for a logistic model)**.** In the case of logistic regression, the posterior is not available in closed form. Consider $Y_i \sim \text{Bernoulli}(\theta_i)$ where $\text{logit}(\theta_i) = \log\{\theta_i/(1-\theta_i)\} = X_i^\top\beta$ with $\beta \sim N(\beta_0, \sigma_\beta^2)$. In this case, the numerator in (3.1) is proportional to

$$L(\beta \mid y, x)\,\pi(\beta) \propto \left[\prod_{i=1}^{n} \frac{\exp(y_i x_i^\top \beta)}{1+\exp(x_i^\top \beta)}\right] \exp\left\{\frac{(\beta-\beta_0)^2}{2\sigma_\beta^2}\right\}.$$

Unfortunately, this does not correspond to the kernel of any well-known distribution, and the integrals in (3.1) and (3.2) cannot be computed in closed form. Options to get around this problem include (i) approximating the posterior distribution with a normal distribution (using, e.g., a *Laplace approximation* [68]) or (ii) using simulation techniques to obtain (approximate) samples from the posterior. For the latter, Markov chain Monte Carlo (MCMC) methods provide an approach to obtain a sample from the posterior distribution without the need to compute the normalizing constant in (3.1). For the above logistic regression model in particular (and the related probit regression), the posterior can be sampled using various *data augmentation* approaches [69, 70, 71]. Details are in Section 3.5.

3.2 Prior Distributions and Identifiability Problems in Causal Inference and Missing Data

The prior distribution, $\pi(\theta)$, quantifies *a priori* knowledge about the parameters in the model. The prior can be constructed based on historical data, expert opinion, or chosen to represent relative ignorance. In situations with finite-dimensional parameters, the data eventually overwhelms the prior and inference about parameters of interest is completely determined by the data (through the likelihood) via the *Bernstein-von Mises* (BVM) theorem [72]; however, in settings with infinite-dimensional parameters (e.g., distribution functions), as considered here, such results often do not hold. In the context of Bayesian nonparametric inference, we need to specify priors on distributions of the observed data (details on priors for distributions are in Chapters 5, 6, and 7).

3.2.1 *Priors in General*

There are many different strategies for specifying priors. When previous (historical) data or scientific knowledge is available, it is desirable to incorporate this information into the prior [73]. To facilitate computation of (or sampling from) the posterior, this information is often used to specify the hyperparameters within a conjugate prior (or a *conditionally conjugate* prior as described in Section 3.3.2) for the parameters.

Priors can also be used to provide estimation stability in settings where there are too many parameters to estimate reliably; in this case, priors might be specified so that many of the parameters are encouraged to be similar. Such priors are often called *shrinkage* priors. They are commonly used in random effects models, where the random effects might be assumed to follow a normal distribution with mean zero and unknown variance. Shrinkage priors are implicitly used in some of the BNP models we will introduce in later chapters.

It is not uncommon to have minimal information about particular parameters in the model. In such settings, diffuse priors are often specified (also called "noninformative," "default," or "flat"). For location parameters (means, regression parameters, etc.), a normal distribution with a large variance is often specified; this prior is essentially flat (or constant) over the parameter values supported by the likelihood. As the variance tends to infinity, this results in an *improper* prior (i.e., a prior that is not integrable). In such cases, it is essential to ensure the resulting posterior is not also improper. For variance parameters, it is common, as recommended in [74], to choose flat priors for the standard deviation over finite ranges (or half normal priors with a large variance). However, for computational simplicity, conjugate inverse gamma priors are often used for the variance.

We will introduce other priors in the context of causal inference and missing data in subsequent chapters. Next, we discuss priors for unidentified parameters in causal inference and missing data.

3.2.2 *Priors for Unidentified Parameters*

In both causal inference and missing data settings, inference requires assumptions that cannot be "checked" by the data (as discussed in Chapters 1 and 2), and, as such, inferences about certain parameters may not become more precise as more data is collected. Such a parameter, of which the estimand of interest is likely a function, will be completely determined by the prior. In what follows, we give an example in the context of the causal and missing data assumptions introduced in the previous two chapters.

Example 3.2.1 (*Prior for uncheckable assumptions in missing data problems*)**.** Consider a bivariate response (Y_1, Y_2) with only missingness in Y_2. Then the missing at random (MAR) assumption implies

$$f(y_2 \mid y_1, r=0, \omega) = f(y_2 \mid y_1, r=1, \omega),$$

where ω denotes the collection of all model parameters. Recall this assumption is uncheckable from the observed data as, for the left hand side, Y_2 is never observed if $R=0$. We can use the pattern mixture factorization from Chapter 2 to assess the sensitivity of inferences to deviations from MAR. For simplicity, assume $[Y_2 \mid Y_1 = y, R=1, \omega]$ is a normal distribution with mean $\beta_0 + \beta_1 y$ and variance σ^2. Assume also that $[Y_2 \mid Y_1 = y, R=0, \omega]$ is a normal distribution with the same variance and the same mean structure, $\alpha_0 + \alpha_1 y$, but with different parameters for the mean. Under MAR:

$$\alpha_j = \beta_j : j = 0, 1.$$

MAR not holding can be expressed as $\alpha_j = \beta_j + \Delta_j$; because α_j is not identified from the observed data, neither is Δ_j. The MAR assumption implicitly assumes Δ_j has the prior $\Pi(\Delta_j) = I(\Delta_j = 0)$; this is very informative and reflects no uncertainty. To quantify uncertainty about MAR and/or deviations from it, we can put a non-degenerate informative prior on Δ_j possibly centered away from zero. Note that, because Δ_j does not appear in the likelihood, its prior and posterior will be identical provided that Δ_j is a priori independent of the remaining parameters in the model.

Suppose $E(Y_2 \mid \omega)$ is the estimand of interest. Specify $f(y_1 \mid R=r, \mu^{(r)})$ as a density with mean $\mu^{(r)}$ and write $\Pr(R_1 = 1 \mid \omega) = p$. Then $E(Y_2 \mid \omega)$ is the following function of Δ_j:

$$E(Y_2 \mid \beta, \Delta, \mu, p) = \sum_r \int_y E(Y_2 \mid Y_1 = y, R=r, \beta, \Delta)\, f(y \mid R=r, \mu^{(r)})\, dy\, p^r (1-p)^{1-r},$$

where

$$E(Y_2 \mid Y_1 = y, R=1, \beta, \Delta) = \beta_0 + \beta_1 y$$

and

$$E(Y_2 \mid Y_1 = y, R=0, \beta, \Delta) = (\beta_0 + \Delta_0) + (\beta_1 + \Delta_1) y.$$

The impact of Δ_j becomes larger as p becomes smaller. The estimand $E(Y_2 \mid \omega)$ is a function of both the unidentified parameter Δ and the identified parameters (β, μ, p); for posterior inference, we would specify priors for all these parameters; however, the posterior for Δ would not get tighter as the sample size increases unless it is a priori dependent on any of the other parameters (but in this case, the posterior would still not converge to a point mass).

Example 3.2.2 (*Prior for uncheckable assumptions in causal problems*)**.** Consider a point treatment setting under the *ignorability* assumption of Chapter 1. Recall that this assumption is equivalent to the statement

$$[Y(1) \mid A=0, L] \equiv [Y(1) \mid A=1, L]$$

and is uncheckable from the observed data. Similar to the missing data example, assume $[Y(1) \mid A=1, L]$ is a normal distribution with mean $\beta_0 + \beta_1 L$. If we assume

$[Y(1) \mid A = 0, L]$ is also a normal distribution with the same mean structure $\alpha_0 + \alpha_1 L$ but possibly different parameters, then under ignorability we have $\alpha_j = \beta_j : j = 0, 1$. Hence, ignorability can be expressed as $\alpha_j = \beta_j + \Delta_j$ where Δ_j is not identifiable from the observed data. To quantify our prior uncertainty in how far the model deviates from ignorability, we can put an informative prior on Δ_j (note, again, that the prior and posterior will be identical to each other due to the lack of identifiability). The ignorability assumption implicitly assumes Δ_j has the prior $\Pi(\Delta_j) = I(\Delta_j = 0)$. The impact of Δ_j on the cdf of the potential outcome, $Y(1)$, $F_{Y(1)}(y)$ (which is of interest for quantile causal effects) can be expressed as

$$F_{Y(1)}(y) = \int [F_{Y(1)}(y \mid A = 1, L = \ell, \beta) \Pr(A = 1 \mid L = \ell) \\ + F_{Y(1)}(y \mid A = 0, L = \ell, \beta, \Delta_0, \Delta_1) \Pr(A = 0 \mid L = \ell)] dF(\ell),$$

where

$$E[Y(1) \mid A = 1, L = \ell, \beta] = \beta_0 + \beta_1 \ell$$

and

$$E[Y(1) | A = 0, L = \ell, \beta, \Delta_0, \Delta_1] = (\beta_0 + \Delta_0) + (\beta_1 + \Delta_1)\ell.$$

So again with this example, the estimand of interest is a function of the unidentified parameters, (Δ_0, Δ_1).

3.2.3 Priors for the Distribution of Covariates

For causal inference (with confounders) and missing data (with auxiliary covariates), to conduct inferences unconditional on the covariates, it is necessary to put a prior on their distribution $f(\ell)$. When these distributions are not modeled explicitly, the empirical distribution is often used as an estimate (with no prior and/or an implicit degenerate prior). The Bayesian bootstrap can be used to incorporate uncertainty in this distribution in practice. We provide details in the following.

The empirical distribution (implicitly) estimates the distribution of confounders (or auxiliary covariates) with a multinomial distribution with fixed weight $1/n$ for each observed set of confounders (so the support of $f(x)$ is assumed to be just the n observed sets of confounders). The empirical distribution of the confounders can be represented as $f_n(\ell) = \sum_{i=1}^n \varpi_i \delta_{\ell_i}$, where δ_{ℓ_i} is a degenerate distribution at ℓ_i and $\{\ell_1, \ldots, \ell_n\}$ are the observed values of the L_i's. The weights $\varpi_i = 1/n$ satisfy $\varpi_i \geq 0$ and $\sum_{i=1}^n \varpi_i = 1$. The Bayesian bootstrap is similar to using the empirical distribution, except that the weights $\varpi = (\varpi_1, \ldots, \varpi_n)$ are considered unknown parameters (as opposed to being fixed at $1/n$) and given a non-informative prior proportional to $\Pi_{i=1}^n \varpi_i^{-1}$. The resulting posterior for ϖ is Dirichlet$(1, \ldots, 1)$. Given the simple form for the posterior and the finite support of the distribution of the covariates in this case, integration over the distribution of the covariates only involves computing a weighted average of the observed covariate sets for each sample (of weights) from the Bayesian bootstrap (i.e., the Dirichlet posterior of the weights). Of course, if there is missingness in the auxiliary covariates or confounders themselves, this approach will not work as the approach would not fill in values for incomplete covariate sets.

3.3 Computation of the Posterior

As noted in Section 3.1, it is atypical that the posterior distribution can be computed in closed form. Fortunately in the late 1980s, various approaches were introduced to obtain a sample from the posterior distribution where the normalizing constant (i.e., the denominator of (3.1)) does not need to be computed. Most of these approaches use MCMC to approximately sample from the desired distribution.

3.3.1 An Overview of Markov Chain Monte Carlo

The goal of MCMC is to approximately sample from a target distribution $p(\theta)$ by constructing a *Markov chain* $\theta^{(1)}, \theta^{(2)}, \ldots$ with the property that $\theta^{(k)}$ converges, in some sense, to a sample from $p(\theta)$. For us, $p(\theta)$ will be a posterior distribution $\pi(\theta \mid y)$, although this need not be the case in general. MCMC differs from directly taking iid samples from $p(\theta)$ in two ways:

1. The samples $\theta^{(k)}$ are not identically distributed.
2. The samples $\theta^{(k)}$ are not independent, but instead form a Markov chain such that $[\theta^{(k)} \mid \theta^{(1)}, \ldots, \theta^{(k-1)}] \sim T(\cdot \mid \theta^{(k-1)})$. The function $T(\theta \mid \theta')$ is known as the *Markov transition function* (MTF). A consequence of this is that $\theta^{(k)}$ is independent of $(\theta^{(1)}, \ldots, \theta^{(k-2)})$ given $\theta^{(k-1)}$.

Under mild regularity conditions, $\theta^{(k)}$ will converge to a sample from $p(\theta)$ provided that $p(\theta)$ is the *stationary distribution* of $T(\theta \mid \theta')$ in the sense that

$$p(\theta) = \int p(\theta') T(\theta \mid \theta') \, d\theta'.$$

The payoff of using MCMC is that it is very easy to construct an MTF with $p(\theta)$ as its stationary distribution and, moreover, this can be done even when $p(\theta)$ is known only up to a normalizing constant (as is typically the case when $p(\theta)$ is a posterior distribution); the downside is that, because the samples are not iid samples from $p(\theta)$, it may be that the $\theta^{(k)}$'s converge very slowly to samples from $p(\theta)$ or that the $\theta^{(k)}$'s are extremely highly correlated so that many samples from the Markov chain are required.

Two classes of MCMC sampling algorithms are *Gibbs sampling* [75] and *Hamiltonian Monte Carlo* (HMC) (see Chapter 5 in [76]). Various softwares are available to implement these algorithms in the programming language `R`; two particularly popular ones are `JAGS` for Gibbs sampling (using the `rjags` package) and `Stan` for HMC (using the `rstan` package).

3.3.2 Gibbs Sampling

Consider a model parameterized by a $p \times 1$ vector θ. The standard *Gibbs sampler* samples from each component θ_j of $\theta = (\theta_1, \ldots, \theta_p)$ conditional on the other parameters, $\theta_{-j} = (\theta_1, \ldots, \theta_{j-1}, \theta_{j+1}, \ldots, \theta_p)$ and the data y. In particular, the steps are given in the following algorithm.

Algorithm 3.1 Gibbs sampling algorithm

For $k = 1, \ldots, B+S$, do the following:

1. Sample $\theta_1^{(k)} \sim \pi(\theta_1 \mid \theta_{-1}^{(k-1)}, y)$
2. Sample $\theta_2^{(k)} \sim \pi(\theta_2 \mid \theta_1^{(k)}, \theta_3^{(k-1)}, \ldots, \theta_p^{(k-1)}, y)$.

$\vdots$

p. Sample $\theta_p^{(k)} \sim \pi(\theta_p \mid \theta_{-p}^{(k)}, y)$.

Then, return the samples $(\theta^{(B+1)}, \ldots, \theta^{(B+S)})$ as an approximate sample from the posterior.

Under mild regularity conditions, samples generated in this fashion converge to the posterior distribution. The steps return a (dependent, approximate) sample $\theta^{(B+1)}, \ldots, \theta^{(B+S)}$ from the posterior. The number B denotes the *burn-in* time of the chain, which gives time for the chain to get close to the posterior before saving samples; the number S denotes the total number of samples collected for analysis. We elaborate below.

When conditionally conjugate priors (i.e., conjugate priors for the full conditional distributions sampled in the Gibbs sampler) are used, these conditionals can be sampled directly. Otherwise they can be sampled using a Metropolis-Hastings step (details below) or a slice sampling step [77]. When the individual components $\theta_1, \ldots, \theta_p$ are themselves vectors, this is referred to as a "blocked" Gibbs sampler for θ; for example, for $p = 3$, instead of updating each component individually, we might iterate between updating (θ_1, θ_2) and then θ_3. We look at a simple example of blocked Gibbs sampling next.

Example 3.3.1 (*Gibbs sampling with a normal regression model*)**.** Consider a simple linear regression model with mean and variance,

$$E(Y \mid L = \ell, \beta, \sigma^2) = \beta_0 + \beta_1 \ell$$
$$\text{Var}(Y \mid L = \ell, \beta, \sigma^2) = \sigma^2$$

and priors

$$(\beta_0, \beta_1) \overset{\text{iid}}{\sim} N(0, \sigma_\beta^2),$$
$$1/\sigma^2 \sim \text{Gam}(a_\sigma, b_\sigma).$$

The Gibbs sampler sequentially samples from the full conditional distribution of each parameter (or set of parameters): $[\beta_0, \beta_1 \mid \sigma^2, y, \ell]$ and $[\sigma^2 \mid \beta_0, \beta_1, y, \ell]$. Let $L_i^\star = (1, L_i)$. At iteration k, the algorithm performs the following two steps:

1. $[1/\sigma^{2(k)} \mid \beta^{(k-1)}, y, \ell] \sim \text{Gam}(a_\sigma + \frac{n}{2}, b_\sigma + \frac{1}{2}\Sigma(y_i - L_i^\star \beta^{(k-1)})^2)$
2. $\beta^{(k)} \mid \sigma^{(k)}, y, \ell \sim N(A, V)$,

where $V = (\sum L_i^{\star\top} L_i^\star / \sigma^{2(k)} + (1/\sigma_\beta^2)I)^{-1}$ and $A = V(\sum L_i^{\star\top} y_i / \sigma^{2(k)})$. These steps are then repeated $B+S$ times, with the first B discarded to burn in. Here, both β_0 and β_1 were updated simultaneously; this is a "block" update (as opposed to sampling them separately from univariate normals) discussed above.

3.3.3 *The Metropolis-Hastings Algorithm*

Metropolis-Hastings (MH) algorithms are useful for constructing an MTF $T(\theta' \mid \theta)$ with essentially any desired stationary distribution $p(\theta)$, provided that $p(\theta)$ can be evaluated up to a normalizing constant, i.e., $p(\theta) = p^*(\theta)/\int p^*(\theta)\,d\theta$ where $p^*(\theta)$ is available in closed form [76]. For us, the distribution of interest $p(\theta)$ will be a posterior distribution $\pi(\theta \mid y)$. To specify an MH sampler, we first specify a *proposal distribution* $q(\theta' \mid \theta)$, which need not have $p(\theta)$ as its stationary distribution. The MH algorithm takes this proposal and provides an automatic adjustment that converts $q(\theta' \mid \theta)$'s stationary distribution into $p(\theta)$. The specific steps are given below.

Algorithm 3.2 Metropolis-Hastings

At iteration k of the sampler:

1. Sample $\theta^* \sim q(\cdot \mid \theta^{(k-1)})$.
2. Compute the acceptance probability

$$\alpha = \min\left\{1, \frac{p(\theta^*)\,q(\theta^{(k-1)} \mid \theta^*)}{p(\theta^{(k-1)})\,q(\theta^* \mid \theta^{(k-1)})}\right\}.$$

3. Sample $A \sim \text{Bernoulli}(\alpha)$. If $A = 1$, set $\theta^{(k)} = \theta^*$; otherwise, set $\theta^{(k)} = \theta^{(k-1)}$.

Under very mild conditions on the proposal $q(\theta' \mid \theta)$, it can be shown that samples from the MH algorithm converge in distribution to $p(\theta)$; additionally, because α depends on $p(\theta)$ only through the ratio $p(\theta^*)/p(\theta^{(k-1)})$, we can compute α even if we only know $p(\theta)$ up to an unknown normalizing constant (which is typically the case when performing Bayesian inference).

To specify an MH sampler, we must choose the proposal distribution $q(\theta' \mid \theta)$. This choice is critical, as poor choices of $q(\theta' \mid \theta)$ can produce algorithms that, while technically valid, generate very poor samples (see Section 3.3.6). There are two common approaches to choosing $q(\theta' \mid \theta)$:

- *Independence samplers* take $q(\theta' \mid \theta) = q(\theta')$ as close as possible to $p(\theta)$ and do not use θ in the proposal distribution. The ideal choice $q(\theta') = p(\theta')$ results in an algorithm with acceptance probability of $\alpha = 1$, and, generally speaking, good samplers have α as close to 1 as possible.
- *Random walk samplers* choose $q(\theta' \mid \theta)$ so that θ' is a small modification of θ; for example, we might take $\theta' \sim N(\theta, \tau^2)$ for some choice of τ. For this candidate, the acceptance probability simplifies to $\alpha = \min\{1, p(\theta^*)/p(\theta^{(k-1)})\}$. There is a trade-off in the choice of τ^2 here: if τ^2 is small, then the chain will move only very

slowly, while if τ^2 is too large, then very few proposals will be accepted. Roberts and Gelman [78] recommend choosing τ^2 so that the acceptance probability is around 25%, and show that this is optimal in certain settings.

The MH algorithm is also used in the HMC algorithm (see Section 3.3.5).

Metropolis-within-Gibbs

The MH algorithm can also be used as part of a Gibbs sampling algorithm when one or more of the required full conditionals cannot be sampled from; this approach is referred to generally as the *Metropolis-within-Gibbs* approach. We illustrate this in the context of a parameter vector θ with two components, $\theta = (\theta_1, \theta_2)$, and suppose the conditional distribution $\pi(\theta_2 \mid \theta_1, y)$ is available only up to a constant of proportionality $\pi(\theta_2 \mid \theta_1, y) \propto \pi^*(\theta_2 \mid \theta_1, y)$ (and correspondingly is not straightforward to sample). To address this, we can specify a proposal distribution $q(\theta_2' \mid \theta_2, \theta_1, y)$ and use the MH algorithm as follows:

Algorithm 3.3 Metropolis-within-Gibbs

At iteration k:

1. Generate $\theta_2^* \sim q(\cdot \mid \theta_2^{(k-1)}, \theta_1^{(k)}, y)$.
2. Compute the acceptance probability

$$\alpha = \min\left\{1, \frac{\pi(\theta_2^* \mid \theta_2^{(k-1)}, \theta_1^{(k-1)}, y)\, q(\theta_2^{(k-1)} \mid \theta_2^*, \theta_1^{(k-1)}, y)}{\pi(\theta_2^{(k-1)} \mid \theta_2^*, \theta_1^{(k-1)}, y)\, q(\theta_2^* \mid \theta_2^{(k-1)}, \theta_1^{(k-1)}, y)}.\right\}$$

3. Sample $A \sim \text{Bernoulli}(\alpha)$. If $A = 1$, set $\theta_2^{(k)} = \theta_2^*$; otherwise, set $\theta_2^{(k)} = \theta_2^{(k-1)}$.

The Metropolis-within-Gibbs strategy uses Algorithm 3.3 as a replacement for sampling $\theta_2^{(k)}$ in the Gibbs sampler given by Algorithm 3.1 with $p = 2$. More generally, one can replace *any* of the p steps of Algorithm 3.1 with an MH update without changing the stationary distribution of the Markov chain.

3.3.4 Slice Sampling

An alternative to the MH algorithm is *slice sampling* [79, 80]. For simplicity, again consider a parameter vector θ with two components, $\theta = (\theta_1, \theta_2)$, and suppose the conditional distribution, $\pi(\theta_2 \mid \theta_1, y)$, is not available in closed form (and correspondingly not straightforward to sample). Let $\pi^\star(\theta_2 \mid \theta_1, y)$ denote an unnormalized density we wish to sample from.

The basic idea of slice sampling is as follows. First, introduce an auxiliary variable A such that the joint distribution of (A, θ_2) is uniform over the region $U = \{(\theta_2, A) : 0 < A < \pi^\star(\theta_2 \mid \theta_1, y)\}$; that is, U is the region below the curve defined by $\pi^\star(\theta_2 \mid \theta_1, y)$ in the plane. The marginal (over A) density of θ_2 is

$$\int_0^{\pi^*(\theta_2|\theta_1,y)} \frac{1}{\int \pi^*(\theta_2 \mid \theta_1,y) d\theta_2} dA = \frac{\pi^\star(\theta_2 \mid \theta_1, y)}{\int \pi^\star(\theta_2 \mid \theta_1, y)\, d\theta_2} = \pi(\theta_2 \mid \theta_1, y),$$

which is the density of interest for Gibbs sampling. The idea now is to jointly sample (θ_2, A) and then discard A to obtain a sample of θ_2 from the desired distribution. See Algorithm 3.4 below:

Algorithm 3.4 Slice sampler

1. Draw $A \sim \text{Uniform}(0, \pi^\star(\theta_2^{(k-1)} \mid \theta_1^{(k)}, y))$
2. Draw $\theta_2^{(k)}$ uniformly from the "slice," $S = \{\theta_2 : A < \pi^\star(\theta_2 | \theta_1^{(k)}, y)\}$.
3. Discard A.

The algorithm above is a special case of the Gibbs sampler, as we are first sampling A from its conditional distribution given (θ_1, θ_2) and then sampling θ_2 from its conditional distribution given (θ_1, A).

Often, the slice in Step 2 is hard to determine. To bypass this issue, Neal [79] proposes generic approaches to approximately sampling from the slice while exactly maintaining the correct target distribution. Neal's schemes are generic enough that they can be fully automated, and are implemented (for example) in the `JAGS` software package.

An advantage of slice sampling relative to MH is that slice sampling does not require much user input: the slice sampling scheme described above has no tuning parameters. The more advanced slice sampling algorithms of Neal *do* have tuning parameters, but these tuning parameters primarily affect the time-per-sample rather than the accuracy of the sampler itself. By contrast, the performance of MH algorithms often depends crucially on the choice of the proposal distribution $q(\theta_2 \mid \theta_1, y)$.

Slice sampling gives a very strong general solution to the problem of having full conditionals that cannot be sampled with a Gibbs sampler: with essentially no tuning parameters, it provides an MTF that approaches the stationary distribution very quickly. Unfortunately, slice sampling is most useful for one-dimensional target distributions, and so is usually limited to Gibbs samplers. The HMC algorithm described in the next section, by contrast, is ideally suited to multivariate distributions.

3.3.5 Hamiltonian Monte Carlo

HMC is a different MCMC method than can be used to sample from posterior distribution of continuous parameters, θ in *one block* using Hamiltonian dynamics [81]. Let $p(\theta)$ denote the distribution we wish to sample from. HMC introduces (auxiliary) momentum variables ρ and generates approximate samples from the joint density, $p(\rho, \theta) = p(\theta)p(\rho \mid \theta)$. The distribution of ρ is arbitrary, but is typically chosen to be multivariate normal with mean 0 and covariance matrix Ω and to be independent of θ. The *Hamiltonian* is the negative logarithm of the joint density,

$$\begin{aligned} H(\rho,\theta) &= -\log p(\rho,\theta) \\ &= K(\rho|\theta)+P(\theta), \end{aligned}$$

where $K(\rho|\theta) = -\log(\rho|\theta)$ is called the *kinetic energy* and $P(\theta) = \log p(\theta)$ is called the *potential energy*. The function $H(\rho,\theta)$ is called the *total energy*.

Each sample of θ involves two steps: (i) sample the momentum variables (independent of the previous iteration) from $\rho \sim N(0,\Omega)$ and (ii) obtain a new value θ' by simulating the set of differential equations,

$$\begin{aligned} \frac{d\theta}{dt} &= \frac{\partial K}{\partial \rho} \\ \frac{d\rho}{dt} &= -\frac{\partial P}{\partial \theta}, \end{aligned}$$

for T units of time (with the initial value corresponding to the current value of θ and the sampled ρ). These differential equations are referred to as the *Hamiltonian equations*. This algorithm assumes (as we did above) that $p(\rho \mid \theta) = p(\rho)$. Unfortunately, it is usually not possible to simulate from the Hamiltonian equations exactly, and so an approximation is required. The most common numeric approximation is the *leapfrog integrator*; a Metropolis-Hastings acceptance step is then used to correct for the approximation at the end of each iteration.

The leapfrog integrator discretizes time and takes L steps of size ε as follows:

$$\begin{aligned} \rho &\leftarrow \rho - \frac{\varepsilon}{2}\frac{\partial P}{\partial \theta} \\ \theta &\leftarrow \theta + \varepsilon\Omega\rho \\ \rho &\leftarrow \rho - \frac{\varepsilon}{2}\frac{\partial P}{\partial \theta}, \end{aligned}$$

where the above corresponds to a full-step update of θ sandwiched between two half step updates of ρ. After L steps, denote the parameter values as (ρ',θ'). Given that the leapfrog integrator only approximates the Hamiltonian dynamics, a Metropolis-Hastings step is needed. The acceptance probability α for this step is

$$\alpha = \min\{1, \exp\{H(\rho^p,\theta^p) - H(\rho',\theta')\},$$

where (ρ^p,θ^p) are the previous values of ρ and θ. This step ensures that the HMC scheme targets the correct distribution. It turns out that, if we were able to simulate the Hamiltonian dynamics exactly, the value of the Hamiltonian would not change – we would just end up with a different configuration of the potential and kinetic energy. Hence, for exact HMC, the acceptance probability is always one. If the approximation is sufficiently accurate, α will be close to one, which will result in an efficient sampling algorithm. Large values of ε result in low acceptance probabilities as the approximation will be poor; on the other hand, small values of ε result in a larger number of steps L to simulate the dynamics for T units of time.

HMC is implemented in Stan, which by default as of this writing sets Ω equal to a diagonalized estimate of the covariance matrix of θ computed during a warm-up period. Stan also automatically chooses ε and the number of steps L to achieve a desired acceptance rate (the default is 0.8). Details can be found in the Stan reference manual [82]. The error of the leapfrog integrator is on the order of ε^3 at each step and ε^2 globally [83].

A drawback of HMC, particularly as implemented by Stan, is that it is not easy to incorporate discrete variables. The recommended approach when using Stan is to marginalize out all discrete variables. There has been recent work that adapts HMC to discrete parameter spaces (see, for example, [84]).

3.3.6 *Drawing Inferences from MCMC Output*

There are (at least) two important issues that need to be considered when implementing MCMC algorithms. The first issue is that, if we run a Markov chain for K iterations, the samples $\theta^{(1)}, \ldots, \theta^{(K)}$ are not from the posterior; rather, they start (potentially) far away from the posterior and converge to the posterior only in the limit. To address this, it is common to run the chain for a *burn-in* period (as noted earlier) and discard the first B samples, the idea being that the samples $\theta^{(B+1)}, \ldots, \theta^{(B+S)}$ will be close to the posterior.

Trace plots An appropriate burn-in is typically determined by running multiple chains with different initial values and then plotting how some of the $\theta_j^{(k)}$'s evolve as the chains run; specifically, we plot the iteration of the chain k on the x-axis and the associated sample of one of the parameters $\theta_j^{(k)}$ on the y-axis. Plots of this form are called *trace plots*. For example, for the Gibbs sampler described in Example 3.3.1, we might construct three plots displaying the parameters $\beta_0^{(k)}$, $\beta_1^{(k)}$, and $\sigma^{2(k)}$ against k for (say) four different initializations of the Gibbs sampler. When the $\theta_j^{(k)}$'s from the different chains begin to overlap, this is an indication that the chains have reached the posterior. For example, in Figure 3.1 we can see after the first few iterations the chains are overlapping , which indicates that the four chains have converged quickly. It is common practice to discard the B samples taken before convergence. So, based on Figure 3.1, we would discard the first five or ten iterations.

The Gelman-Rubin Diagnostic As an alternative to visually inspecting trace plots from multiple chains, Gelman and Rubin [85] proposed a statistic to measure the convergence to the stationary distribution. For a univariate parameter θ, define $\theta_j^{(k)}$ to be the kth sample taken on the jth chain, $\bar{\theta}_j$ to be the post-burn-in mean of θ on the jth chain, and $\bar{\theta}$ to be the post-burning mean of θ averaged over all chains. Then the *potential scale factor reduction* ($\hat{R}$) for θ is given by

$$\hat{R} = \sqrt{\frac{\frac{S-1}{S}W + \frac{1}{S}\mathscr{B}}{W}},$$

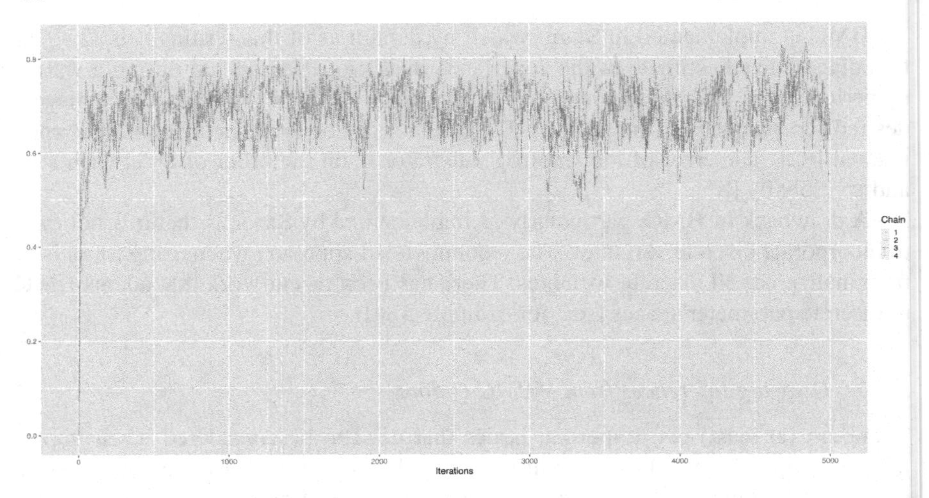

Figure 3.1 *Trace plot of four chains showing very quick convergence (short burn-in).*

where

$$\mathscr{B} = \frac{S}{m-1}\sum_{j=1}^{m}(\bar{\theta}_j - \bar{\theta})^2 \qquad \text{and} \qquad W = \frac{1}{m(S-1)}\sum_{j=1}^{m}\sum_{k=B+1}^{B+S}(\theta_j^{(k)} - \bar{\theta}_j)^2$$

are the between- and within-sequence variances, S is the number of saved samples, and m is the number of chains. The numerator and denominator of $\hat{R}$ are both an estimate of the within-chain variance so that when $\hat{R}$ is close to one the chains are likely sufficiently converged. Gelman and Rubin recommend that samples are collected until the associated $\hat{R}$ for each parameter is below 1.1.

The second issue is that the samples $\theta^{(k)}$ are (unlike standard Monte Carlo sampling) dependent and may be highly correlated with one another; this issue persists even after the chain has reached the stationary distribution. How quickly or slowly the samples become independent is referred to as the *mixing* of the chain. Figures 3.2 and 3.3 demonstrate good mixing (close to independent) and poor mixing (high dependence). Intuitively, if the chain is slowly mixing, then it may take many samples from the chain to obtain the equivalent of a single "effective" sample from the posterior, so that it is necessary to run the chain for many iterations to have the necessary precision to "approximate" posterior means or quantiles [86]. Autocorrelation plots are often used to quantify the strength of the dependence. When the lag k autocorrelation is sufficiently small, e.g., less than .05, then taking every k iteration would result in close to an independent sample (this is referred to as *thinning* the chain). Figure 3.4 shows the dependence dissipating quickly. Figure 3.5 shows the dependence weakening much more slowly and would probably require thinning of at least 20 for approximate independence. We note that, while thinning is not necessary strictly speaking (we cannot expect to *increase* the effective sample size by thinning), in cases where we need to "post-process" the MCMC samples (e.g., Bayesian

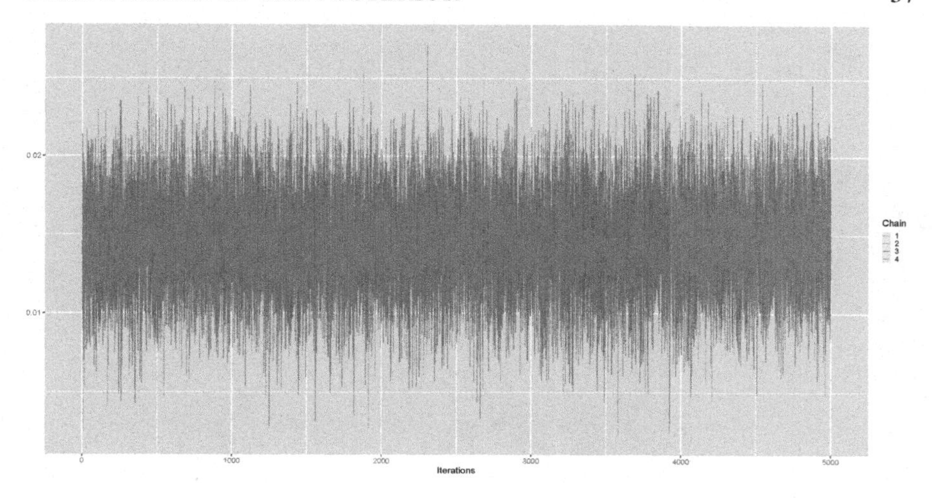

Figure 3.2 *Trace plot of four chains showing very good mixing.*

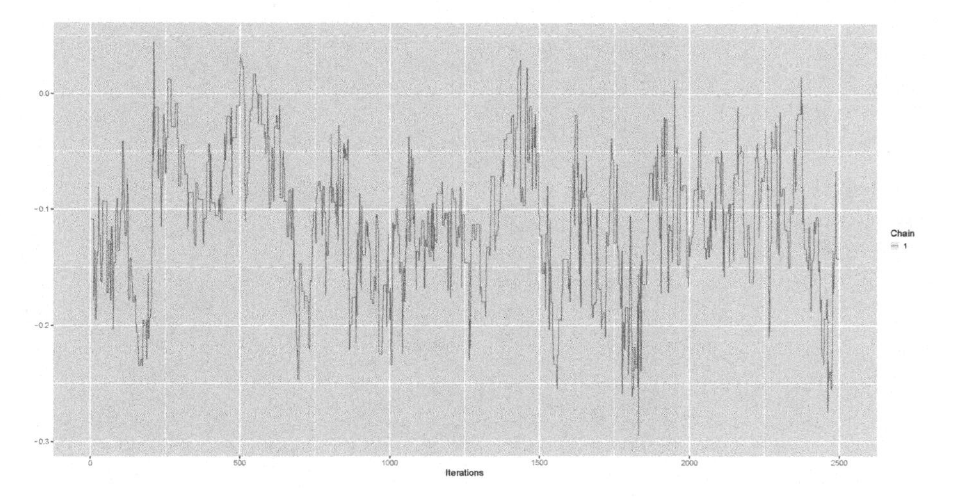

Figure 3.3 *Trace plot of one chain showing very slow mixing (high dependence).*

g-computation, see Example 3.6.1), it will typically be most efficient to use only the thinned samples, as this will dramatically reduce computation time.

In settings with many parameters, the log likelihood (or log posterior) can be used as a global measure of burn-in and mixing (e.g., compute the potential scale factor reduction statistic using the log likelihood, which is a function of $\boldsymbol{\theta}$, as the parameter); this is particularly helpful for the Bayesian nonparametric approaches in Chapters 5-7. A review of how to use MCMC samples for posterior inference

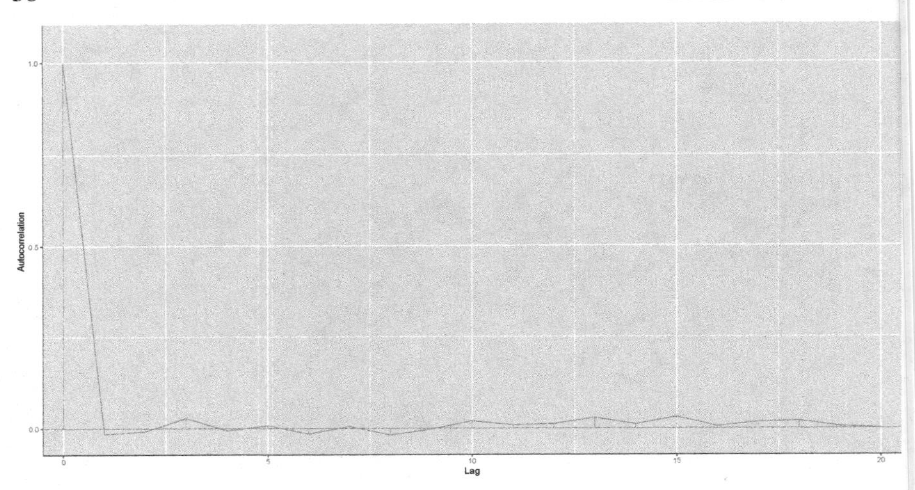

Figure 3.4 *Autocorrelation plot demonstrating very quickly dissipating dependence.*

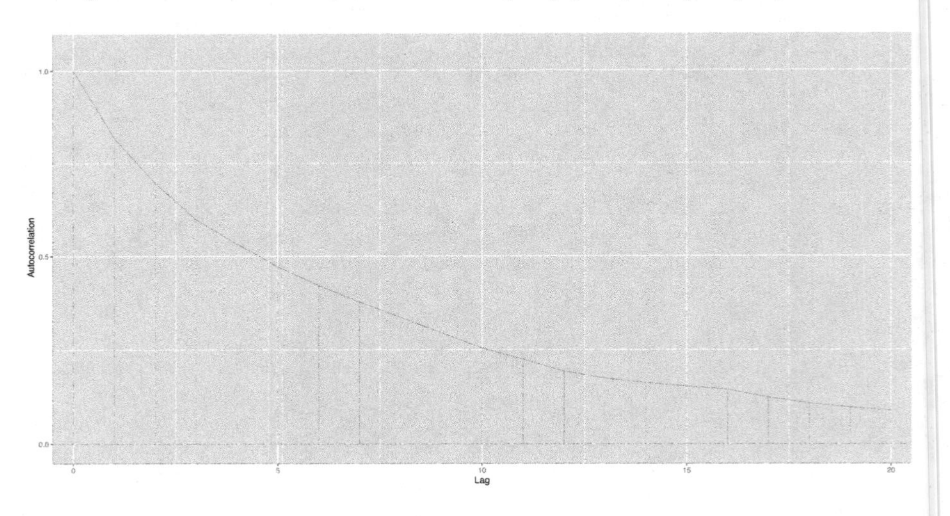

Figure 3.5 *Autocorrelation plot demonstrating very slowly dissipating dependence.*

can be found in Chapter 1 in [76], and a review of the many other diagnostics for convergence of chains can be found in [87] and Chapter 6 in [76].

3.4 Posterior Inference and Model Selection/Checking

MCMC provides approximate samples from the joint posterior distribution of all the parameters. It is typical to want a point estimate and a measure of uncertainty. Typical point estimates used are the posterior mean or the posterior median, which can be computed as the sample mean or median from the posterior sample. Uncertainty is often characterized by an equal tail, $100(1-\alpha)\%$ credible interval. So for a 95% credible interval, in practice, the endpoints can be computed from the 2.5th and 97.5th percentiles of the MCMC (posterior) sample. Alternatively, one can use a *highest posterior density* (HPD) interval, which is the narrowest interval obtaining the desired coverage [88]. These intervals are not restricted to $\alpha/2$ in each tail.

It is also often of interest to perform hypothesis testing. One way to do this in terms of credible intervals is to assess whether the interval covers the null value. The strength of evidence can be quantified via posterior probabilities: for example, if we have the null $H_0 : \theta > 0$, then we can explicitly compute $\Pr(H_0 \mid y)$. It is useful to note that under priors which do not explicitly include a point mass at the null value, the posterior probability of the null value is zero with probability one, and hence one should be careful when using posterior probabilities to assess point nulls. As an example, with a non-point null, consider the causal inference problem with a binary treatment and define $\eta = E\{Y(1) - Y(0)\}$ and assume we want to know whether this difference is larger than zero. We can quantify the strength of this via $\Pr(\eta > 0 \mid y, \ell)$. In "classical" settings, one can often show that this is, in large samples, equivalent to a frequentist p-value for the one-sided hypothesis test $H_0 : \eta > 0$ against $H_0 : \eta \leq 0$, although this may not hold in the nonparametric settings we consider in this book.

Model selection criteria like Bayes factors or the LPML can be also used to test individual parameters and/or different (not necessarily nested) models, which we describe in the following section.

3.4.1 Model Selection

The traditional Bayesian approach for model selection is the Bayes factor (BF). BFs compare the marginal likelihoods for two competing models via their ratio. For each model, the BF ostensibly requires computation of the marginal likelihood, i.e., it requires computation of the denominator of the posterior in (3.1). This is notoriously difficult to compute and is typically not even easy to compute using the samples obtained from MCMC algorithms [89]. Geisser and Eddy [90] proposed an alternative, predictive approach to model selection. In particular, they proposed the *log pseudo marginal likelihood* (LPML), which is the sum of the log *conditional predictive ordinates* (CPOs) $f(y_i \mid y_{-i})$ summed over all units. Specifically, the LPML is defined as

$$\text{LPML} = \sum_{i=1}^{n} \log f(y_i \mid y_{-i})$$

where y_{-i} is the outcome data without unit i, and $f(y_i \mid y_{-i})$ is given by

$$f(y_i \mid y_{-i}) = \int f(y_i \mid y_{-i}, \theta)\, \pi(\theta \mid y_{-i})\, d\theta.$$

When the observations are iid, we have the simplification $f(y_i \mid y_{-i}, \theta) = f(y_i \mid \theta)$. The LPML is a leave-one-out cross-validated variant of the log likelihood, and hence larger values are preferred. While the individual CPOs can be computed by brute force by refitting the model n times, Gelfand and Dey [91] show that CPO_i, and thus LPML, can be easily estimated from a posterior sample $\theta^{(1)}, \ldots, \theta^{(K)}$ as

$$\text{CPO}_i^{-1} \approx \frac{1}{K} \sum_{k=1}^{K} f(y_i \mid \theta^{(k)})^{-1}.$$

This computation can be easily done in `JAGS` or `Stan`. While this estimate is simple, results may be unstable (potentially having infinite variance); the CPOs can be computed automatically and stabilized using the `R` package `loo` [92].

In the context of mixture models (as will be discussed in Chapter 6), we can also use a form of the *deviance information criterion* (DIC) for model comparison [93]:

$$\text{DIC}_3 = -4E_\theta[\log f(O \mid \theta) \mid O] + 2\log \hat{f}(O),$$

where $O = (O_1, \ldots, O_n)$ is the observed data and O_i is the observed data on observation i, $\hat{f}(O) = \prod_{i=1}^{n} \hat{f}(O_i)$ and $\hat{f}(O_i) = \frac{1}{K}\sum_{l=1}^{K} f(O_i \mid \theta^{(K)})$. The terms $\hat{f}(O_i)$ are computed using the K sets of $\theta^{(k)}$ from the MCMC samples. The quantity $\hat{f}(O)$ is the posterior predictive density. DIC_3 can be rewritten as the additive combination of a goodness-of-fit term, $\hat{f}(O)$, and a penalty term, $E_\theta[\log f(O \mid \theta)|O] - \hat{f}(O)$, with the latter characterizing overfitting. We will use this in the causal mediation case study (Chapter 13).

3.4.2 *Model Checking*

An intuitive (and simple) way to check models in the Bayesian framework is to use *posterior predictive checks* (PPCs). PPCs compare datasets simulated from the fitted model to dataset we actually observed, checking for any systematic differences between the two. For a setting with no missingness, the replicated data y_{rep} is sampled from the posterior predictive distribution,

$$f(y_{\text{rep}} \mid y) = \int f(y_{\text{rep}} \mid \theta, y) f(\theta \mid y)\, d\theta.$$

Statistics, $T(y)$ or discrepancies, $T(y, \theta)$ can be used to assess whether the replicated data (under the model) captures the relevant features of the observed data. The strength of evidence can be quantified via posterior predictive probabilities. For example, we can compute the posterior predictive probability that a discrepancy based on the replicated data, $T(y_{\text{rep}}, \theta)$, is larger than the discrepancy based on the observed data, $T(y, \theta)$, as follows:

$$P(T(y_{\text{rep}}, \theta) > T(y, \theta) \mid y) = \iint I\{T(y_{\text{rep}}, \theta) > T(y, \theta)\} f(y_{\text{rep}} \mid \theta, y) f(\theta \mid y)\, d\theta\, dy_{\text{rep}}.$$

In settings with missing data, it has been recommended to replicate the observed data $O = (O_1, \ldots, O_n)$ where $O_i = (y_{ir_i}, r_i)$ [94], which are sampled from

$$f(O_{\text{rep}} \mid O) = \int f(O_{\text{rep}} \mid O, \theta)\, \pi(\theta \mid O).$$

Such an approach allows direct assessment of model fit for the observed data model. An example of a statistic $T(\cdot)$ that is appropriate in this case might be $T(O) = \frac{1}{n}\sum_i r_{iJ}(y_{iJ} - y_{i1})$, which can be used to compare the change from baseline in the observed data to the change in baseline in the replicated data; alternatively, rather than taking the mean, we might take the standard deviation to define $T(\cdot)$. The posterior predictive probability then essentially "counts" how frequently the replicated statistic is smaller/larger than the observed statistic:

$$\int I\{T(O) > T(O_{\text{rep}})\} f(O_{\text{rep}} \mid O)\, dO_{\text{rep}}.$$

In the setting of causal inference, we might instead compare the residual sum of squares, $T(O,\theta) = \sum(y_i - E[y_i \mid a_i, \ell_i, \theta])^2$, between the observed and replicated data.

Note that the choice of statistic/discrepancy is, in general, application/setting specific and often depends on the target of inference [95]. We will explore these posterior predictive checks and other model checks in the case studies (Chapters 11 and 13).

3.5 Data Augmentation

Data augmentation is a convenient tool to facilitate posterior sampling via MCMC. The basic idea is to introduce additional random variables that facilitate posterior sampling but preserve the posterior distribution of interest. We consider three variations that are particularly relevant to our settings: (i) probit regression, (ii) mixture/latent class models, and (iii) augmenting missing data. We start by describing data augmentation for probit regression.

Example 3.5.1 (*Data augmentation with probit regression*). Consider a binary response Y_i with covariates L_i. A simple *probit regression model* takes

$$\Phi\{\Pr(Y = 1 \mid L = \ell, \beta)\} = \ell^\top \beta, \tag{3.3}$$

where $\Phi(\cdot)$ is the standard normal cdf. If we place an improper (flat) prior $\pi(\beta) \propto 1$ on the regression coefficients, then there will not be a simple, analytically tractable posterior for β. However, we can introduce a latent variable Z by considering the following expanded data generating process:

$$\begin{aligned} Z &\sim N(L^\top \beta, 1), \\ Y &= I(Z > 0). \end{aligned}$$

It is easy to show that this model leads to the model (3.3) after marginalizing out Z. Additionally, it can be shown that $[Z \mid Y, \beta]$ follows a *truncated normal distribution*, with

$$Z \sim \begin{cases} N(L^\top \beta, 1) I(Z > 0) & \text{if } Y = 1, \\ N(L^\top \beta, 1) I(Z \leq 0) & \text{if } Y = 0, \end{cases}$$

where $N(\mu, \sigma^2)I(Z \in A)$ denotes the normal distribution with mean μ and variance σ^2 truncated to the set A. We can therefore use the Gibbs sampler (Algorithm 3.1) to sample approximately from the posterior distribution of (Z, β) and discard the sampled Z to get a sample from the posterior of β: Bayesian additive regression trees,

---

Algorithm 3.5 Data augmentation: Probit regression

At the kth iteration,

1. For $i = 1, \ldots, n$, sample Z_i from a truncated normal distribution with mean $L_i^\top \beta$ and variance 1, with the truncation to the positive real line for $Y_i = 1$ and the negative real line for $Y_i = 0$. Then set $\mathbf{Z}^{(k)} = (Z_1, \ldots, Z_n)$.
2. Sample $\beta^{(k)} \mid \mathbf{Z}^{(k)}, \mathbf{Y}, \mathbf{L} \sim N(m, \Omega)$ where $\Omega = (\mathbf{L}^\top \mathbf{L})^{-1}$, $m = \Omega \mathbf{L}^\top \mathbf{Z}$, $\mathbf{Z} = (Z_1, \ldots, Z_n)^\top$, and $\mathbf{L}$ is the design matrix formed from $(L_1, \ldots, L_n)$.

---

a semiparametric regression approach, which we will introduce in Chapter 5, is implemented for binary regression by using the probit link and the above augmentation scheme with the mean $L^\top \beta$ replace by a sum of trees (details in Chapter 5).

Similar approaches can be used for logistic regression, as mentioned earlier. One approach is to approximate the logistic distribution with a t-distribution and replace the truncated normal distributions above with truncated t-distributed latent variables; see [69, 96] for details. Alternatively, we can fit the logistic regression model exactly using the *Pólya-gamma* data augmentation approach proposed in [71].

Next, we describe using data augmentation to sample missing data with a simple example.

Example 3.5.2 (*Data augmentation with ignorable missingness in a bivariate normal distribution*)**.** Assume $(Y_1, Y_2) \sim N(\mu, \Sigma)$, where σ_{jk} is the jk-th element of Σ and with missingness only in Y_2 and assuming ignorability. We specify a normal prior on μ and a *Wishart* prior on Σ^{-1} (both are conditionally conjugate). However, they are not conditionally conjugate for the observed data likelihood,

$$\prod_i f(y_{i1} \mid \mu_1, \sigma_{11})^{I\{R_i=0\}} f(y_{i1}, y_{i2} \mid \mu, \Sigma)^{I\{R_i=1\}}.$$

So, at the kth iteration, we replace the two-step standard Gibbs sampler with the three steps given in Algorithm 3.6.

Finally, we describe the use of latent indicators, which will be particularly useful for the Bayesian nonparametric approaches in Chapter 6 in a simple finite mixture/latent class model.

Example 3.5.3 (*Data augmentation with finite mixture model*)**.** Consider a generalization of a simple linear regression model to a latent class regression model. Suppose there are $c = 1, \ldots, K$ classes (i.e., mixture components). Within the cth class suppose that

$$[Y \mid L = \ell, C = c; \beta^c, \sigma^2] \sim N(\beta_0^c + \beta_1^c \ell, \sigma^2), \qquad c = 1, \ldots, K.$$

Algorithm 3.6 Data augmentation: Ignorable missingness in a bivariate normal

At iteration k,

1. For all i with $R_{i2} = 0$, sample $Y_{i2}^{(k)} \mid \mu^{(k-1)}, \Sigma^{(k-1)}, y_{i1} \sim N(m_i, s^2)$ where $m_i = \mu_2^{(k-1)} + \sigma_{12}^{(k-1)}(y_{i1} - \mu_1^{(k-1)})/\sigma_{11}^{(k-1)}$ and $s^2 = \sigma_{22}^{(k-1)} - (\sigma_{12}^{(k-1)})^2/\sigma_{11}^{(k-1)}$.
2. Sample $\mu^{(k)}$ from its full-conditional (a normal distribution) given $\Sigma^{(k-1)}$ and the (augmented) complete data $y^{(k)}$.
3. Sample $\Sigma^{-1(k)}$ from its full-conditional (a Wishart distribution) given $\mu^{(k)}$ and the (augmented) complete data $y^{(k)}$.

Here, $y^{(k)}$ is composed of the observed data and the imputed $y_{i2}^{(k)}$ (for $r_{i2} = 0$) at iteration k.

Suppose further that the distribution of class membership is

$$[C \mid L = \ell, \xi] \sim \text{Categorical}_K(\xi),$$

where ξ is a probability vector [69]. We typically assume normal priors for the within-class regression coefficients, β^c. The marginal distribution of the outcome (over classes) is a mixture of K normal distributions,

$$\begin{aligned} & f(y \mid L = \ell, \xi, \{\beta^c : c = 1, \ldots, K\}, \sigma^2) \\ & = \sum_{c=1}^{K} \xi_c(\ell) f(y \mid L = \ell, C = c, \beta^c, \sigma^2). \end{aligned}$$

Updating β^c with a Gibbs sampling algorithm is more difficult here than in the $K = 1$ case, as its full-conditional does not correspond to a standard distribution. However, data augmentation can be used to facilitate sampling as follows. Define C_i to be the class of subject i, which takes values from $\{1, \ldots, K\}$. We augment the sampler at each iteration by sampling $C_i : i = 1, \ldots, n$ from its conditional distribution. In particular, we now also sample the C_i's from $\{1, \ldots, K\}$ with probability of $[C_i = c]$ given by

$$\Pr(C_i = c \mid y_i, L_i = \ell, \beta^c, \sigma^2) \propto \xi_c(\ell) f(y_i \mid L_i = \ell, C_i = c, \beta^c, \sigma^2).$$

Conditional on the C_i's, we now have independent samples from the K different subpopulations so that $f(\beta^c \mid \sigma^2, y, \ell, C)$ is now a normal distribution (cf. Example 3.3.1) (using only the individuals $i : C_i = c$).

3.6 Bayesian g-Computation

After we have collected samples $\theta^{(1)}, \ldots, \theta^{(K)}$ from the posterior, our next task is to convert these samples into samples of the causal parameters of interest. For example, given θ in an observational study, we might need to compute the average causal effect of some treatment (see Chapter 1). Under ignorability, this is given by

$$\begin{aligned}\text{ACE}(\theta) &= E\{Y(1) - Y(0) \mid \theta\} \\ &= \int y f(Y_i = y \mid A = 1, L = \ell, \theta) f(\ell \mid \theta) \, dy \, d\ell \\ &\quad - \int y f(Y_i = y \mid A = 0, L = \ell, \theta) f(\ell \mid \theta) \, dy \, d\ell.\end{aligned}$$

In general, and especially for nonparametric models, there is no reason for the integrals above to be available in closed form, and hence effects like $\text{ACE}(\theta)$ will not be simple functions of θ. Because of this, we will often require an extra layer of Monte Carlo integration beyond MCMC to compute parameters like $\text{ACE}(\theta)$ (see Example 3.6.1 for more details on the ACE in particular). In the case of causal inference, this corresponds to simulating from the g-formula given in Section 1.2. We refer to this approach as *Bayesian g-computation* (or just g-computation).

In the examples below, we simulate a total of N_L potential outcomes for each $\theta^{(k)}$ and use these to approximate the effects of interest. We note at the outset that N_L is entirely in the control of the analyst, with larger values of N_L corresponding to better approximations but longer computation times. While this approach is expensive computationally, we note that (fortunately) the Monte Carlo for each $\theta^{(k)}$ can be done in parallel [97].

Example 3.6.1 (*Bayesian g-computation for population-level causal inference*)**.** Expanding on the above example, to obtain population-level causal effects, it is necessary to standardize the conditional distributions with the desired population distribution of confounders (which is often the combined, across treatment and control, group population). For example, to compute the ACE in the population, under ignorability, we need to compute

$$E\{Y(a) \mid \theta\} = \int E(Y \mid A = a, \ell, \theta) f(\ell \mid \theta) \, d\ell, \qquad \text{for } a = 0, 1.$$

Suppose we have a posterior sample of the parameters of the distribution $[Y \mid A = a, \ell, \theta]$ for $a = 0, 1$. For each posterior sample of θ, we sample N_L realizations of L from its unconditional distribution, $f(\ell \mid \theta)$; if only the conditional distribution of $[Y \mid L]$ is modeled, then often the empirical distribution or the Bayesian bootstrap is used. Using Monte Carlo integration, we compute

$$E\{Y(a) \mid \theta\} \approx N_L^{-1} \sum_{i=1}^{N_L} E(Y \mid A = a, \ell^{(i)}, \theta).$$

Note that this step can be done in parallel for each sampled $\theta^{(k)}$ as mentioned above. To obtain a point estimate of the treatment effect, we can then average these over the K posterior samples of θ as

$$E\{Y(a)\} \approx K^{-1} \sum_{k=1}^{K} E\{Y(a) \mid \theta^{(k)}\},$$

where the sum is over the posterior samples of θ. This is a Monte Carlo estimate of $E\{Y(a)\}$ for $a = 0, 1$. Note that the same procedure can be done to compute the distribution function of $Y(a)$, $F_{Y(a)}(y \mid \theta)$ by replacing the expectations $E(Y \mid A = a, \ell, \theta)$ above with $F_{Y(a)}(y \mid A = a, L = \ell, \theta)$.

Example 3.6.2 (*Bayesian G-computation for causal mediation).* To obtain population-level causal mediation effects, it is again necessary to standardize the conditional distributions with the desired population distribution of confounders (which is often combined across the treatment and control group populations). But it is also necessary to sample from the proper mediator distribution. Suppose we have a posterior sample of the parameters of the distributions $[Y \mid A = a, \ell, \theta]$ and $[M \mid A = a, \ell, \theta]$ for $a = 0, 1$. To obtain the potential outcomes needed to compute direct and indirect effects under sequential ignorability, we need to use Algorithm 3.7 for each posterior sample of θ.

Algorithm 3.7 Bayesian g-computation for computing $E[Y\{a, M(a')\}]$

At iteration k:

1. Sample N_L realizations $L = \ell^{(i)}$ from its unconditional distribution $f(\ell \mid \theta^{(k)})$.
2. Sample N_L realizations of the mediator $M = m^{(i)}$ from $f(M \mid A = a', L = \ell^{(i)}, \theta^{(k)})$.
3. Approximate

$$E[Y\{a, M(a')\} \mid \theta^{(k)}] \approx N_L^{-1} \sum_{i=1}^{N_L} E(Y \mid A = a, M = m^{(i)}, L = \ell^{(i)}, \theta^{(k)}).$$

Our approximate samples of $E[Y\{a, M(a')\} \mid \theta^{(k)}]$ can then be used for inference for the natural direct/indirect effects. Again note that this can be parallelized across the $k = 1, \ldots, K$ collected samples. We can obtain point estimates by averaging over the posterior distribution as

$$E[Y\{a, M(a')\}] \approx K^{-1} \sum_{k=1}^{K} E[Y\{a, M(a')\} \mid \theta^{(k)}].$$

Bayesian G-computation is also sometimes needed for posterior inference for non-ignorable missingness, which requires a model for (y_r, r) as discussed in Chapter 2.

Example 3.6.3 (*Bayesian G-computation for non-ignorable missingness).* In the setting of longitudinal data with non-ignorable missingness, g-computation is often needed to compute marginal quantities (e.g., marginal expectations) [98, 99]. Consider a longitudinal response with J measurement times, with outcome $Y = (Y_1, \ldots, Y_J)$ and missing indicators $R = (R_1, \ldots, R_J)$. We need to the evaluate the following integral to compute the average outcome at the last time point:

$$E(Y_J \mid \theta) = \sum_{r \in \{0,1\}^J} \int E(Y_J \mid \overline{Y}_{J-1} = \bar{y}_{J-1}, r) f(\bar{y}_{J-1}, r \mid \theta)\, dy.$$

where $\bar{y}_{J-1} = (y_1, \ldots, y_{J-1})$ and $r = (r_1, \ldots, r_J)$. Note that sometimes this can be done in closed form. Otherwise, this can be computed using a similar approach to the population-level causal inference above, i.e., by simulating realizations of $(\overline{Y}_{J-1}, \bar{R}_J)$ from the density $f(\bar{y}_{J-1}, r \mid \theta^{(k)})$. See Chapter 4, Section 4.3.5, for details.

In some situations, it might be computationally burdensome to take N_L large; in such cases, one might wish to estimate (or correct) the Monte Carlo error resulting from taking N_L to be finite. We describe an approach to do this in the next example.

Example 3.6.4 (*Accelerated g-Computation (AGC)*). Let $\eta = E(Y_J \mid \theta)$ in Example 3.6.3. Using g-computation, we obtain samples $\widehat{\eta}_1, \ldots \widehat{\eta}_K$ where each $\widehat{\eta}_k$ is a Monte Carlo estimate of the parameter η at iteration k. We might use the approximation $\eta \approx N_L^{-1} \sum_i Y_{iJ}^{\star}$, where the $Y_{iJ}^{\star}$'s are $i = 1, \ldots, N_L$ samples generated according to $f(y, r \mid \theta)$, and, implicitly, using Bayesian g-computation as in previous examples assumes that the Monte Carlo error in $\widehat{\eta}_k$ is negligible. Unfortunately, taking N_L large can be computationally prohibitive. For example, taking $N_L = 100N$ in the mixture modeling example of Chapter 11 is as computationally expensive as fitting the mixture model in the first place, and this is compounded by the fact that the Monte Carlo integration must be run for every setting of a sensitivity parameter ξ. On the other hand, taking N_L too small risks inefficient inferences. Because the Monte Carlo accuracy increases like $\sqrt{N_L}$, we must have N_L grow faster than N for the Monte Carlo error to be of lower order than the posterior standard error of η.

When the posterior distribution of η is approximately normal, it is possible to use *accelerated g-computation* (AGC) [100] to correct the Monte Carlo error exactly, allowing for N_L to be taken as low as N. Let $\widetilde{\eta} = E(\eta \mid O)$ and $V = \text{Var}(\eta \mid O)$, where O denotes the collection of observed data $\{(R_i, Y_{iR_i}) : i = 1, \ldots, n\}$. Then, conditional on $\theta^{(k)}$ at the kth iteration of the chain, by the central limit theorem, we have $\widehat{\eta}_k \sim N(\eta^{(k)}, s_k^2/N_L)$, where $s_k^2 = \text{Var}(Y_{iJ} \mid \theta^{(k)})$. Using the iterated expectation and variance formulas, we can estimate

$$\widetilde{\eta} \approx K^{-1} \sum_k \widehat{\eta}^{(k)} \qquad \text{and} \qquad V \approx \frac{\sum_k (\widehat{\eta}_k - \widetilde{\eta})^2}{K-1} - \frac{1}{K} \sum_k \hat{s}_k^2$$

where $\hat{s}_k^2 = \frac{\sum_i (Y_{k,iJ}^{\star} - \widehat{\eta}_k)^2}{N_L - 1}$ is an estimate of s_k^2. As $K \to \infty$, we have

$$\frac{1}{K} \sum_k \widehat{\eta}_k \to \widetilde{\eta} \quad \text{and} \quad \frac{\sum_k (\widehat{\eta}_k - \widetilde{\eta})^2}{K-1} - \frac{1}{K} \sum_k \hat{s}_k^2 \to V \quad \text{almost-surely given } O.$$

Hence we can compute the approximate posterior $\eta \sim N(\widetilde{\eta}, V)$.

Use of these approximations essentially eliminates the computational burden of g-computation, at the cost of requiring approximate normality. Our experience is that such approximations are quite accurate for the models considered in this monograph. Even if approximate normality does not hold, however, we note that calculating the estimated Monte Carlo contribution to the posterior variance $K^{-1} \sum_k \hat{s}_k^2$ can be useful for determining if a sufficiently large N_L has been chosen. We illustrate AGC in Chapters 11 and 14. We can also use AGC for all the g-computations in the monograph by just outputting the appropriate $\hat{s}_k^2$'s.

3.7 Summary

We have reviewed essential concepts in Bayesian inference for missing data and causal inference that are needed for Bayesian nonparametric approaches to causal inference and missing data. In the next chapter, we provide an extensive discussion of identifiability and the use of sensitivity parameters to capture deviations from uncheckable causal and missing data assumptions.

Chapter 4

Identifiability and sensitivity analysis

DOI: 10.1201/9780429324222-4

Causal inference and missing data methods rely on untestable assumptions (such as the MAR assumption for missing data or the ignorability assumption for observational studies) in order to identify the effects of interest. Because of the inherently uncheckable nature of these assumptions, the best one can do is (i) think carefully about which (if any) assumptions are plausible for a given problem and (ii) assess the robustness of inferences to these assumptions by performing a sensitivity analysis. In this chapter we take a more in-depth look at specifying untestable identifying assumptions and performing a sensitivity analysis to assess them. We start by discussing strategies to calibrate sensitivity parameters. We then introduce identifying restrictions for various causal and missing data settings.

4.1 Calibration of Sensitivity Parameters

To calibrate sensitivity parameters, as introduced in the Examples from Chapter 3, we begin by specifying a baseline or *anchoring* assumption, such as MAR or ignorability. We then embed that assumption within a larger family of assumptions, indexed by a substantively meaningful sensitivity parameter ξ – for example, we will see settings where ξ corresponds to (i) a location-shift parameter corresponding to the effect of missingness or (ii) a shift on the scale of log-odds of an outcome being missing depending on a possibly missing value. After doing this, we must decide on a plausible range of values of ξ to investigate. There are three general approaches we will consider:.

1. We can perform a *tipping point* analysis [101] to identify the values of ξ that cause our substantive inference to change. If the tipping point region is far away from the values of ξ that we deem substantively plausible, then we can conclude that our analysis is robust to violations of our identifying assumptions. We illustrate this in the case study in Chapter 11.
2. We can calibrate the sensitivity parameters based on observed data, with the assumption that the sensitivity parameter can be bounded based on observed data summaries (such as the "proportion of variability explained" by unmeasured confounders, see Section 4.2.2) or standard deviations of "equivalent" quantities in the observed data (see Chapter 11). The latter strategy is advocated in [52].
3. By working with a subject matter expert, we can attempt to construct a realistic informative prior $\pi(\xi)$ for ξ. By incorporating this prior (which, due to non-identifiability, will usually also be the posterior of ξ), we can arrive at a single inference that combines all plausible assumptions in a principled fashion (see [98] and Chapter 8 of [52]).

In any of these cases, we emphasize that collaboration with subject matter experts is essential. The first two strategies have the advantage of not requiring subject-matter input about ξ prior to fitting the model, and we do not have to engage in a possibly complicated elicitation process. The third strategy has the advantage that we reduce the range of possible inference to a single inference that averages over our uncertainty in ξ.

We can also consider "default" priors, in particular for the second strategy above. Given a range of plausible values of the sensitivity parameters, we can consider various priors that weight the values in the interval. Often, one end of the interval (or the middle) corresponds to the anchoring restriction. A common "default" choice is a uniform prior over the interval. However, this can be viewed as potentially putting too much weight near the anchoring restriction; a simple way to place more weight on deviations from the anchoring assumption is to use, for example, a triangular prior with a mode at the opposite end of the interval from the anchoring assumption. A triangular prior allows for differential weighting in a default way [102]. Specifically, the (triangular shaped) prior has a mode at one end of the interval (equal to 2 divided by the length of the interval) and a value of zero at the opposite end. The prior allows for a default differential weighting favoring the desired end of the interval.

In what follows, we introduce several identifying restrictions with embedded sensitivity parameters and discuss how to calibrate them using the approaches described above.

4.2 Identifiability Restrictions for Causal Inference and Sensitivity Analysis

4.2.1 *Sensitivity to the Ignorability Assumption for Causal Inference with a Point Treatment*

One of the key assumptions for identifying the average causal effect in the point treatment setting is ignorability: (see Section 1.1.2): $[Y(a) \perp\!\!\!\perp A \mid L]$. Here we show one approach to carrying out a sensitivity analysis for possible violations of the ignorability assumption, which is related to the approach introduced in [103].

As shown in Section 1.1.2, under SUTVA, ignorability, and positivity, we can identify the counterfactual mean $E\{Y(a)\}$ as

$$E\{Y(a)\} = \int E\{Y(a) \mid L = \ell\}\, dF_L(\ell) \tag{4.1}$$

$$= \int E\{Y(a) \mid L = \ell, A = a\}\, dF_L(\ell) \tag{4.2}$$

$$= \int E(Y \mid L = \ell, A = a)\, dF_L(\ell) \tag{4.3}$$

where (4.2) holds because of ignorability and (4.3) holds because of consistency.

If we are not confident in the ignorability assumption, we can consider either weakening it or accounting for uncertainty about it. What allowed us to go from (4.1) to (4.2) is the fact that under ignorability we have

$$E\{Y(a) \mid L = \ell\} = E\{Y(a) \mid L = \ell, A = 1\} = E\{Y(a) \mid L = \ell, A = 0\}.$$

However, without ignorability, the difference $\Delta_a(\ell) = E\{Y(a) \mid L = \ell, A = 1\} - E\{Y(a) \mid L = \ell, A = 0\}$ may not be 0. Denote by Ψ the average causal effect $\Psi = E\{Y(1) - Y(0)\}$. Then the contrast in standardized means can be written as the true causal effect plus a bias term:

$$\int E(Y \mid L = \ell, A = 1)\, dF_L(\ell) - \int E(Y \mid L = \ell, A = 0)\, dF_L(\ell) = \Psi - \xi,$$

where $\xi = \int [\Delta_0(\ell)e(\ell) + \Delta_1(\ell)\{1-e(\ell)\}]\, dF_L(\ell)$. The form of the above expression can be seen more clearly by writing

$$\begin{aligned} E\{Y(a) \mid L=\ell\} &= \sum_{a'} E\{Y(a) \mid A=a', L=\ell\} \Pr(A=a' \mid L=\ell) \\ &= \sum_{a'} E\{Y(a) \mid A=a', L=\ell\} \Pr(A=a' \mid L=\ell) \\ &\quad + E\{Y(a) \mid A=a, L=\ell\}(1-\Pr(A=a \mid L=\ell)) \\ &\quad - E\{Y(a) \mid A=a, L=\ell\}(1-\Pr(A=a \mid L=\ell)) \\ &= E\{Y(a) \mid A=a, L=\ell\} + (1-\Pr(A=a \mid L=\ell)) \\ &\quad \times [E\{Y(a) \mid A=1-a, L=\ell\} - E\{Y(a) \mid A=a, L=\ell\}]. \end{aligned}$$

For $a=1$, the first term is just $E[Y \mid A=1, L=\ell]$ and the second term is $1-e(\ell)$ times $-\Delta_1(\ell)$.

The ignorability assumption has the implication that $\xi = 0$; dropping this assumption, we can replace ignorability with a *direct* assumption about the value of ξ, with different values of ξ yielding different (but, in terms of the observed data, equally valid) conclusions about Ψ. In this case, ξ is a sensitivity parameter. Sensitivity analysis will typically involve trade-offs between (i) allowing a realistic range of types of violations of ignorability that could happen and (ii) keeping the sensitivity parameters both small in number and interpretable. For example, if we specified $\Delta_a(\ell)$ in a way that is a complex function of a and ℓ, then there would be no realistic way to carry out a sensitivity analysis. Alternatively, a simple functional form for $\Delta_a(\ell)$ with (say) one to three interpretable parameters allows for a sensitivity analysis that can be understood by subject matter experts.

Simplifying and calibrating Δ

Below we illustrate how one might impose structure on $\Delta_a(\ell)$ to facilitate a sensitivity analysis. Suppose we simplify the problem by assuming that people who actually received treatment had potential outcome $Y(a)$ that was Δ units different, on average, from people who did not receive treatment. Suppose the amount Δ does not depend on either a or the confounders L; that is, $\Delta = \Delta_1(\ell) = \Delta_0(\ell)$. Such a scenario could arise if there was an unmeasured confounder, independent of $L <$, that leads to healthier people being more likely to receive treatment. This could be viewed as a worst-case scenario, because our observed L's tell us nothing about the unmeasured confounder. So, although we have simplified the problem, we did so in such a way that could be viewed as conservative. This form also simplifies ξ, as we have

$$\xi = \int \Delta e(\ell) + \Delta\{1-e(\ell)\}\, dF_L = \Delta.$$

Next, we would like to specify a reasonable range for deviations from ignorability via Δ (recall that ignorability implies that $\Delta = 0$). We can do this by calibrating Δ based on a summary of the observed data (Strategy 2 from Section 4.1). Denote by R^2 the total amount of variability in Y that is explained by L. We might assume that unmeasured confounding would account for less than $k\%$ of the remaining variance, i.e., $|\Delta| < \sqrt{\text{Var}(Y)(1-R^2)k}$. We can then specify a sign for Δ (do we expect, if there is unmeasured confounding, that bias will be in the positive or negative direction) and a distribution for k (e.g., if we thought the bound to be 20% of the variability, and we expect the bias to be positive, we might specify a uniform distribution (or a triangular distribution) on $(0, 0.2)$).

Posterior inference for Ψ can then proceed by taking posterior draws from the usual difference in standardized means to compute Ψ under ignorability, taking draws of the sensitivity parameter $\Delta = \xi$, and then adjusting the value of Ψ under ignorability by subtracting ξ. In the case study in Chapter 13, we present an example of this type of sensitivity analysis.

4.2.2 *Sensitivity to the Sequential Ignorability Assumption for Causal Mediation*

The previous subsection explored sensitivity to the ignorability assumption from Section 1.1.2. Thus, sensitivity for the first part of sequential ignorability (1.5) can use those approaches. In this section, we discuss a strategy to assess sensitivity to violations of (1.6), the second part of the sequential ignorability assumption.

Under the sequential ignorability assumptions, (1.5) is guaranteed to hold in settings of a randomized experiment; this occurs in the case study in Chapter 13. However, (1.6) is based on an untestable assumption that there are no unobserved confounding relationships between the mediator and the outcome. Although a few approaches to sensitivity analysis have been proposed [41, 104, 105], they are restricted to settings, such as linear (or generalized linear) models, and do not directly apply to the BNP models introduced in Chapter 6 and Chapter 7. We propose a strategy applicable to these models next.

Similar to the sensitivity to ignorability in the previous section, a convenient starting point for a sensitivity analysis is to start with the proof of identification under the given assumptions. Here, to introduce a method for sensitivity to (1.6), we restate the identification results from [41],

$$
\begin{aligned}
&E[Y\{1, M(0)\} \mid L = \ell] \\
&= \int E\{Y(1,m) \mid M(0) = m, A = 0, L = \ell\}\; dF_{M(0)|A=0,L=\ell}(m) \\
&= \int E\{Y(1,m) \mid A = 0, L = \ell\}\; dF_{M(0)|A=0,L=\ell}(m) && (4.4) \\
&= \int E\{Y(1,m) \mid A = 1, L = \ell\}\; dF_{M(0)|A=0,L=\ell}(m) && (4.5) \\
&= \int E\{Y(1,m) \mid M(1) = m, A = 1, L = \ell\}\; dF_{M(0)|A=0,L=\ell}(m) && (4.6) \\
&= \int E(Y \mid M = m, A = 1, L = \ell)\; dF_{M|A=0,L=\ell}(m)
\end{aligned}
$$

where both (4.4) and (4.6) follow from (1.6). Equality (4.5) follows from (1.5).

Now, assume instead that (1.6) does not hold. Let

$$g_0(m,\ell) = E\{Y(1,m) \mid M(0) = m, A = 0, L = \ell\} - E\{Y(1,m) \mid A = 0, L = \ell\}, \quad \text{and}$$
$$g_1(m,\ell) = E\{Y(1,m) \mid M(1) = m, A = 1, L = \ell\} - E\{Y(1,m) \mid A = 1, L = \ell\}.$$

Using g_0 and g_1, we can reexpress $E[Y\{1,M(0)\} \mid L = \ell]$ without using (1.6) as

$$\begin{aligned}
&E[Y\{1,M(0) \mid L = \ell\}] \\
&= \int E\{Y(1,m) \mid M(0) = m, A = 0, L = \ell\}\, dF_{M(0)|A=0,L=\ell}(m) \\
&= \int \Big\{g_0(m,\ell) + E\{Y(1,m) \mid A = 0, L = \ell\}\Big\}\, dF_{M(0)|A=0,L=\ell}(m) \\
&= \int \Big\{g_0(m,\ell) + E\{Y(1,m) \mid A = 1, L = \ell\}\Big\}\, dF_{M(0)|A=0,L=\ell}(m) \\
&= \int \Big\{g_0(m,\ell) - g_1(m,\ell) + E\{Y(1,m) \mid M(1) = m, A = 1, L = \ell\}\Big\} \\
&dF_{M(0)|A=0,L=\ell}(m) \\
&= \int \Big\{g_0(m,\ell) - g_1(m,\ell) + E\{Y \mid M = m, A = 1, L = \ell\}\Big\}\, dF_{M|A=0,L=\ell}(m).
\end{aligned}$$

Thus, $g_0(m,\ell)$ and $g_1(m,\ell)$ are sensitivity parameters embedded with in (1.6); recall that (1.6) implies that both $g_0(m,\ell)$ and $g_1(m,\ell)$ are 0 for all values of m and ℓ.

To help calibrate the sensitivity parameters, under a randomized experiment where (1.5) holds, we first note that, from (1.5), we have

$$Y(a',m) \perp\!\!\!\perp A \mid L = \ell, \tag{4.7}$$
$$Y(a',m) \perp\!\!\!\perp A \mid M(a) = m', L = \ell. \tag{4.8}$$

Using (4.7) and (4.8), $g_0(m,\ell)$ and $g_1(m,\ell)$ can be simplified to

$$\begin{aligned}
g_0(m,\ell) &= E\{Y(1,m) \mid M(0) = m, A = 0, L = \ell\} - E\{Y(1,m) \mid A = 0, L = \ell\} \\
&= E\{Y(1,m) \mid M(0) = m, L = \ell\} - E\{Y(1,m) \mid L = \ell\}, \qquad (4.9) \\
g_1(m,\ell) &= E\{Y(1,m) \mid M(1) = m, A = 1, L = \ell\} - E\{Y(1,m) \mid A = 1, L = \ell\} \\
&= E\{Y(1,m) \mid M(1) = m, L = \ell\} - E\{Y(1,m) \mid L = \ell\}. \qquad (4.10)
\end{aligned}$$

Similar to the previous section, we can calibrate $g_0(m,\ell)$ and $g_1(m,\ell)$ using the observed data by computing the total amount of variability in the outcomes under $A = 0$ and $A = 1$ explained by L. For $g_1(m,\ell)$ in (4.10), we estimate the coefficient of determination among the treated ($A = 1$) with ℓ as covariates (but not $M(1)$) and denote this as R_1. Then, we might then assume a bound on $g_1(m,\ell)$ of the form

$$|g_1(m,\ell)| \leq \sqrt{\text{Var}(Y \mid A = 1) \times (1 - R_1) \times k_1}$$

where the square root of k_1 is now the percent of total variance not explained by L. So we expect any unmeasured confounding that invalidates (1.6) to have an impact bounded by the variance of $Y|A = 1$ multiplied by $100 \times k_1\%$ of the variability unexplained by the measured confounders. Similarly, for (4.9), we might assume a bound

$$|g_0(m,\ell)| \leq \sqrt{\text{Var}(Y \mid A=1) \times (1-R_1) \times k_0}.$$

Since we expect that $M(0)$ does not explain as much of the variance of $Y(1,m)$ as $M(1)$, we might add the further restriction $k_0 \leq k_1$. Thus, we have two sensitivity parameters (k_0,k_1) bounded in a region of the unit square.

To accommodate these sensitivity parameters, Step 3 of the g-computation algorithm in Example 3.6.2 needs to be adjusted by adding the appropriate offsets as follows:

Modified Step 3 Sample (k_0,k_1) from their prior. Then, approximate

$$E[Y\{a,M(a')\} \mid \theta,\eta] \approx N_L^{-1}\sum_{i=1}^{N_L} E(Y \mid A=a, M=m^{(i)}, L=\ell^{(i)}, \theta) + g_a(m,\ell) - g_{a'}(m,\ell).$$

An alternative strategy for introducing sensitivity parameters for the sequential ignorability assumptions is given in the case study in Chapter 14.

4.2.3 *Monotonicity Assumptions for Principal Stratification*

As discussed in Chapter 1, principal stratification is an approach to causal inference with post-treatment variables. An example of a principal stratification-based estimand is the survival average causal effect (SACE), which often uses a monotonicity assumption for identification [50]. Define $S(a)$ to be the potential survival outcome under treatment a. The (deterministic) monotonicity assumption states that $S(1) \geq S(0)$, i.e.,

$$\Pr\{S(1)=1 \mid S(0)=1\} = 1.$$

Intuitively, this states that any individual who died while receiving the treatment would also have died if they had received the control.

To quantify uncertainty about this assumption, it can be embedded within a stochastic monotonicity assumption [106] via a sensitivity parameter

$$\Pr\{S(1)=1 \mid S(0)=1\} = \Pr\{S(1)=1\} + \rho\left[\min\left\{1, \frac{\Pr\{S(1)=1\}}{\Pr\{S(0)=1\}}\right\} - \Pr\{S(1)=1\}\right].$$

The stochastic monotonicity assumption introduces a sensitivity parameter ρ that controls how likely an individual is to survive under the treatment given that they survived under the control. When a degenerate prior $\rho \equiv 1$ is used (and $\Pr\{S(1)=1\} > \Pr\{S(0)=1\}$), we obtain the deterministic monotonicity assumption results. Uncertainty about ρ, can be encoded by placing a non-degenerate prior over $[0,1]$ (e.g., a uniform or triangular prior).

4.3 Monotone Restrictions for Missing Data and Sensitivity Analysis

In this section, we consider alternatives to, and generalizations of, MAR when missingness is monotone (see Chapter 2). By considering different assumptions, we can gauge how robust inferences are to the occurrence of informative missingness. The motivation for considering monotone missingness is that, in clinical trials, missingness is often associated with patient dropout, wherein we observe the outcome Y_{ij} at the times up-to-and-including the dropout time $\overline{Y}_{iS_i} = (Y_{i1}, \ldots, Y_{iS_i})$, but do not observe the outcome after dropout $\widetilde{Y}_{iS_i} = (Y_{i,S_i+1}, \ldots, Y_{iJ})$. Here, $S_i = \max\{j : R_{ij} = 1\}$ is called the *follow-up time*. For the sake of concreteness, we will consider the schizophrenia clinical trial (SCT) described in Example 2.1.1 throughout this section. Recall that in the SCT Y_{ij} corresponds to a numeric measure of the severity of symptoms (the Positive and Negative Syndrome Scale (PANSS) score).

A helpful observation when missingness is monotone is that MAR, while introduced as a restriction on the selection model, has a dual representation in terms of pattern mixture models. MAR has the intuitive interpretation that missingness at time j depends only on the values of the outcome prior to time j and possibly covariates. In terms of the selection model, we have

$$\Pr(S_i = s \mid Y_i = y, \omega) = \Pr(S_i = s \mid \overline{Y}_{is} = \overline{y}_s, \omega). \tag{4.11}$$

Molenberghs et al. [107] showed that, under monotone missingness, MAR can be expressed in terms of a pattern mixture model (see Section 2.6.2) with the restriction

$$f(y_j \mid \overline{y}_{j-1}, S_i = k, \omega) = f(y_j \mid \overline{y}_{j-1}, S_i \geq j, \omega) \quad \text{for all } 1 \leq k < j. \tag{4.12}$$

When formulated in this fashion, MAR is called the *available case missing value* (ACMV) restriction. Pattern mixture representations like (4.12) are convenient, as they only impose restrictions on the extrapolation distribution (Section 2.6.2) and hence we can introduce sensitivity parameters without affecting the fit of the model to the data.

Throughout the remainder of this section, we develop generalizations and extensions of (4.12) and we explain how to interpret them. We show how these generalizations can be used to perform a sensitivity analysis in the case study in Chapter 11.

4.3.1 *Pattern Mixture Alternatives to MAR*

One approach to assessing sensitivity to MAR is to vary the identifying assumption "discretely" by modifying (4.12). We can understand (4.12) as providing a "donor distribution" for the subjects with missing observations: a missing Y_{ij} is imputed using a comparable group of Y_{ij}'s with $S_i \geq j$, implying that we believe that for predictive purposes these two groups are similar.

To provide motivation for the alternative restrictions, we ask the following question: why should we match $S_i = k$ (those who drop out at time t_j) with $S_i \geq j$ (those who drop out after time j) rather than, for example, $S_i = j$ (those who drop out at exactly time j)? Intuitively, the event $[S_i = k]$ is more similar to the event $[S_i = j]$ than $[S_i \geq j]$, so borrowing from $[S_i = j]$ is arguably more natural; for example, an individual who is known to have dropped out at (say) time t_3 should be more similar to someone who dropped out at time t_4 than someone who dropped out at some time *after* t_4. This leads us naturally to the following alternate restriction.

Definition 4.3.1. We say that the *nearest case missing value* (NCMV) [108] restriction is satisfied if

$$f(y_j \mid \bar{y}_{j-1}, S_i = k, \omega) = f(y_j \mid \bar{y}_{j-1}, S_i = j, \omega) \quad \text{for all } 1 \leq k < j. \tag{4.13}$$

Like ACMV, NCMV places a restriction directly on the extrapolation distribution, and so switching from ACMV to NCMV does not affect the fit of the model to the data (although it will change the estimated treatment effects).

Once one has the idea of considering different donor distributions, many possibilities open up for defining alternatives to MAR. For example, one might take the donor class to consist only of subjects who are fully observed.

Definition 4.3.2. We say that the *complete case missing value* (CCMV) [109] restriction is satisfied if

$$f(y_j \mid \bar{y}_{j-1}, S_i = k, \omega) = f(y_j \mid \bar{y}_{j-1}, S_i = J, \omega) \quad \text{for all } 1 \leq k < j. \tag{4.14}$$

We generally find NCMV to be a more natural assumption than CCMV because the event $[S_i = k]$ seems more comparable to $[S_i = j]$ than $[S_i = J]$, i.e., we would expect outcomes among people who dropped out to be more similar to one another than either of them is to someone who completed the study.

Taking this idea to its most general form, one can define the class of *interior family restrictions* [108, 110], which takes

$$f(y_j \mid \bar{y}_{j-1}, S_i = k, \omega) = \sum_{\ell=j}^{J} \varpi_{\ell,j}(\bar{y}_{j-1}; \omega) f(y_j \mid \bar{y}_{j-1}, S_i = \ell, \omega),$$

where the $\varpi_{\ell,j}(\cdot;\omega)$'s are arbitrary functions satisfying $\varpi_{\ell,j}(\bar{y}_{j-1};\omega) \geq 0$ and $\sum_{\ell=j}^{J} \varpi_{\ell,j}(\bar{y}_{j-1};\omega) = 1$. The interior family contains several special cases of interest:

- $\varpi_{j,j}(\bar{y}_{j-1}) = 1$ corresponds to NCMV.
- $\varpi_{j,J}(\bar{y}_{j-1}) = 1$ corresponds to CCMV.
- $\varpi_{\ell,j}(\bar{y}_{j-1}) = f(S_i = \ell \mid S_i \geq \ell, \bar{y}_{j-1}, \omega)$ corresponds to ACMV.

From the perspective of an anchoring distribution, if we consider ACMV, the above interior family generalizes ACMV with the weights being the sensitivity parameters. Alternatively, given a collection of assumptions, one approach to sensitivity analysis is to simply conduct inference under each assumption and see if any of the substantive conclusions are significantly affected. Our experience is that this approach is usually not powerful enough, as inferences are *usually* fairly robust if one stays in the interior family; in the cases where there is a substantive difference, however, it can be very illuminating to explore features of the data, that lead to the differences post hoc. For an example of this approach, see the case study in Chapter 12. The use of restrictions in the interior family should usually be combined with some approach for introducing continuous sensitivity parameters to the restrictions, as we will discuss in Section 4.3.3.

4.3.2 The Non-Future Dependence Assumption

A particularly useful identifying restriction that is amenable to the introduction of sensitivity parameters is the *non-future dependence* (NFD) restriction [110]. This assumption states

$$\Pr(S_i = s \mid Y_i = y, \omega) = \Pr(S_i = s \mid \overline{Y}_{i(s+1)} = \overline{y}_{s+1}), \tag{4.15}$$

for all possible values of the complete data (y, s). Heuristically, NFD states that the event $[S_i = s]$ depends on the "past" values of the response $(\overline{Y}_s)$ and the "present" (possibly unobserved) value of the response Y_{s+1}; for this reason [110], refer to models satisfying NFD as having *proper time dependence*. MAR is a special case of NFD, which assumes further that missingness depends on the past (but not the present); in fact, MAR is precisely at the intersection of NFD and the interior family. While there are plausible causal mechanisms that do not correspond to models with proper time dependence (such as those involving latent variables), NFD represents a natural generalization of MAR/ACMV.

Equation (4.15) describes the NFD assumption in terms of the selection mechanism. For our purposes, it is more convenient to work with a pattern-mixture representation, as pattern mixture models tell us directly how to impute missing data. NFD has the pattern mixture representation

$$f(y_j \mid \overline{y}_{j-1}, s = k, \omega) = f(y_j \mid \overline{y}_{j-1}, s \geq j-1, \omega) \qquad \text{for all } 1 \leq k < j-1. \tag{4.16}$$

We refer to the pattern mixture representation of NFD as the *non-future missing value* (NFMV) restriction [110].

4.3.3 Completing the NFD Specification

The NFD assumption by itself is insufficient to identify the complete data distribution $f(y, s \mid \omega)$. This is immediate from the fact that NFD is implied by MAR (which *is* sufficient) but is not equivalent to it. The missing piece required to identify an NFD model is a specification of $f(y_j \mid \overline{y}_{j-1}, s = j-1, \omega)$ (i.e., the distribution of the outcome immediately after dropout), which (4.16) does not place any restriction on.

A schematic description of NFD is given in Table 4.1. This table shows how the outcome distributions $f(y_j \mid \overline{y}_{j-1}, s = k, \omega)$ are identified. Below the dividing line, we have $k \geq j$ and the conditional density is identified from the data as none of the outcomes are missing. Above the line, we have $k < j$ and the distribution is not identified. When $k < j-1$, the NFD assumption tells us how to fill in the unidentified distribution, whereas for $k = j-1$ the NFD assumption does not state how to fill in the unidentified distribution. Completing the NFD assumption is equivalent to filling in the ?'s in Table 4.1.

One strategy to complete NFD is to notice that the additional restriction

$$f(y_j \mid \overline{y}_{j-1}, s = j-1, \omega) = f(y_j \mid \overline{y}_{j-1}, s \geq j, \omega)$$

Table 4.1 *Schematic representation of NFD when $J = 4$. Distributions above the dividing line are not identified by the observed data (dependence on ω is suppressed). Subscripting by j denotes conditioning on the event $S = j$, while subscripting by $\geq j$ denotes conditioning on the event $S \geq j$.*

	$j = 1$	$j = 2$	$j = 3$	$j = 4$
$S = 1$	$f_1(y_1)$	?	$f_{\geq 2}(y_3 \mid \bar{y}_2)$	$f_{\geq 3}(y_4 \mid \bar{y}_3)$
$S = 2$	$f_2(y_1)$	$f_2(y_2 \mid y_1)$	?	$f_{\geq 3}(y_4 \mid \bar{y}_3)$
$S = 3$	$f_3(y_1)$	$f_3(y_2 \mid y_1)$	$f_3(y_3 \mid \bar{y}_2)$	?
$S = 4$	$f_4(y_1)$	$f_4(y_2 \mid y_1)$	$f_4(y_3 \mid \bar{y}_2)$	$f_4(y_4 \mid \bar{y}_3)$

turns NFD into the MAR assumption; equivalently, MAR corresponds to the pair of assumptions

$$[Y_{ij} \mid \overline{Y}_{i,j-1}, S_i = j-1, \omega] \stackrel{d}{=} [Y_{ij} \mid \overline{Y}_{i,j-1}, S_i \geq j, \omega] \qquad \text{and} \tag{4.17}$$

$$[Y_{ij} \mid \overline{Y}_{i,j-1}, S_i = k, \omega] \stackrel{d}{=} [Y_{ij} \mid \overline{Y}_{i,j-1}, S_i \geq j-1, \omega] \quad \text{for all } 1 \leq k < j-1, \tag{4.18}$$

while (4.18) alone defines NFD. Using this observation, we can construct a subfamily of NFD which "passes through" MAR. For continuous outcomes, one approach is to use a family of location transformations by replacing (4.17) with

$$[Y_{ij} \mid \overline{Y}_{i,j-1}, S_i = j-1, \omega] \stackrel{d}{=} [Y_{ij} + \xi_j \mid \overline{Y}_{i,j-1}, S_i \geq j, \omega]. \tag{4.19}$$

The parameters $\xi = (\xi_2, \ldots, \xi_J)$ are sensitivity parameters, with MAR corresponding to $\xi = (0, \ldots, 0)$. Thus, we have embedded MAR within NFD via sensitivity parameters, ξ. Importantly, ξ has an interpretation which is easily understood by practitioners:

> *Suppose that individual A and individual B are observed to be identical up to time $j-1$ in terms of their response history and covariates, but that A drops out ($S_i = j-1$) while B does not ($S_i \geq j$). Then, on average, A will have a response Y_{ij} that is ξ_j units higher at time j than B; other than this difference, the distribution of the responses is the same.*

In total, this family has $J-1$ sensitivity parameters, and it can be difficult to consider the impact of varying all 15 parameters simultaneously. For example, for the SCT, we have 15 sensitivity parameters: five parameters $\{\xi_j : j = 2, \ldots, 6\}$ for each of the three treatment arms. To address this, one might reduce the number of sensitivity parameters by (say) taking a single $\xi_j \equiv \xi$ for each treatment as we did in Section 4.2.1 and calibrate these sensitivity parameters using observed data summaries as discussed in Sections 4.2.1 and 4.2.2. The case study in Chapter 11 illustrates calibrating the sensitivity parameters using observed data summaries as well as the tipping point strategy.

The strategy of shifting Y_{ij} by a fixed amount ξ_j can be extended to outcomes of other types through the use of a nonlinear link function. For example, when Y_{ij} is binary, we can modify (4.19) by setting

$$\begin{aligned}&\operatorname{logit}\Pr(Y_{ij}=1 \mid \overline{Y}_{i,j-1}, S_i = j-1, \omega)\\&\quad = \operatorname{logit}\Pr(Y_{ij}=1 \mid \overline{Y}_{i,j-1}, S_i \geq j, \omega) + \xi.\end{aligned} \tag{4.20}$$

An elicitation strategy for this setting can be found in the second case study in Chapter 10 of [52].

If it is easier for clinicians to think in terms of the effect of the outcome on missingness rather than vice versa, then using Bayes rule we can reformulate the above restriction in terms of the selection model as

$$\begin{aligned}&\operatorname{Odds}(S_i = j-1 \mid Y_{ij}=1, \overline{Y}_{i,j-1}, S_i \geq j-1, \omega)\\&\quad = \operatorname{Odds}(S_i = j-1 \mid Y_{ij}=0, \overline{Y}_{i,j-1}, S_i \geq j-1, \omega)\, e^{\xi},\end{aligned}$$

so that ξ can also be interpreted as the effect of the (unobserved) value of Y_{ij} on the probability that subject i will be missing at time t_j (i.e., $S_i = j-1$) or not. Hence we have the following interpretation.

> *Suppose that individual A and individual B are observed to be identical up to time $j-1$ in terms of their response history and covariates, but that $Y_{ij}=1$ for individual A and $Y_{ij}=0$ for individual B. Then the odds of individual A having dropout time $[S_i = j-1]$ are e^{ξ} times as big as the odds of individual B.*

It is also possible to elicit sensitivity parameters on the scale of relative risks rather than odds ratios as well; see [98] for an illustration and a discussion of this.

4.3.4 *g-Computation for Interior Family Restrictions*

We describe now how to perform g-computation for any identifying restriction in the interior family. For concreteness, suppose that we are interested in the change from baseline $\eta = E(Y_{iJ} - Y_{i1} \mid \omega)$. For a given value of ω, we estimate η using the following steps.

1. Using MCMC, collect a sample of ω's from the posterior distribution.
2. For each value of ω, sample a large number of pairs $(Y_i^\star, S_i^\star)$ for $i = 1, \ldots, N_L$. This is done in two steps for each i:
   (a) Sample new observed data $(\overline{Y}^\star_{iS_i^\star}, S_i^\star)$ from the model $f(\overline{Y}_{iS_i} = \overline{y}_s, S_i = s \mid \omega)$.
   (b) For $j = s+1, \ldots, J$, do the following:
      i. Sample $\ell \sim \text{Categorical}\{\varpi_{jj}(\overline{Y}^\star_{i,j-1}), \ldots, \varpi_{Jj}(\overline{Y}^\star_{i,j-1})\}$.
      ii. Sample $Y^\star_{ij} \sim f(y_j \mid \overline{Y}_{i,j-1} = Y^\star_{i,j-1}, S_i = \ell, \omega)$.
3. Approximate $\eta \approx \frac{1}{N_L}\sum_{i=1}^{N_L} Y^\star_{iJ} - Y^\star_{i1}$ with simulation standard error given by $\{\widehat{\text{Var}}(Y^\star_{iJ} - Y^\star_{i1})/N_L\}^{1/2}$ where $\widehat{\text{Var}}(Y^\star_{iJ} - Y^\star_{i1})$ denotes the empirical variance of the $(Y^\star_{iJ} - Y^\star_{i1})$'s.

This algorithm applies immediately to the ACMV, CCMV, and NCMV restrictions, using their respective definitions of the $\varpi_{\ell j}$'s. By returning the simulation standard error, we allow for the possibility of applying the AGC algorithm (described in Chapter 3, Section 3.6.4) to remove the simulation error.

4.3.5 g-Computation for the NFD Restriction

The algorithm for performing g-computation under NFD is very similar to the algorithm for interior family restrictions given in Section 4.3.4. The only change we need to make is to replace steps (2bi) and (2bii) with the following steps.

(i) Sample $\ell \sim \text{Categorical}\{\varpi_{j-1,j}(\overline{Y}^{\star}_{i,j-1}),\ldots,\varpi_{Jj}(\overline{Y}^{\star}_{i,j-1})\}$ where $\varpi_{\ell j}(\overline{y}_{j-1}) = f(S_i = \ell \mid S_i \geq j-1, \overline{Y}_{i,j-1} = \overline{y}_{j-1})$.

(ii) Sample $Y^{\star}_{ij} \sim f(y_j \mid \overline{Y}_{i,j-1} = Y^{\star}_{i,j-1}, S_i = \ell, \omega)$.

Under NFD, we are additionally required to specify the unidentified distribution $f(y_j \mid \overline{y}_{j-1}, S_i = j-1, \omega)$, as this is not identified from the observed data. In the case of the location shift, this distribution can be sampled by sampling $Y^{\star}_{ij} \sim f(y_j \mid \overline{Y}^{\star}_{i,j-1}, S_i \geq j, \omega)$ and shifting the result by ξ, while in the binary case we instead sample $Y^{\star}_{ij}$ according to (4.20).

4.3.6 Differential Reasons for Dropout

In addition to observing the follow-up time, we often have auxiliary information in the form of the *reason* for the dropout. This information is important because, intuitively, patients who are removed from a clinical trial due to (say) progression of symptoms should be imputed differently than individuals who are removed from the trial due to (say) pregnancy. To take this information into account, we can divide reasons for missingness into informative reasons (withdrawal of consent, lack of efficacy, disease progression, and physician decision) and non-informative reasons (pregnancy, side effects, and accidental protocol violation).

It is important to account for the reason for dropout in the analysis because of how it interacts with the sensitivity analysis. The more non-MAR missingness there is, the more influence the sensitivity parameters have on the inference, decreasing the robustness of our final conclusions. By taking into account the missingness reason, we reduce the impact of individuals who are missing for "MAR reasons," making our inferences more robust to variations in the sensitivity parameters. We illustrate this in the case study in Chapter 11.

4.4 Non-monotone Restrictions for Missing Data and Sensitivity Analysis

Compared to the setting of monotone missingness, where missingness is due primarily to dropout, the problem of developing identifying restrictions for non-monotone missingness is more difficult. There are several issues unique to the setting of non-monotone missingness that make the problem particularly complex.

(i) Unlike the monotone setting where we can identify MAR with the available case missing value (ACMV) restriction, it is difficult to write MAR in the non-monotone setting in the pattern mixture form in a manner that is practically useful. This makes it difficult to anchor an analysis to the MAR assumption.

(ii) Additionally, it has been argued that MAR does not correspond to a plausible causal mechanism for generating non-monotone missingness in longitudinal settings [111, 112], making the use of MAR as a benchmark for comparison itself dubious.

In this section we develop some alternative identifying assumptions that can be used to identify and compute treatment effects and describe how interpretable sensitivity parameters can be introduced.

For clarity, we will put the identifying restrictions in context by linking them to the *Breast Cancer Prevention Trial* (BCPT), which was introduced in Chapter 2 and will be analyzed in Chapter 12.

Example 4.4.1 (*Missingness in BCPT*). The BCPT was a large-scale clinical trial with the goal of determining the efficacy and side effects of the drug tamoxifen when used as a treatment for breast cancer in women. As tamoxifen may be used as a prophylactic, it is important to consider both its efficacy in preventing breast cancer and its effect on the quality of life of women; in particular, there was some concern that tamoxifen increased the risk of clinical depression.

Our primary interest is in the relationship between tamoxifen and depression (as measured by the PANSS). Over the course of the study, there was substantial missingness, and there is potential non-ignorability due to the fact that if a woman is depressed at time t_j then they will be less likely to have their depression measured at that time, even after accounting for their observed depression at other times.

4.4.1 The Partial Missing at Random Assumption

We can achieve something like MAR when the aspect of the missingness that is informative about the response is the dropout time. Let $g(R_i)$ denote a generic *coarsening* of the missingness indicator. We say that the missingness is *partially missing at random* (PMAR; [113]) if we have

$$f(R_i = r \mid Y_i = y, g(R_i) = g(r), \omega) = f(R_i = r \mid Y_{ir} = y_r, g(R_i) = g(r), \omega),$$

for all observed values of y_r and values of y consistent with y_r. PMAR is completely analogous to MAR, except that we additionally condition the missing data mechanism on $g(R_i)$ rather than just Y_i. An immediate consequence is that, as with MAR, the likelihood function can be factored in a particularly simple fashion:

$$L_{\text{obs}}(\omega_y, \psi_r, \psi_g) = L_{y,g}(\omega_y, \psi_g)\, L_r(\psi_r)$$

where $\psi = (\psi_r, \psi_g)$ parameterize $f(R_i = r \mid Y_i = y, g(R_i) = g(r))$ and $f(g(R_i) = g(r))$, respectively, and

$$L_{y,g}(\omega_y, \psi_g) = \prod_{i=1}^{n} f(Y_{ir_i} = y_{r_i}, g(R_i) = g(r_i) \mid \omega_y, \psi_g), \qquad \text{and}$$

$$L_r(\psi_r) = \prod_{i=1}^{n} f(R_i = r_i \mid Y_{ir_i} = y_{r_i}, g(R_i) = g(r_i) \mid \psi_r).$$

Proof. Suppressing dependence on the parameters and i, we have

$$\begin{aligned} f(Y=y_r,R=r) &= \int f(Y=y,g(R)=g(r))\,f(R=r\mid Y=y,g(R)=g(r))\,dy_{-r} \\ &= \int f(Y=y,g(R)=g(r))\,f(R=r\mid Y=y_r,g(R)=g(r))\,dy_{-r} \\ &= f(Y=y_r,g(R)=g(r))\,f(R=r\mid Y=y_r,g(R)=g(r)). \end{aligned}$$

□

If we additionally assume that (ω_y, ψ_g) is a priori independent of ψ_r, then we say that the missing data mechanism is *partially ignorable (PI) given* $g(R)$. We can use the PI assumption to revert back to the setting of monotone missingness by taking $g(R_i) = S_i = \max\{j : R_{ij} = 1\}$ to be the dropout time. Rather than being free from modeling the entire missing data mechanism (the case with ignorability), PI frees us from needing to specify a model for $f(R_i = r_i \mid Y_i = y_i, g(R_i) = g(r_i), \psi_r)$. Essentially all strategies for identifying the causal estimands of interest and introducing sensitivity parameters now carry over from the monotone setting.

Computationally, inference under PI allows for the same algorithms to be used as under monotone missingness, as PI identifies the distribution $f(\bar{y}_j \mid s = j, \omega)$ when $g(r) = \max\{j : r_j = 1\}$ is the follow-up time. Hence we can implement the g-computation algorithm for MAR missingness directly.

Example 4.4.2 (*Partial MAR in BCPT*). The partial MAR assumption was used by [98] to analyze the Breast Cancer Prevention Trial (BCPT) data. This is sensible if it is judged that depression and missingness are related only through the effect of depression on the dropout time (S_i), but is not predictive of any intermittent missingness given the dropout time.

4.4.2 Generic Non-Monotone Restrictions

PMAR may not be appropriate when there is no natural coarsening $g(r)$ to apply it to, or (in the case of reducing the dropout time) if we believe the intermittent missing values may be predictive of their missing outcome. There have been several recent proposals for addressing non-monotone missingness in ways that do not invoke PMAR. Like our modifications to MAR under monotone missingness, the assumptions we discuss are best interpreted as specifying an appropriate imputation (or "donor") distribution for $f(Y_j = y_j \mid Y_r = y_r, R = r)$ based on an alternative pattern $[R = r^\star]$ for which Y_j is identified. We consider the following assumptions, which are also considered in [114]. Each of these assumptions can be viewed as non-monotone extensions of the assumptions considered in [115].

Nearest Identified Pattern Suppose we believe that, all other observed quantities being equal, missingness at time t_j is not predictive of depression at time t_j. We can then impose the *nearest identified pattern* (NIP) restriction [55]

$$f(y_j \mid R_i = r, y_r, \omega) = f(y_j \mid R_i = r_j^\star, y_r, \omega) \tag{4.21}$$

where $r_j^\star$ is equal to r, but with the jth component fixed at 1. That is, we identify the pattern $R_i = r$ by making the smallest possible change to r, which identifies the distribution.

Sequential Explainability Suppose we believe that the observed depression levels prior to time j are sufficient to predict whether or not a subject will be measured at time j, while the outcome at time j is not predictive. We can then impose the *sequential explainability* (SE) restriction [112]

$$f(y_j \mid \overline{o}_{j-1}, R_{ij} = 0, \omega) = f(y_j \mid \overline{o}_{j-1}, R_{ij} = 1, \omega). \tag{4.22}$$

Here, $\overline{O}_{i,j-1}$ denotes the *observed data up to time* $j-1$, i.e., $\overline{O}_{i,j-1} = (Y_{ik}, R_{ik} : k \leq j-1, R_{ik} = 1)$.

The nearest identified pattern is "minimal" in the sense that the pattern $[R = r_j^\star]$ that we borrow from is the one that makes the smallest possible modification to r while still identifying the distribution of Y_j when $r_{ij} = 0$. We note that neither NIP nor SE identify the full joint distribution $[Y_{-r} \mid Y_r, R = r]$, as it leaves the joint distribution of $[(Y_j, Y_k) \mid R = r]$ unidentified when $r_{ij} = r_{ik} = 0$. When our interest lies in the mean $E(Y_j \mid \omega)$, or the contrast $E(Y_j - Y_1 \mid \omega)$ when Y_1 is always observed, then the lack of identification of the joint is not concerning, as we will still be able to carry out our primary inferential goals.

Full identification of the distribution of Y can be obtained using one of the following assumptions.

Pairwise Missing at Random Suppose we believe that the distribution of the missing responses can be reasonably approximated using an equivalent subject who was observed at all times. We can then posit the pairwise missing at random (pairwise MAR) [116] restriction

$$f(y_{-r} \mid R_i = r, y_r, \omega) = f(y_{-r} \mid R_i = \mathbf{1}, y_r, \omega). \tag{4.23}$$

Pairwise MAR is a generalization of CCMV to non-monotone missingness and is also "minimal" in a sense: it identifies $[Y_{-r} \mid Y_r, R = r]$ with the *single* pattern $R = \mathbf{1}$ which identifies the joint distribution of Y. In many cases, however, we might think that the *complete case* pattern $[R = \mathbf{1}]$ is quite different from patterns with missingness and may seek alternative restrictions.

Itemwise Conditional Independence Suppose we believe that the distribution of a missing response Y_{ij} can be reasonably approximated as being conditionally independent of R_{ij} given both the missingness indicators and outcomes (both missing and not) at all other times. We can then posit the *itemwise conditional independence* (ICIN) assumption

$$\begin{aligned} &f(y_j \mid Y_{-j} = y_{-j}, R_{i,-j} = r_{-j}, R_j = 0, \omega) \\ &\quad = f(y_j \mid Y_{-j} = y_{-j}, R_{i,-j} = r_{-j}, R_j = 1, \omega). \end{aligned}$$

Unlike the other restrictions discussed so far, it is not at all trivial to show that the ICIN restriction defines a valid identifying restriction, as the law $[Y_j \mid Y_{-j} =$

$y_{-j}, R_{i,-j} = r_{-j}, R_j = 1]$ conditions on the value of both missing and observed data. Despite this, the ICIN assumption can be shown to nonparametrically identify the joint distribution of $[Y, R; \omega]$ [54, 56].

ICIN is similar to NIP in that both identify Y_{ij} when $R_{ij} = 0$ by looking at the closest missingness pattern such that $R_{ij} = 1$. The difference is that ICIN conditions on the *complete* data $Y_{i,-j}$ whereas NIP conditions on the *observed* data $\{Y_{ik} : k \neq j, r_k = 1\}$. Interestingly, ICIN can also be shown to be equivalent to NCMV when missingness is monotone [54]. The downside of ICIN as an identifying restriction is that it is not, in our experience, easy to implement using the g-formula.

General comments on these restrictions

On a first read, it can be difficult to get a feel for how these assumptions differ from one another mathematically; much less come to a conclusion regarding which assumption best reflects our substantive beliefs about the form of the missing data mechanism. In practice, the different assumptions often lead to similar conclusions about the effects of interest. When they *do* produce different conclusions, however, they can point us toward the features of the missingness process that are driving our conclusions. We will see precisely this benefit when we analyze the BCPT in Chapter 12.

Unlike the other identifying restrictions we have considered for longitudinal data, SE and NIP *do not identify the joint distribution of* Y_i. The marginal for each Y_{ij} is, however, identified, with both (4.22) and (4.21) stating explicitly how to simulate from the marginal distribution of Y_{ij} whenever $R_{ij} = 0$. For this reason, assuming that $R_{i1} \equiv 1$, we are limited to estimating features of the joint distribution of (Y_{i1}, Y_{ij}) such as the change-from-baseline $E(Y_{ij} - Y_{i1} \mid \omega)$.

4.4.3 *Computation of Treatment Effects under Non-Monotone Missingness*

For most of the identifying assumptions discussed above, it is fairly simple to implement the g-formula via Monte Carlo integration. For example, Algorithm 4.1 gives an implementation of the NIP assumption (4.21).

Algorithm 4.1 returns both the Monte Carlo g-formula estimate of $\eta = E(Y_{iJ} - Y_{i1} \mid \omega)$ as well as an estimate of the Monte Carlo variance $s^2_{\hat{\eta}}$ that would be needed for the AGC algorithm introduced in Section 3.6.4 if desired. Algorithm 4.1 is very straight forward to implement, being simpler even than algorithms for monotone missingness in Section 4.3.

As another illustration, we give an algorithm for computing η under the pairwise MAR assumption in Algorithm 4.2. Algorithm 4.2 is basically identical to Algorithm 4.1 except that the distribution used to sample the missing values differs at step 2.b.ii. Unlike NIP, which samples $Y^\star_{iJ}$ under different patterns for each sample, pairwise MAR only uses the fully observed pattern $R_i = \mathbf{1}$.

Algorithm 4.1 Monte Carlo implementation of the g-formula for computing $\eta = E(Y_{iJ} - Y_{i1} \mid \omega)$ under the NIP assumption

1. Update ω by sampling from the posterior distribution.
2. For $i = 1, \ldots, N_L$:
   (a) Sample $(Y^{\star}_{iR^{\star}_i}, R^{\star}_i) \sim f(y_r, r \mid \omega)$.
   (b) If $R^{\star}_{iJ} = 0$:
      i. Set $r = r^{\star} = R^{\star}_i$ then set $r^{\star}_J = 1$.
      ii. Sample $Y^{\star}_{iJ} \sim f(y_J \mid Y_{ir} = Y^{\star}_{ir}, R_i = r^{\star})$.
3. Set $\widehat{\eta} = (N_L)^{-1} \sum_{i=1}^{N_L} (Y^{\star}_{iJ} - Y^{\star}_{i1})$ and $s^2_{\widehat{\eta}} = \{N_L(N_L - 1)\}^{-1} \sum_i (Y^{\star}_{iJ} - Y^{\star}_{i1} - \widehat{\eta})^2$.
4. Return $(\widehat{\eta}, s^2_{\widehat{\eta}})$.

Algorithm 4.2 Monte Carlo implementation of the g-formula for computing $\eta = E(Y_{iJ} - Y_{i1} \mid \omega)$ under the pairwise MAR

1. Update ω by sampling from the posterior distribution.
2. For $i = 1, \ldots, N_L$:
   (a) Sample $(Y^{\star}_{iR^{\star}_i}, R^{\star}_i) \sim f(y_r, r \mid \omega)$.
   (b) If $R^{\star}_{iJ} \neq 1$:
      i. Set $r = R^{\star}_i$.
      ii. Sample $Y^{\star}_{i,-r} \sim f(y_{-r} \mid Y_{ir} = Y^{\star}_{ir}, R_i = \mathbf{1})$.
3. Set $\widehat{\eta} = (N_L)^{-1} \sum_{i=1}^{N_L} (Y^{\star}_{iJ} - Y^{\star}_{i1})$ and $s^2_{\widehat{\eta}} = \{N_L(N_L - 1)\}^{-1} \sum_i (Y^{\star}_{iJ} - Y^{\star}_{i1} - \widehat{\eta})^2$.
4. Return $(\widehat{\eta}, s^2_{\widehat{\eta}})$.

4.4.4 *Strategies for Introducing Sensitivity Parameters*

We can introduce sensitivity parameters for non-monotone missingness in much the same way that we did under monotone missingness. For continuous data, we can introduce location-shift parameters to model the relationship between missingness and the outcome. The following example anchors the sensitivity analysis to the nearest identified pattern restriction.

Example 4.4.3 (*Sensitivity analysis for a continuous outcome in the BCPT*)**.** Suppose in the BCPT that Y_{ij} is a continuous measure of depression at time t_j and that we observe $R_i = r$ and $Y_{ir} = y_r$ with $R_{iJ} = 0$ (i.e., Y_{iJ} is missing). Suppose we believe that, given all observed data, the depression level at time t_J is expected to be ξ_r units higher if an individual is missing than if an otherwise identical individual were observed. Then we can set $[Y_{iJ} \mid R_i = r, Y_{ir} = y_r] \stackrel{d}{=} [Y_{iJ} + \xi_r \mid R_i = r^{\star}_J, Y_{ir} = y_r]$. Setting $\xi_r = 0$ corresponds to the NIP restriction (4.21).

Location-shift sensitivity parameters are particularly convenient to use when the estimand of interest is a mean parameter. For instance, Example 4.4.3 has

$$E(Y_{iJ} \mid \omega, \xi) = E(Y_{iJ} \mid \omega, \xi = \mathbf{0}) + \sum_{r:r_J=0} \xi_r \Pr(R_i = r \mid \omega). \tag{4.24}$$

The equation (4.24) can be used to change the value of ξ_r without needing to perform Monte Carlo integration multiple times. When $\xi_r \equiv \xi$ is a constant across r, we get the further simplification $\sum_{r:r_J=0} \xi \Pr(R_i = r) = \xi \Pr(R_{iJ} = 0)$.

For binary data, we once again can apply the location-shift idea on the logit scale to shift the probability of success.

Example 4.4.4 (*Sensitivity analysis for a binary outcome in the BCPT*). Suppose in the BCPT that Y_{ij} is a binary measure of depression (1 for depressed and 0 otherwise) and that we observe $R_i = r$ and $Y_{ir} = y_r$ with $R_{iJ} = 0$ (i.e., Y_{iJ} is missing). Suppose we believe that, given all observed data, the *odds* of depression at time t_J are e^{ξ_r} times larger if an individual is missing than if an otherwise identical individual were observed. Then we can set

$$\begin{aligned} &\Pr(Y_{iJ} = 1 \mid R_i = r, Y_{ir} = y_r, \omega) \\ &\quad = \operatorname{expit}\{\xi_r + \operatorname{logit}\Pr(Y_{iJ} = 1 \mid R_i = r_J^\star, Y_{ir} = y_r, \omega)\}. \end{aligned} \tag{4.25}$$

Due to the well-known symmetry property of the odds ratio

$$\frac{\text{Odds}(A = 1 \mid B = 1)}{\text{Odds}(A = 1 \mid B = 0)} = \frac{\text{Odds}(B = 1 \mid A = 1)}{\text{Odds}(B = 1 \mid A = 0)}$$

the assumption (4.25) also has an interpretation in terms of the selection model:

$$\begin{aligned} &\Pr(R_i = r \mid Y_r, Y_{iJ} = 1, R_i \in \{r, r_J^\star\}, \omega) \\ &\quad = \operatorname{expit}\{\xi_r + \operatorname{logit}\Pr(R_i = r \mid Y_r, Y_{iJ} = 0, R_i \in \{r, r_J^\star\}, \omega)\}. \end{aligned} \tag{4.26}$$

Note that assumption (4.26) makes sense regardless of whether the response is binary or continuous, and leads to the class of *exponentially tilted* NIP models

$$f(y_J \mid y_r, r, \omega) = \frac{f(y_J \mid y_r, r_j^\star, \omega)\, e^{\xi_r(y_J)}}{E\{e^{\xi_r(Y_{iJ})} \mid Y_r = y_r, R_i = r, \omega\}},$$

where $\xi_r(y_J)$ represents the effect of y_J on the log-odds of $R_i = r$ and is a non-identified user-specified function of y_J. Setting $\xi_r(y_J) = 0$ corresponds to the NIP restriction (4.21). This exponential tilting strategy can generally be used to embed sensitivity parameters into any of the anchoring assumptions we have discussed so far.

Example 4.4.5 (*More sensitivity analysis in the BCPT*). Both the location-shift and logit-shift approaches can be used with the other identifying restrictions for non-monotone missingness discussed here. For example, an analogous version of the location-shift approach for pairwise MAR is to take $[Y_{iJ} \mid R_i = r, Y_{ir} = y_r, \omega] =$

$[Y_{iJ} + \xi_r \mid R_i = \mathbf{1}, Y_{ir} = y_r, \omega]$, and an analogous version of the logit-shift approach takes

$$P(Y_{iJ} = 1 \mid R_i = r, Y_{ir} = y_r, \omega) \\ = \text{expit}\{\xi_r + \text{logit}\, P(Y_{iJ} = 1 \mid R_i = \mathbf{1}, Y_{ir} = y_r, \omega)\}.$$

Calibration of these sensitivity parameters and priors can proceed as in the monotone setting. However, in general, there is nothing requiring us to use a simple location shift for continuous data; location-scale shifts, or other more exotic transformations, are equally easy to implement but are less interpretable by clinicians. Algorithm 4.3 shows how to implement the location-shift model to obtain a sensitivity analysis anchored at pairwise MAR when $\xi_r \equiv \mathbf{0}$. In Chapter 12, we apply the logit-shift approach to the BCPT and provide an associated algorithm.

Algorithm 4.3 Monte Carlo implementation of the g-formula for computing $\eta = E(Y_{iJ} - Y_{i1} \mid \omega)$ under a location-shifted pairwise MAR assumption.

1. Update ω by sampling from the posterior distribution.
2. For $i = 1, \ldots, N_L$:
   (a) Sample $(Y^\star_{iR^\star_i}, R^\star_i) \sim f(y_r, r \mid \omega)$.
   (b) If $R^\star_{iJ} \neq 1$:
      i. Set $r = R^\star_i$.
      ii. Sample $Z \sim f(y_j \mid Y_{ir} = Y^\star_{ir}, R_i = \mathbf{1})$ and set $Y^\star_{iJ} = Z + \xi_r$.
3. Set $\widehat{\eta} = (N_L)^{-1} \sum_{i=1}^{N_L} (Y^\star_{iJ} - Y^\star_{i1})$ and $s^2_{\widehat{\eta}} = \{N_L(N_L - 1)\}^{-1} \sum_i (Y^\star_{iJ} - Y^\star_{i1} - \widehat{\eta})^2$.
4. Return $(\widehat{\eta}, s^2_{\widehat{\eta}})$.

4.5 Summary

In this chapter, we introduced strategies for sensitivity analysis, including introducing and calibrating sensitivity parameters for a variety of uncheckable assumptions for causal inference including ignorability, sequential ignorability, and monotonicity, and for a variety of uncheckable assumptions for monotone and non-monotone missing data. We also demonstrated how to incorporate them into g-computation algorithms. We will illustrate these strategies in the case studies in Chapters 10-15.

Part II

Bayesian nonparametrics for causal inference and missing data

Chapter 5

Bayesian decision trees and their ensembles

DOI: 10.1201/9780429324222-5

5.1 Motivation: The Need for Priors on Functions

Consider the problem of inferring the average causal effect of a binary exposure $\Delta = E\{Y(1) - Y(0)\}$. As discussed in Chapter 1, Section 1.1.2, given the Stable Unit Treatment Value (SUTVA), consistency, ignorability, and positivity assumptions, Δ can be identified from the observed data in two different ways. First, we can estimate the *regression* function

$$\mu_a(x) = E\{Y(a) \mid X = x\} = E(Y \mid A = a, X = x).$$

Given an estimate of $\widehat{\mu}_a(x)$ of this function, a frequentist estimate of Δ is

$$\widehat{\Delta}_\mu = N^{-1} \sum_{i=1}^{N} \{\widehat{\mu}_1(X_i) - \widehat{\mu}_0(X_i)\} = \int \{\widehat{\mu}_1(x) - \widehat{\mu}_0(x)\}\, \mathbb{F}_N(dx)$$

where $\mathbb{F}_N$ is the empirical distribution of the X_i's. Alternatively, from Proposition 1.3.1 we can use the *propensity score* $e(x) = \Pr(A = 1 \mid X = x)$ to identify Δ as

$$\Delta = E\left\{\frac{AY(1)}{e(X)} - \frac{(1-A)Y(0)}{1-e(X)}\right\} = E\left\{Y \frac{A - e(X)}{e(X)\{1 - e(X)\}}\right\}.$$

Given an estimate of the propensity score $\widehat{e}(x)$, we can construct the estimator

$$\widehat{\Delta}_e = N^{-1} \sum_{i=1}^{N} Y_i \frac{A_i - \widehat{e}(X_i)}{\widehat{e}(X_i)\{1 - \widehat{e}(X_i)\}}.$$

The estimates $\widehat{\Delta}_\mu$ and $\widehat{\Delta}_e$, while fundamentally different, both require estimating *functions* ($\mu_a(\cdot)$ or $e(\cdot)$). Other approaches to estimating Δ exist, such as doubly robust estimating equations [117, 118], or conditional density regression of Y on the propensity score $e(X)$ as illustrated in Chapter 8; however, all of these methods require estimating at least one, and possibly both, of $\mu_a(x)$ and $e(x)$.

If interest lies in the average causal effect Δ, then the functions $\mu_a(x)$ and $e(x)$ are not themselves of direct interest. The parametric Bayesian approach is to specify models such as

$$\mu_a(x) = \beta_0 + x^\top \beta \qquad \text{or} \qquad e(x) = \{1 + \exp(-\gamma_0 - x^\top \gamma)\}^{-1}. \tag{5.1}$$

There are two reasons one might favor models like (5.1).

Reason 1: It is easier to interpret the regression coefficients $(\beta_0, \beta, \gamma_0, \gamma)$ than it is to interpret the fit of nonparametric models.

Reason 2: Parametric models with few parameters tend to yield low-variance estimates of Δ.

If we are strictly concerned with estimating Δ, Reason 1 is not compelling because we are not interested in interpreting the parameters of $\mu_a(x)$ and $e(x)$; we can interpret Δ directly, regardless of how $\mu_a(x)$ and $e(x)$ are estimated. In the absence of

Reason 1, Reason 2 is also not compelling because there are less crude ways to stabilize estimators of Δ than artificially restricting attention to parametric models. On the other hand, while models like (5.1) might result in lower variance for Δ, they also risk increasing *bias* due to model misspecification. Our philosophy is that, by using Bayesian nonparametrics, we allow the data to decide for itself an appropriate degree of model complexity.

The situation is similar for missing data problems. Consider estimating the mean response $\eta = E(Y)$ when Y is subject to missingness. Under the ignorability and positivity assumptions described in Chapter 2, we require estimating either the regression function $\mu(x) = E(Y \mid R = 1, X = x) = E(Y \mid X = x)$ or the missingness probability $e(x) = \Pr(R = 1 \mid X = x)$. Given estimates $\widehat{\mu}(x)$ and $\widehat{e}(x)$, we again can construct an estimator $\widehat{\eta}$ of η using either the regression estimator $\widehat{\eta} = N^{-1}\sum_i \widehat{\mu}(X_i)$ or inverse probability weighting $\widehat{\eta} = N^{-1}\sum_i R_i Y_i / \widehat{e}(X_i)$. Models like (5.1) can again be used for these purposes, but suffer the same drawbacks as in the causal inference setting.

A recent trend in the causal inference literature has been to apply machine learning to estimate $\mu_a(x)$ and $e(x)$ [11, 119]. This chapter is concerned, in particular, with applying *Bayesian additive regression trees* (BARTs) for these purposes. We favor BARTs due to its ease of use and state-of-the-art results in causal inference problems [120]. By virtue of being Bayesian, BARTs also provide uncertainty estimates, such as credible intervals, automatically. These uncertainty estimates have been shown to be competitive with other machine learning methods even by frequentist criteria.

5.1.1 *Nonparametric Binary Regression and Semiparametric Gaussian Regression*

In this chapter our main focus will be on the problems of (i) nonparametric binary regression and (ii) semiparametric Gaussian regression. BARTs are adaptable to many other problems, including survival analysis and conditional distribution estimation [121, 122], but are ideally suited to these two problems in particular. Nonparametric binary regression considers the problem of estimating $g(x)$ in the model

$$[Y_i \mid X_i = x] \sim \text{Bernoulli}[h\{g(x)\}],$$

where $g(x)$ is an arbitrary function and $h(\mu)$ is a known *link function*, such as the standard normal cdf $h(\mu) = \Phi(\mu)$. Binary responses are of interest because estimating $e(x)$ when the treatment is binary is equivalent to performing binary regression, and the response $Y_i(a)$ may also be binary itself.

The semiparametric Gaussian response model takes

$$[Y_i \mid X_i = x] \sim N\{g(x), \sigma^2\}, \tag{5.2}$$

where $g(x)$ is again modeled nonparametrically. This model is convenient to work with when Y_i is continuous. The working assumption that the error distribution is homoskedastic and normal is typically made to simplify computations. There are various ways of weakening this assumption [122, 123, 124]; however, we note that BART is somewhat robust to the normality assumption, at least when only the average causal effect is of interest [125].

5.1.2 *Running Example: Medical Expenditure Data*

Throughout this chapter, we will illustrate concepts on data from the Medical Expenditure Panel Survey (MEPS); we will revisit this dataset in the case study in Chapter 14 as well. MEPS is a large, ongoing collection of surveys of individuals, families, employers, and medical providers. We take the outcome of interest to be the total medical expenditures of an individual during a given year. Covariates include age, sex, race, income, martial status, education, and whether an individual smokes. Our goal is to determine whether this data supports the hypothesis that individuals who smoke have higher medical expenditures, and whether there is evidence that this relationship is causal. In this case, the exposure A_i is whether an individual smokes and X_i is the collection of potential confounders. It turns out that the log medical expenditures are well modeled by a normal distribution, so that $Y_i \sim N\{g(X_i), \sigma^2\}$ approximately.

5.2 From Basis Expansions to Tree Ensembles

In this section we gradually build up to BARTs by considering several familiar approaches and discussing where they break down. We focus on the regression model

$$Y_i(a) = \mu_a(X_i) + \varepsilon_i(a),$$

where $\varepsilon_i(a) \sim N(0, \sigma_a^2)$. Our goal is to construct a flexible estimate of $\mu_a(\cdot)$.

Temporarily, we will assume that X_i is univariate. A fully parametric approach might use a first-order approximation $\widehat{\mu}_a(x) = \beta_0 + \beta_1 x$. If $\mu_a(x)$ is well approximated by a linear function in x, then this approach is very attractive. When this is not the case, we might replace the linear approximation with a *basis function expansion* $\widehat{\mu}_a(x) = \widehat{\beta}_0 + \sum_{b=1}^{B} \widehat{\beta}_b \psi_b(x)$. For example, we might approximate $\mu_a(x)$ with a polynomial $\widehat{\mu}_a(x) = \sum_{b=0}^{B} \widehat{\beta}_b x^b$ of sufficiently high order. Other expansions, which may have advantages over polynomials, are also possible. A *cubic spline* basis takes

$$\psi_b(x) = (c_{b0} + c_{b1}x + c_{b2}x^2 + c_{b3}x^3) I(\xi_{b1} \le x < \xi_{b2})$$

with the coefficients c_{bj} constrained so that $\widehat{\mu}_a(x)$ is continuous and twice differentiable. A *natural cubic spline* basis [126] is given in Figure 5.1 with $B = 10$ basis functions. Natural cubic splines are compactly supported, which allows them (unlike polynomial the basis) to capture local behavior in $\mu_a(x)$ without causing large global changes.

Example 5.2.1 (*MEPS*). We consider the relationship between the medical expenditure of an individual (measured on the log scale) and age. Intuitively this relationship should be increasing, but may not be linear. The left panel of Figure 5.2 gives the posterior mean and 95% credible bands for the regression

$$\log(\texttt{expenditure}_i) = \mu(\texttt{age}_i) + \varepsilon_i, \tag{5.3}$$

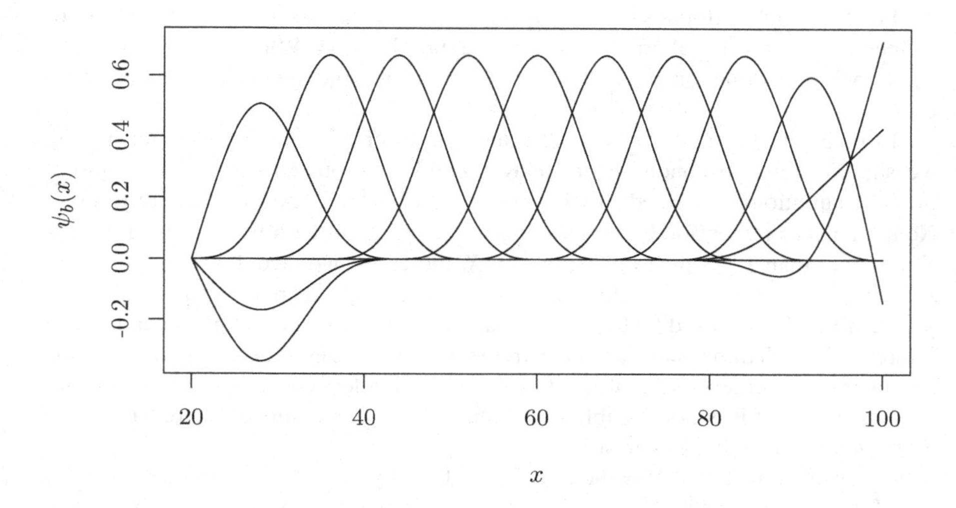

Figure 5.1 *An example of a natural cubic spline basis with $B = 10$ basis functions.*

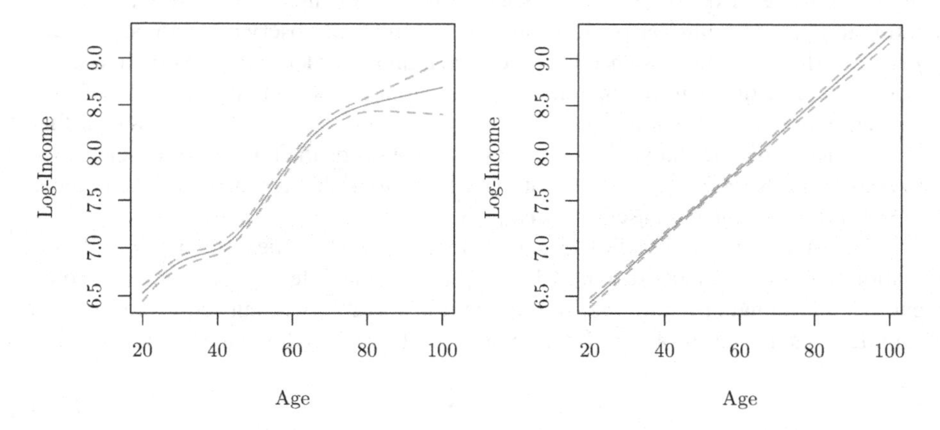

Figure 5.2 *Left: regression of log-income on age using the natural cubic splines model* (5.3). *Right: regression of log-income on age using a simple linear regression.*

where $\varepsilon_i \stackrel{\text{iid}}{\sim} N(0, \sigma^2)$ and $\mu(\texttt{age})$ is modeled as a natural cubic spline with five basis functions. We see that natural cubic splines allow for a flexible relationship between the variables and allow us to account for the fact that the relationship is non linear – because the natural cubic spline basis is nested in the linear model, this non-linearity is easily confirmed via a likelihood ratio test. The right panel displays the fit of a linear regression of $\log(\texttt{expenditure}_i)$ on $\texttt{age}_i$ for comparison, which misses curvature in the relationship; in particular, the slope of $\mu(\texttt{age})$ decreases for larger values of `age`.

For a single confounding variable X_i, basis expansions using polynomials or splines are a powerful alternative to linear approximation. While we give up some parsimony by estimating $\mu_a(x)$ flexibly, we gain robustness to model misspecification.

In practice, it is rare that we are interested in only a single covariate; in fact, we should prefer to condition on many possible confounders so that the ignorability assumptions described in Chapter 1 and Chapter 2 are more likely to hold. Nonparametric regression becomes substantially more difficult when X_i is of dimension higher than 1. As the dimension p of X_i increases, we face the *curse of dimensionality*: the accuracy with which we can estimate $\mu_a(x)$ decreases exponentially in p. The difficulties caused by the curse of dimensionality for missing data and causal inference in high dimensions can be extreme, with estimates of Δ failing to obtain the parametric convergence rate of $N^{-1/2}$ [65]. At a high level, the curse of dimensionality implies that it is not feasible to estimate $\mu_a(x)$ in a completely nonparametric way, even if p is only of modest size.

One way of understanding the difficulty is to consider the estimation rate of $\mu_a(x)$ as a function of the dimension of x [65]. In parametric models, one expects that as the number of observations N tends to infinity, we will obtain a parametric rate of convergence of order $\sigma_a\sqrt{p/N}$, where p is the dimension of x; equivalently, to estimate $\mu_a(x)$ to within error δ, we need the number of observations to be of order $p\sigma_a^2/\delta^2$. Hence, N does not need to be of much higher order than p to obtain a small error. The situation with nonparametric estimation is substantially more bleak; we instead expect an estimation rate for $\mu_a(x)$ of roughly $\sigma_a N^{-2/(4+p)}$, provided that $\mu_a(x)$ is twice differentiable. This implies that one needs the number of observations to be of order $N = (\sigma_a/\delta)^{2+p/2}$ to estimate $\mu_a(x)$ to within accuracy δ. The number of observations now increases *exponentially* in p.

The curse of dimensionality does not completely rule out flexible methods of estimation. Rather, we must use a model for $\mu_a(x)$ which, while not parametric, imposes enough additional structure that the problem is solvable without requiring massive sample sizes. For example, the linear model can be replaced by an *additive* model

$$\mu_a(x) = \alpha + \sum_{d=1}^{p} g_d(x_d), \tag{5.4}$$

where the functions $\{g_d(x_d) : d = 1, \ldots, p\}$ are estimated nonparametrically. Models of this form are referred to as *generalized additive models*, or GAMs. GAMs generalize the linear model (5.1) by allowing the effects of the confounders to be nonlinear. An approach to fitting GAMs, which we will revisit later, is to estimate the $g_d(\cdot)$'s using *backfitting*. A backfitting algorithm for fitting a GAM is given in Algorithm 5.1.

A limitation of GAMs is that they do not allow for interaction effects among the covariates. One possible alternative to the BART models we discuss here is to use *Gaussian process* models, which are discussed in Chapter 7. In principle, one can extend GAMs to allow for low-order dependencies in the predictors. However, it is difficult to do this due to the fact that there are $\binom{p}{2}$ possible interaction effects, with the problem becoming even more challenging for higher order interactions. If we

Algorithm 5.1 Backfitting algorithm for fitting $g_1, \dots, g_p$ in model (5.4)

Repeat the following until convergence:

1. For $j = 1, \dots, p$:
   (a) Compute the backfit residuals
   $$R_i^{(j)} = Y_i - \alpha - \sum_{d \neq j} g_d(x_d)$$
   for $i = 1, \dots, N$
   (b) Update $g_j(x_j)$ by regressing $R_i^{(j)}$ on X_{ij} using any nonparametric regression technique.
2. Compute $R_i^{(0)} = Y_i - \sum_d g_d(x_d)$ for $i = 1, \dots, N$ and set $\alpha = N^{-1} \sum_{i=1}^{N} R_i^{(0)}$.

believe that there is a possibility for interaction effects, it seems that we must use a model that is capable of actively searching over the space of possible interactions to find a small number of low-order interactions. This suggests a model of the form

$$\mu_a(x) = \alpha + \sum_{b=1}^{B} g_b(x) \tag{5.5}$$

where each of the functions $g_b(x)$ depends on a small number of the covariates. That is, each of the $g_b(x)$'s is a *sparse* function of the predictors. As we will see, the BART model is of exactly this form, with each of the $g_b(x)$'s being a *decision tree*. Models of the form (5.5) appeal strongly to intuition – in many causal inference/missing data problems, we expect that functions are roughly additive and that any interactions that are present will be of low order. BART excels in this regime, making it an excellent choice in practice.

5.3 Bayesian Additive Regression Trees

5.3.1 Decision Trees

The BART models we are working toward use *decision trees* as a fundamental building block. Decision trees are powerful tools for constructing flexible models for $\mu_a(x)$ and $e(x)$. While we will rarely use a single decision tree in isolation, it is useful to understand how individual decision trees are structured.

Decision trees themselves are fairly self-explanatory. In Figure 5.3 we show a decision tree for predicting medical expenditures (on the log scale). This tree was constructed using the CART algorithm, which we discuss in Section 5.3.6. Each node of the decision tree gives the prediction of log expenditure at that node, the number of individuals associated with that node (n), and the percentage of individuals associated with that node. Starting from the node labeled (1) in Figure 5.3, we check the decision rule associated with that node. In the case of (1), this rule checks whether the individual has an age less than 53; if so, the individual is associated with the node

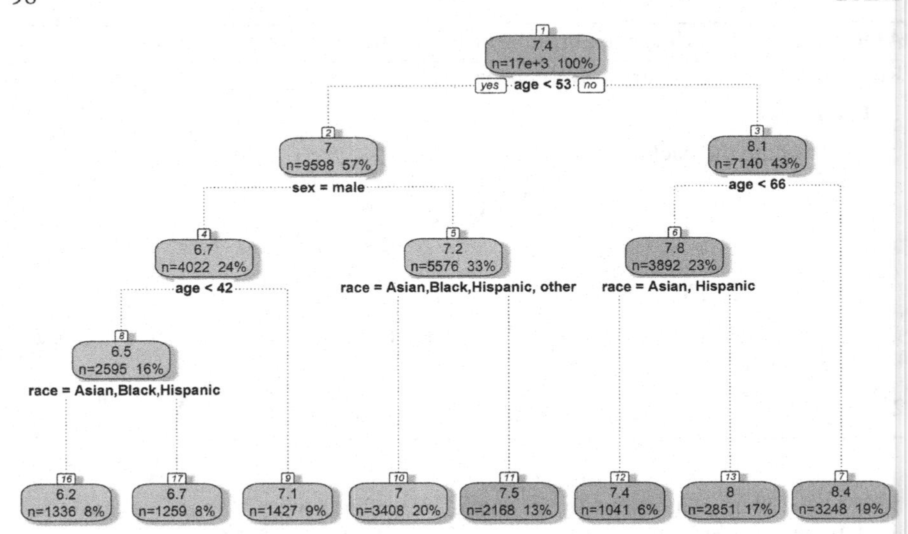

Figure 5.3 *Decision tree grown using the CART algorithm on the MEPS data with log-expenditure as the response.*

labeled (2); otherwise the individual is associated with the node labeled (3). This process iterates until we reach the bottom of the tree. The final predictions are given at the bottom of the tree, and the tree is organized in such a way that the smaller predictions are sent down the left branch of each node; because of this, the leaf node predictions are nearly monotonically increasing from left to right. We see that age is highly predictive of log expenditure, with younger individuals having smaller expenditures on average. We also see that men and non-White individuals tend to have lower expenditures, although this relationship varies slightly by age and sex.

To help understand the salient features of decision trees, it is worth comparing decision tree approaches with linear models, where $\mu_a(x) = x^\top \beta_a$. Linear models are also highly interpretable, with the β's corresponding to the effect of shifting covariates by a unit. Whereas decision trees recover subgroups of the population where the response is predicted to be the same, linear models recover the effect of covariates *in isolation*. By nature, a single decision tree represents one large interaction effect, whereas a linear model captures many separate non-interactive effects.

Formally, a binary tree consists of nodes η_o where o is a (potentially empty) string of the symbols L (for left) and R (for right). The empty string $o = \emptyset$ corresponds to the root of the tree. The root node $\eta_\emptyset$ in Figure 5.4 has two child nodes η_L and η_R. We refer to any node which has a child node as a *branch node*, and any node which has no children as a *leaf node*. In Figure 5.4, η_R is a leaf node while η_L is a branch node, with children η_{LL} and η_{LR}. We write $\mathscr{B}$ for the collection of branch nodes and $\mathscr{L}$ for the collection of leaf nodes. In Figure 5.4, $\mathscr{B} = \{\eta_\emptyset, \eta_L\}$ and $\mathscr{L} = \{\eta_R, \eta_{LL}, \eta_{LR}\}$. The *depth* of a node η_o, i.e., how far down the tree we must travel to each η_o, is equal to the length of o.

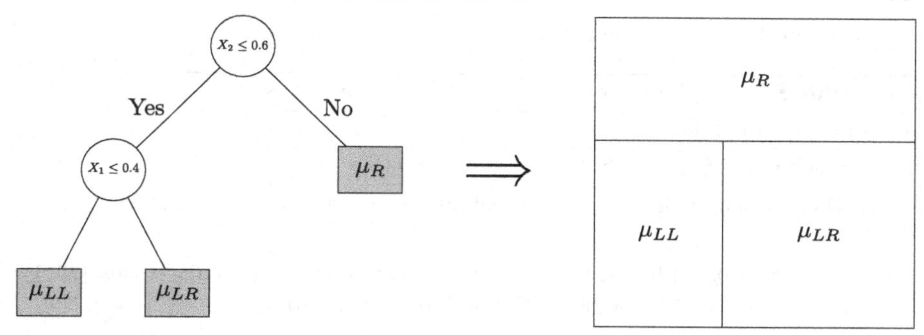

Figure 5.4 *Schematic of a decision tree and the associated partition.*

Associated with each branch node $\eta_o \in \mathscr{B}$ is a *splitting direction* j_o. If X_{j_o} is a continuous or ordinal variable, then we also associate with η_o a *cutpoint* C_o. If X_{j_o} is a categorical variable taking values in $\{1,\ldots,K\}$, then we instead associate with η_o a set $\mathscr{C}_o \subseteq \{1,\ldots,K\}$ and have x take the left path at η_o if $x \in \mathscr{C}_o$ and the right path otherwise. If $\eta_o \in \mathscr{L}$, then we associate a prediction μ_o with the node. In Figure 5.4, for example, we have $C_L = 0.4$ and $j_L = 1$; the prediction associated with all X_i's such that $X_{i2} > 0.6$ is given by μ_R and so forth. Figure 5.4 also illustrates how a decision tree induces a partition of the predictor space such that $\mu_a(x)$ is piecewise constant on the partition.

Using a decision tree, we can define a model for $\mu_a(x)$ and $e(x)$. Let $\mathscr{T}$ denote a decision tree with leaf node parameters $\mathscr{M}$. We let $\text{Tree}(x; \mathscr{T}, \mathscr{M})$ return the value μ_o if x is associated with the leaf node η_o. Then a model for the semiparametric Gaussian regression model (5.2) might take $Y_i(a) \sim N\{\text{Tree}(x; \mathscr{T}_a, \mathscr{M}_a), \sigma^2\}$ and a model for the propensity score might take $e(x) = \text{Tree}(x; \mathscr{T}_e, \mathscr{M}_e)$. There are many different approaches to constructing a decision tree $\mathscr{T}$ and obtaining the leaf node predictions $\mathscr{M}$. We will focus on the Bayesian approach, but we also discuss other approaches in Section 5.3.6.

5.3.2 *Priors over Decision Trees*

The Bayesian nonparametric approach requires us to specify a prior for both the decision tree $\mathscr{T}$ and the leaf node predictions $\mathscr{M}$ given $\mathscr{T}$. We denote these priors $\pi_{\mathscr{T}}$ and $\pi_{\mathscr{M}}$. The most common prior used for $\mathscr{T}$ is a *branching process* [127]. A sample $\mathscr{T} \sim \pi_{\mathscr{T}}$ from this branching process prior is obtained by iteratively deciding, for all of the nodes at each depth, whether the node will be a branch or a leaf. Given the tree structure (consisting of the shape of the tree, the splitting rules j_o, and cutpoints C_o and $\mathscr{C}_o$), we then sample $\mu_0 \stackrel{\text{iid}}{\sim} \pi_\mu$ for some π_μ, which is usually chosen to be a conditionally conjugate prior.

An algorithm for sampling from this prior is given in Algorithm 5.2.

Algorithm 5.2 Algorithm for Generating $\mathscr{T} \sim \pi_{\mathscr{T}}$ and $\mathscr{M} \sim \pi_{\mathscr{M}}(\cdot \mid \mathscr{T})$

1. Set the depth of $\mathscr{T}$ to $d \leftarrow 0$.
2. For each node η_o of depth d:
   (a) Make η_o a branch node with probability $\gamma/(1+d)^{\beta}$; otherwise, make η_o a leaf node.
   (b) If all possible splitting rules for η_o result in either η_{oL} or η_{oR} being empty, make η_o a leaf node; otherwise, proceed to Step 2(c).
   (c) Select a coordinate to construct a split on by sampling $j_o \sim \text{Categorical}(s)$.
   (d) If X_{j_o} is quantitative, select a cutpoint value C_o by sampling an index i from the collection of X_i's that lead to η_o. Then set $C_o = X_{ij_o}$. If X_{j_o} is qualitative, sample $\mathscr{C}_o$ uniformly from the set of subsets of possible values for X_{j_o}.
   (e) If either η_{oL} or η_{oR} is empty, return to Step 2(c).
3. If all nodes of depth d are leaf nodes, proceed to Step 4. Otherwise, set $d \leftarrow d+1$ and return to Step 2.
4. For $\eta_o \in \mathscr{L}$, sample $\mu_o \stackrel{\text{iid}}{\sim} \pi_{\mu}$.

We emphasize that Algorithm 5.2 *does not* describe how trees are learned from the data. Rather, it is an algorithm for sampling from the *prior* distribution. Algorithm 5.2 is guaranteed to terminate with no more than n leaf nodes.

There are several hyperparameters in the tree growing prior $\pi_{\mathscr{T}}$. In particular, we must specify s, γ, and β. Both γ and β control the overall shape of the tree. The parameter β acts to control the maximal depth of the tree, with larger values of β leading to the branching process terminating quickly. The parameter γ controls the probability that the tree extends beyond depth 0. A common choice is to set γ large (say, 0.95) to encourage at least one split. When considering a single decision tree, β parameterizes the overall complexity of the trees we will end up with, and is perhaps best treated as a tuning parameter. For BART (discussed in the following section), however, the depth of the decision tree is not the sole determinant of the model complexity. Reasonable practical performance for BART can be obtained by fixing $\beta = 2$ or 1, and results are generally robust to the selection of β. Selection of s is covered in detail in Section 5.3.4.

The choice of π_{μ} will vary by the structure of the problem. For the model (5.2) with $Y_i \sim N\{\text{Tree}(x; \mathscr{T}, \mathscr{M}), \sigma^2\}$, the choice $\mu_o \sim N(0, \sigma_{\mu}^2)$ is conditionally conjugate and leads to tractable computations. Similarly, if $Y_i \sim \text{Bernoulli}\{\text{Tree}(x; \mathscr{T}, \mathscr{M})\}$, we can take $\mu_o \sim \text{Beta}(a_{\mu}, b_{\mu})$, which is conjugate to binomial sampling.

5.3.3 Ensembles of Decision Trees

When used in isolation, there are several deficiencies of decision trees that make them suboptimal for our purposes. The first deficiency is that decision trees are *unstable* –

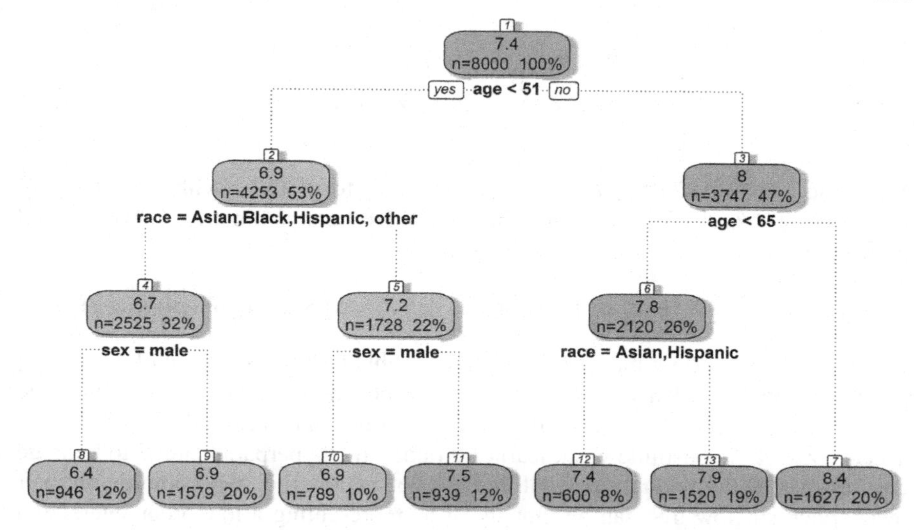

Figure 5.5 *Tree built by subsampling 8000 individuals without replacement from the MEPS dataset and applying the CART algorithm.*

two moderately large samples from the same population can often produce quite different estimates of the tree structure. For example, Figure 5.5 shows a tree structure similar to the tree from Figure 5.3, but was obtained by subsampling roughly half of the dataset without replacement. We see that, while similar, the individual predictions assigned to particular observations can vary substantially from tree to tree; a 65-year-old Hispanic individual, for example, has a prediction which changes from 8.4 to 7.4, while a 66-year-old Hispanic does not have their prediction change at all. Moreover, the changes between Figure 5.5 and Figure 5.3 are relatively mild compared to what one sees in smaller datasets, where a change at the root node of the decision tree can have drastic consequences on the predictions in the leaves [128, Chapter 8].

A second problem is that decision trees do not produce smooth estimates of $\widehat{\mu}_a(x)$ and essentially estimate the response x independently for each leaf node ℓ (conditional on $\mathscr{T}$). Intuitively, we might expect that the values μ_{LL} and μ_{LR} in Figure 5.4 should be closer in value than μ_{LL} and μ_R because they have a nearer common ancestor. Because decision trees as described in Section 5.3.1 do not enforce this, the estimates of $\mu_a(x)$ can be extremely non-smooth in x. This is also seen in Figure 5.5, where we again note that changing age from 65 years old to 66 years old produces a large change in the prediction.

To address these problems, [125] introduced *Bayesian additive regression trees* (BART). The Bayesian additive regression trees framework approximates an unknown function $g(x)$ using shallow decision trees as basis functions. That is, they set

$$g(x) = \sum_{m=1}^{M} \text{Tree}(x; \mathscr{T}_m, \mathscr{M}_m). \tag{5.6}$$

We write $g \sim \text{BART}(\pi_{\mathscr{T}}, \pi_{\mathscr{M}})$ if $g(x)$ has the form (5.6) with $(\mathscr{T}_m, \mathscr{M}_m) \stackrel{\text{iid}}{\sim} \pi_{\mathscr{T}}(\mathscr{T}_m)\, \pi_{\mathscr{M}}(\mathscr{M}_m \mid \mathscr{T}_m)$. We can then consider semiparametric Gaussian and nonparametric probit regressions of the form

$$[Y_i \mid X_i = x] \sim N\{g(x), \sigma^2\} \qquad \text{or} \qquad [Y_i \mid X_i = x] \sim \text{Bernoulli}\{\Phi(g(x))\}.$$

The approach of adding many trees together in this fashion is inspired by *boosting* algorithms; in the parlance of machine learning, boosting is an approach to combine many *weak learners* into a single *strong learner* [129]. In this case, the tree functions $\text{Tree}(x; \mathscr{T}_m, \mathscr{M}_m)$ constitute weak learners. Taking the hyperparameter β to be large ensures that the individual trees will tend to be shallow (i.e., not complex on their own). Each shallow tree can be thought of as representing a low-order interaction effect in the data, so that the BART framework is a particular implementation of the low-order additive model structure described in Equation (5.5). When the trees are aggregated together, we obtain the strong learner $g(x)$, which is highly flexible but also lacking in high-order interactions.

Bayesian additive regression trees produce estimates that are both smoother than single-tree model predictions and more stable. Figure 5.6 shows the fit of a single decision tree to the MEPS dataset with BART. While both fits lack smoothness, the BART fit is smoother in the sense that there are a large number of small discontinuities rather than a small number of large discontinuities. An even smoother alternative to both of these possibilities is the spline estimate in Figure 5.2, although this is not straight forward to generalize to larger numbers of predictors in a way that preserves the sparse-additivity feature of BART.

5.3.4 Prior Specification for Bayesian Additive Regression Trees

A benefit of BART relative to other Bayesian nonparametric/semiparametric regression techniques (such as the Gaussian process regression techniques discussed in Chapter 7) is that it is easy to specify *default priors* that work well in many situations. Due to an abundance of software packages that automatically implement these priors, practitioners need not burden themselves with prior specification. In this section, we describe a prior for the semiparametric regression model

$$Y_i = g(X_i) + \varepsilon_i, \qquad \varepsilon_i \sim N(0, \sigma^2).$$

As a preprocessing step, the response Y_i is first standardized in some fashion. One common approach [125] is to standardize the observed Y_i's so that $\min(Y_i) = -0.5$ and $\max(Y_i) = 0.5$. Most software packages use the branching process prior described in Algorithm 5.2 with default values $\gamma = 0.95$ and $\beta = 2$; various authors modify step (d) [130], but we find this to have few practical consequences.

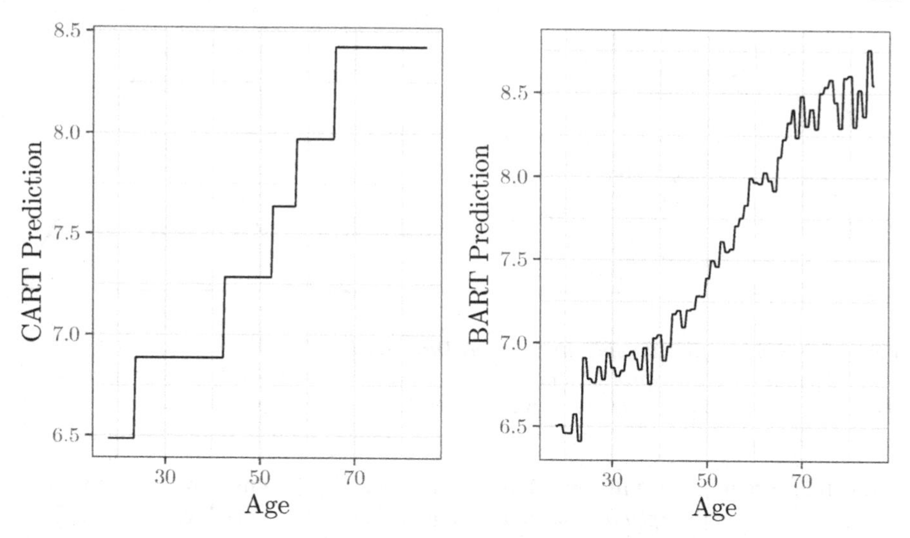

Figure 5.6 *Left: an estimate of the relationship between age and log expenditure for the MEPS dataset obtained from a decision tree. Right: the same relationship, but estimated using the posterior mean of the BART model.*

Let $\mathscr{L}_m$ and $\mathscr{B}_m$ denote the leaf and branch nodes of $\mathscr{T}_m$, and let $\eta_o^{(m)}$ denote a node of $\mathscr{T}_m$. For $\eta_o^{(m)} \in \mathscr{L}_m$, a conditionally conjugate prior is used for the predicted value $\mu_o^{(m)} \sim N(0, \sigma_\mu^2)$. The remaining parameters in the model are the number of trees M, the error variance σ^2, and the variance of the leaf node parameters σ_μ^2. We specify σ_μ indirectly by specifying $\mathrm{Var}\{g(x)\} = \tau_\mu^2 = M\sigma_\mu^2$. This suggests that σ_μ^2 should be inversely proportional to M, so that the number of trees can be varied independently of the signal-to-noise ratio. A common default choice is $\sigma_\mu = 0.5/(k\sqrt{M})$, so that the prior for $g(x)$ assigns mass near both $\min(Y_i)$ and $\max(Y_i)$. The parameter k controls how far out in the tails $\min(Y_i)$ and $\max(Y_i)$ are in the prior; the choice $k = 2$, for example, encodes the belief that $g(x) \in \{-0.5, 0.5\}$ with probability 95% a priori.

The prior on the error variance σ^2 is chosen so that it is a "rough, data-based, overestimate of σ^2" [125]. First, a pilot estimate $\widehat{\sigma}$ is obtained for σ, either by fitting a linear model $Y_i = X_i^\top \beta + \varepsilon_i$ or as the empirical standard deviation of the Y_i's. Next, a prior for σ based on $\widehat{\sigma}$ is specified. A common choice is to use an inverse-gamma prior $\sigma^2 \sim \mathrm{InvGam}(\alpha_\sigma, \beta_\sigma)$ with $(\alpha_\sigma, \beta_\sigma)$ chosen so that $[\sigma < \widehat{\sigma}]$ holds with some specified probability q; the original BART paper [125] considers $q \in \{0.75, 0.9, 0.99\}$, so that the prior expresses a preference for values of σ that are smaller than what one can predict from a linear model. The overall default prior recommended for σ^2 takes $q = 0.9$ and $\alpha_\sigma = 1.5$.

The parameter s in the prior $j_o \sim \mathrm{Categorical}(s)$ can also be given a prior. The role of this hyperprior is to filter out irrelevant predictors; if s_j is very small in the

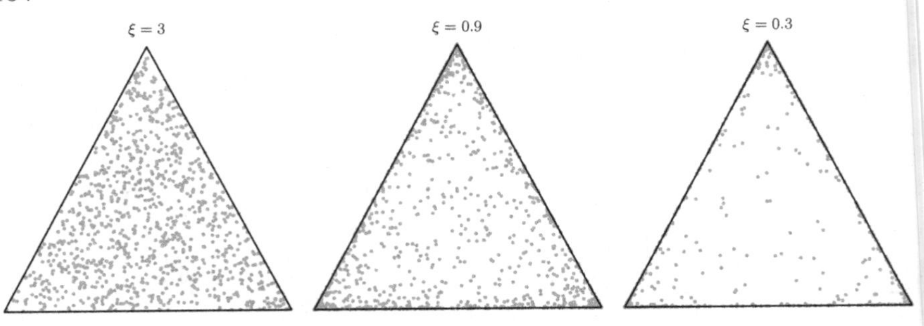

Figure 5.7 *Samples from the sparsity-inducing Dirichlet distribution with* $\xi \in \{3, 0.9, 0.3\}$ *and* $p = 3$. *Samples in* $\mathbb{R}^3$ *are projected down to the two-dimensional simplex, with the vertices of the simplex corresponding to the points* $s = (1,0,0), (0,1,0)$, *and* $(0,0,1)$.

prior, then predictor j will most likely not appear in the ensemble. A convenient choice of prior is the *sparsity-inducing Dirichlet* prior $s \sim \text{Dirichlet}(\xi/p, \ldots, \xi/p)$. To develop intuition about this prior, Figure 5.7 displays 1000 samples from the sparsity-inducing Dirichlet prior with $p = 3$ and $\xi \in \{3, 0.9, 0.3\}$, projected onto the two-dimensional simplex. The vertices in Figure 5.7 correspond to the 1-sparse vectors $s = (1,0,0), (0,1,0)$, and $(0,0,1)$, while the edges correspond to 2-sparse vectors of the form $s = (s_1, s_2, 0), (s_1, 0, s_3)$, and $(0, s_2, s_3)$, and the interior points correspond to dense s vectors. We see that as ξ decreases the prior becomes increasingly concentrated on sparse vectors. In causal inference problems, this allows the analyst to be aggressive in including potential confounders, as irrelevant confounders can be filtered out in a data-adaptive fashion. It can be shown that, for large p and conditional on there being B splitting rules in the ensemble, the number of unique predictors Q included in the tree ensemble is approximately distributed as $(Q-1) \sim \text{Poisson}(\theta)$ where $\theta = \xi \sum_{b=0}^{B-1} (\xi + b)^{-1}$. Using this, the parameter ξ can be chosen to target a specified number of predictors to be included in the ensemble. Alternatively, a hyperprior on ξ can be used to allow the data to adaptively determine an appropriate amount of sparsity.

Prior specification for the nonparametric binary regression problem follows a similar approach. Suppose that we specify $Y_i \sim \text{Bernoulli}[\Phi\{g(X_i)\}]$ where $\Phi(\mu)$ is the distribution function of a standard normal random variable. Then we have the following *latent variable* model of [69] introduced in Chapter 3 (Section 3.5)

$$Y_i = I(Z_i \geq 0), \qquad Z_i \sim N\{g(X_i), 1\}. \tag{5.7}$$

We can then follow the prior specification given above, with the caveat that the Z_i's are not standardized to be between -0.5 and 0.5 and the variance σ^2 is fixed at 1. For this setting, we modify the default choice $\sigma_\mu = 3/(k\sqrt{M})$ so that if, for example, $k = 2$, then the prior assigns 95% mass to the event $g(x) \in (-3, 3)$.

5.3.5 *Posterior Computation for Bayesian Additive Regression Trees*

Computation for BART models is performed by (approximate) sampling from the posterior using Markov chain Monte Carlo.

Computation for a Single Tree

We first describe how to perform inference for a *single* decision tree in the model

$$[Y_i \mid X_i = x] \sim f\{y \mid \text{Tree}(x; \mathscr{T}, \mathscr{M})\}$$

where $f(y \mid \mu)$ is a parametric model for Y_i. Let $\mathscr{D} = \{(X_i, Y_i) : i = 1, \ldots, N\}$ denote the observed data. This model (with the number of trees set to $M = 1$) is referred to as the *Bayesian CART* (BCART) [127]. Our goal is to construct a Markov chain $(\mathscr{T}^{(1)}, \mathscr{M}^{(1)}), (\mathscr{T}^{(2)}, \mathscr{M}^{(2)}), \ldots$ that converges to $\pi(\mathscr{T}, \mathscr{M} \mid \mathscr{D})$. We do this with the following pair of updates:

1. Sample $\mathscr{T}^{(t)} \sim a(\mathscr{T}^{(t)} \mid \mathscr{T}^{(t-1)})$, where $a(\mathscr{T}' \mid \mathscr{T})$ leaves the distribution $\pi(\mathscr{T} \mid \mathscr{D})$ invariant.
2. Sample $\mathscr{M}^{(t)} \sim \pi(\mathscr{M} \mid \mathscr{T}^{(t)}, \mathscr{D})$.

A transition $a(\mathscr{T}^{(t)} \mid \mathscr{T}^{(t-1)})$ can be constructed from the *Metropolis-Hastings* algorithm, provided that the integrated likelihood

$$L(\mathscr{T}) = \prod_{\eta_o \in \mathscr{L}} \int \prod_{i \rightsquigarrow \eta_o} f(Y_i \mid \mu)\, \pi_\mu(\mu)\, d\mu$$

can be computed in closed form; here, the notation $i \rightsquigarrow \eta_o$ means that X_i leads to the leaf node η_o of the tree. For the semiparametric Gaussian model with $\mu_o \sim N(0, \sigma_\mu^2)$, for example, we can compute $L(\mathscr{T})$ as

$$L(\mathscr{T}) = (2\pi\sigma^2)^{-n/2} \prod_{\eta_o \in \mathscr{L}} \sqrt{\frac{\sigma_\mu^{-2}}{\sigma_\mu^{-2} + N_o}} \exp\left\{-\frac{1}{2}\left(\frac{S_o^2}{\sigma^2} + \frac{N_o \bar{Y}_o^2}{\sigma^2 + N_o \sigma_\mu^2}\right)\right\}, \quad (5.8)$$

where N_o is the number of observations that lead to η_o, $\bar{Y}_o$ is the sample mean of the observations that lead to η_o, and $S_o^2 = \sum_{i \rightsquigarrow \eta_o} (Y_i - \bar{Y}_o)^2$ is the residual sum-of-squares for η_o. Other parametric models with conjugate priors lead to different expressions for $L(\mathscr{T})$.

Using $L(\mathscr{T})$, we can compute the marginal posterior density $\pi(\mathscr{T} \mid \mathscr{D})$ up to a normalizing constant as $\pi(\mathscr{T}) L(\mathscr{T})$. Given a *transition kernel* $q(\mathscr{T}' \mid \mathscr{T})$ satisfying minimal requirements, the Metropolis-Hastings algorithm provides a transition $a(\mathscr{T}' \mid \mathscr{T})$ that leaves the target distribution invariant. A sample from $a(\mathscr{T}' \mid \mathscr{T})$ is obtained by first sampling $\mathscr{T}^\star \sim q(\mathscr{T}' \mid \mathscr{T})$ and then setting $\mathscr{T}' = \mathscr{T}^\star$ with probability

$$A = \min\left\{\frac{L(\mathscr{T}^\star)\, \pi_{\mathscr{T}}(\mathscr{T}^\star)\, q(\mathscr{T}^\star \mid \mathscr{T})}{L(\mathscr{T})\, \pi_{\mathscr{T}}(\mathscr{T})\, q(\mathscr{T} \mid \mathscr{T}^\star)}, 1\right\},$$

and setting $\mathscr{T}' = \mathscr{T}$ otherwise. Implementations of BART/BCART differ in their choice of the transition kernel $q(\mathscr{T}' \mid \mathscr{T})$. All packages we are aware of take $q(\mathscr{T}' \mid \mathscr{T})$ to be a mixture distribution over possible modifications to $\mathscr{T}$ [127]. The two most important transitions are the `BIRTH` and `DEATH` proposals, which propose adding or deleting a pair of leaf nodes. The `BIRTH` step first randomly selects a leaf node $\eta_o \in \mathscr{L}$ uniformly from $\mathscr{L}$ and then converts this η_o to a branch node by adding the nodes η_{oL} and η_{oR} to $\mathscr{L}$. The `DEATH` proposal is the inverse of the `BIRTH` move: it randomly selects a leaf node pair (η_{oL}, η_{oR}) from $\mathscr{L}$ and deletes them, which converts η_o from a branch to a leaf node. `BIRTH` and `DEATH` are dual to one another in the sense that if $q(\mathscr{T}^\star \mid \mathscr{T})$ represents a `BIRTH` move then $q(\mathscr{T} \mid \mathscr{T}^\star)$ represents a `DEATH` move. Other potential transitions include the following:

- `CHANGE`: pick a branch node η_o at random and reassign its splitting rule according to the prior for $j_o \sim s$ and C_o.
- `PRIOR`: Sample $\mathscr{T}'$ from the prior distribution.
- The `ROTATE` and `PERTURB` moves: These transitions are generalizations of the `SWAP` and `CHANGE` transitions, but are substantially more involved. Interested readers are referred to [131].

When fitting the Bayesian CART, some variant of all of these moves should be used. Because BART uses many *shallow* trees, however, not all of these are as useful as they are for Bayesian CART. Some implementations, for example, use only the `BIRTH` and `DEATH` steps; the intuition behind needing only `BIRTH` and `DEATH` steps is that they are sufficient to explore the space of shallow trees on their own. We have found using only `BIRTH` and `DEATH` moves to be inadvisable, however. In practice, we have found combinations of the `BIRTH`, `DEATH`, and `PRIOR` moves to be highly effective for BART, and the `CHANGE/PERTURB` moves to also be helpful, while the `ROTATE` proposal is less useful.

Bayesian backfitting for BART

To implement BART, we extend the single-tree procedure using *Bayesian backfitting* (cf. Algorithm 5.1). Consider the model $Y_i = \sum_{m=1}^{M} \text{Tree}(X_i; \mathscr{T}_m, \mathscr{M}_m) + \varepsilon_i$. Bayesian backfitting iteratively updates the parameters $(\mathscr{T}_m, \mathscr{M}_m)$ by computing the *backfit residuals* $R_i^{(m)} = Y_i - \sum_{j \neq m} \text{Tree}(X_i; \mathscr{T}_j, \mathscr{M}_j)$. This reduces the problem to updating a single decision tree from the model $R_i^{(m)} \sim N\{\text{Tree}(X_i; \mathscr{T}_m, \mathscr{M}_m), \sigma^2\}$. Based on this, we can compute a conditional posterior distribution

$$
\begin{aligned}
&\pi\{\mathscr{T}_m \mid \mathscr{D}, (\mathscr{T}_j, \mathscr{M}_j : j \neq m)\} \\
&\quad \propto \pi_{\mathscr{T}}(\mathscr{T}_m) \prod_{\eta_o^{(m)} \in \mathscr{L}_m} \int \prod_{i \rightsquigarrow \eta_o} N(R_i^{(m)} \mid \mu, \sigma^2)\, N(\mu \mid 0, \sigma_\mu^2)\, d\mu \qquad (5.9) \\
&\quad = \pi_{\mathscr{T}}(\mathscr{T}_m) L_m(\mathscr{T}_m)
\end{aligned}
$$

where $L_m(\mathscr{T}_m)$ is defined by (5.8) with the $R_i^{(m)}$'s in the role of the Y_i's. We now can perform the same update for the single-tree BCART model, with the same choice of $a(\mathscr{T}' \mid \mathscr{T})$ to update $(\mathscr{T}_m, \mathscr{M}_m)$, but with the $R_i^{(m)}$'s used instead of the Y_i's. This is valid because (5.9) shows that this is equivalent to making an update that leaves $\pi\{\mathscr{T}_m \mid \mathscr{D}, (\mathscr{T}_j, \mathscr{M}_j : j \neq m)\}$ invariant and then sampling $\mathscr{M}_m$ from its full conditional. This update is then performed in turn for $m = 1, \ldots, M$. A single iteration of this Markov chain is given in Algorithm 5.3.

Algorithm 5.3 One iteration of the Bayesian backfitting algorithm to update $\mathscr{T}_1, \ldots, \mathscr{T}_M$

For each $j = 1, \ldots, m$:

1. Compute the backfit residuals $R_i^{(m)} = Y_i - \sum_{j \neq m} \text{Tree}(X_i; \mathscr{T}_j, \mathscr{M}_j)$.
2. Sample a candidate tree from the transition kernel $\mathscr{T}' \sim q(\mathscr{T}' \mid \mathscr{T}_m)$.
3. Compute $L_m(\mathscr{T}_m)$ and $L_m(\mathscr{T}')$, where $L_m(\mathscr{T})$ is expression (5.8) evaluated at $[Y_i = R_i^{(m)}]$.
4. Sample $U \sim \text{Uniform}(0,1)$ and compute the acceptance probability

$$A = \min\left\{\frac{L_m(\mathscr{T}')\,\pi_{\mathscr{T}}(\mathscr{T}')\,q(\mathscr{T}_m \mid \mathscr{T}')}{L_m(\mathscr{T}_m)\,\pi_{\mathscr{T}}(\mathscr{T}_m)\,q(\mathscr{T}' \mid \mathscr{T}_m)}, 1\right\}.$$

5. If $U \leq A$, set $\mathscr{T}_m = \mathscr{T}'$. Otherwise, leave $\mathscr{T}_m$ unchanged.
6. For each $\eta_o^{(m)} \in \mathscr{L}_m$, sample

$$\mu_o^{(m)} \sim N\{\widehat{\mu}_o^{(m)}, 1/(a_o^{(m)})\},$$

where

$$a_o^{(m)} = N_o^{(m)}\sigma^{-2} + \sigma_\mu^{-2} \qquad \text{and} \qquad \widehat{\mu}_o^{(m)} = \frac{\sigma^{-2}\sum_{i:X_i \rightsquigarrow \eta_o^{(m)}} R_i^{(m)}}{a_o^{(m)}}.$$

For both practitioners and researchers, we do not recommend implementing the Metropolis-Hastings algorithms for updating the tree parameters. Implementing BART efficiently is a difficult programming task, and expertise in algorithms and data structures is required to make BART practical. Instead, we recommend the use of existing software. Variants of BART for regression, classification, survival analysis, and many other problems are available in various software packages; examples include the `BART`, `bartMachine`, and `dbarts` packages in `R`. For researchers who wish to use BART as a building block in larger models, the `SoftBart` package allows for separate BART objects to be used in a larger `R` program.

We can extend the Bayesian backfitting approach to the nonparametric binary regression model by using the latent variable representation of the probit model:

$$Z_i(a) \sim N\{g_a(X_i), 1\}, \qquad Y_i(a) = I\{Z_i(a) \geq 0\}.$$

The probit BART model can then be implemented by adding an additional step that samples

$$Z_i(a) \sim \begin{cases} N\{g_a(X_i), 1\} I\{Z_i(a) > 0\} & \text{if } Y_i(a) = 1, \\ N\{g_a(X_i), 1\} I\{Z_i(a) \leq 0\} & \text{if } Y_i(a) = 0, \end{cases}$$

as in Example 3.5.

5.3.6 Non-Bayesian Approaches

For context, we briefly contrast BART with other *algorithmic* approaches to building ensembles of decision trees. These methods have been primarily developed in the machine learning community.

Decision trees have a long history in both statistics and machine learning, with key works being [132] and [133]. A popular algorithm for constructing decision trees is the CART algorithm introduced by [133], which was used to produce the decision trees displayed in Figure 5.3 and Figure 5.5. The CART algorithm builds trees in a greedy fashion, starting from an empty root node. A splitting coordinate j_o and cutpoint C_o (or set $\mathscr{C}_o$) are chosen to maximize a measure of *purity* of the child nodes. For example, if we are modeling the response Y_i in a semiparametric fashion with a model $f(y \mid \mu)$, then the rule might be chosen so that the total likelihood of the tree $\prod_{\eta_o \in \mathscr{L}} \prod_{i \rightsquigarrow \eta_o} f(Y_i \mid \widehat{\mu}_o)$ is maximized. This process then iterates across all leaves until a stopping criteria (such as a minimal sample size for each node) is reached. While not typically phrased in these terms, the CART algorithm can be thought of as a heuristic approach to optimize the posterior density

$$(\mathscr{T}_{\text{CART}}, \mathscr{M}_{\text{CART}}) = \arg\max_{\mathscr{T}, \mathscr{M}} \pi(\mathscr{T}, \mathscr{M} \mid \mathscr{D}),$$

with the stopping-rule constraints and regularization incorporated implicitly into the prior $\pi(\mathscr{T}, \mathscr{M})$. Because finding the optimal tree requires searching over the massive space of all possible trees, solving this optimization problem is computationally infeasible. CART instead iteratively solves a simpler problem: given $(\mathscr{T}, \mathscr{M})$, find the $(\mathscr{T}', \mathscr{M}')$ that maximizes the posterior density, subject to the constraint that $\mathscr{T}'$ is obtained from $\mathscr{T}$ by converting a leaf node into a branch node with two children.

As noted earlier, decision trees do not give optimal performance when used in isolation, and it is much better to use an ensemble of many decision trees. The most common techniques for doing this are *random forests* and *boosted decision trees*, both of which have been used for causal inference in the frequentist framework. Boosting is similar in spirit to BART, using an ensemble of the form $g(x) = \sum_{m=1}^{M} \text{Tree}(x; \mathscr{T}_m, \mathscr{M}_m)$ [129]. Whereas BART averages over many collections of trees, typical variants of boosting use a *single* collection of M decision trees. In certain cases, boosting can also be thought of as a greedy algorithm for optimizing a posterior density, or more generically as minimizing a measure of empirical risk. For example, the L_2-boosting algorithm [134] iteratively selects $(\mathscr{T}_m, \mathscr{M}_m)$ to minimize the residual error

$$(\mathscr{T}_m, \mathscr{M}_m) = \arg\min_{\mathscr{T},\mathscr{M}} \sum_{i=1}^{N} \{R_i^{(m)} - \text{Tree}(X_i; \mathscr{T}, \mathscr{M})\}^2 + \Lambda(\mathscr{T}, \mathscr{M}), \tag{5.10}$$

where $R_i^{(m)} = Y_i - \sum_{j=1}^{m-1} \text{Tree}(X_i; \mathscr{T}_j, \mathscr{M}_j)$ is the same backfit residual used in the Bayesian backfitting algorithm and $\Lambda(\mathscr{T}, \mathscr{M})$ is a penalty term. The L_2-boosting algorithm described by (5.10) is particularly simple and is highly reminiscent of BART; other boosting algorithms, such as AdaBoost or LogitBoost [135, 136], are not as directly analogous, but are similar in the sense that one iteratively optimizes $(\mathscr{T}_m, \mathscr{M}_m)$ by fitting the tree to residual-like quantities $R_i^{(m)}$. Like BART, the trees in boosting are typically restricted to be shallow (say, of depth 2), which provides enough simplification that it may be computationally feasible to exactly optimize $(\mathscr{T}_m, \mathscr{M}_m)$. Of course, it is still computationally infeasible to optimize $\sum_i \{R_i^{(m)} - \sum_m \text{Tree}(X_i; \mathscr{T}_m, \mathscr{M}_m)\}^2 + \sum_m \Lambda(\mathscr{T}_m, \mathscr{M}_m)$ jointly for all trees; hence, boosting is still a greedy algorithm. For a generic description of boosting, see [137].

Random forests [138] are qualitatively different from BART and boosting. Let $\mathscr{T}_{\mathscr{D}}$ and $\mathscr{M}_{\mathscr{D}}$ denote the result of applying the CART algorithm to a dataset $\mathscr{D}$. We first describe a simpler algorithm known as *bootstrap aggregating* or *bagging*. We replace $\text{Tree}(x; \mathscr{T}_{\mathscr{D}}, \mathscr{M}_{\mathscr{D}})$ with an expectation

$$\mu(x) = E\{\text{Tree}(x; \mathscr{T}_{\mathscr{D}^\star}, \mathscr{M}_{\mathscr{D}^\star})\} \tag{5.11}$$

with the expectation taken with respect to some distribution $\mathscr{D}^\star \sim \mathbb{G}$. Using the bias/variance decomposition, if $\mathbb{G}$ is the true distribution of the data, then this estimator will have the same bias as the original estimator but a smaller variance [139]. There are two problems with this estimator: we do not know the true distribution of the data, and even if we did, we would not be able to compute the expectation. Bagging approximates the distribution of the data with the empirical distribution of the data and approximates the expectation by applying the CART algorithm to many bootstrap samples of the data and averaging them together.

The random forests algorithm is similar to bagging but with additional randomization of the CART algorithm. This randomization gives more diversity in the trees we obtain than bagging alone would – for example, we would like the trees in Figure 5.3 and Figure 5.5 to be more different than they are. Random forests can be thought of as augmenting the data $\mathscr{D}^\star$ in (5.11) with random variables $U^\star$ that randomize the CART algorithm to increase diversity in trees. A common randomization is to limit which variables j can be selecting as a splitting coordinate j_o. That is, at each split, we randomly select a subset of $\{1, \dots, p\}$ (of size, say, $p/2$) to select j_o.

For reference, we provide fits of random forests and L_2-boosting to the MEPS dataset, regressing log expenditure on age, in Figure 5.8 (random forests being the same as bagging since there is only one predictor). We see that the results are quite similar, with random forests being noisier. The salient features of random forests and boosting are quite different in higher dimensions. Whereas boosting combines many *weak, dependent* learners, random forests combine *strong learners* which are ideally

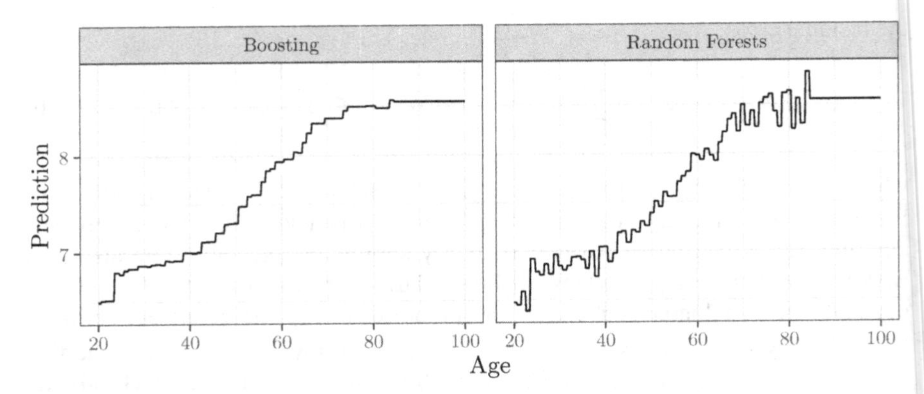

Figure 5.8 *Prediction of log-income as a function of age for the MEPS dataset, obtained using L_2-boosting (left) and random forests (right).*

as different as possible. In boosting, care must be taken to ensure that one avoids overfitting by either limiting the number of trees or using regularization; conversely, adding more trees to a random forest gives a better approximation to (5.11), with increasing the number of trees only having the positive feature of reducing Monte Carlo error.

Both boosting and random forests have variants that can be applied to causal inference. From our perspective, one big advantage of BART is that it gives uncertainty estimates automatically, whereas the algorithmic approaches do not. For more detail on these approaches in a causal setting, see [140] or [141].

5.4 Bayesian Additive Regression Trees Applied to Causal Inference

5.4.1 *Estimating the Outcome Regression Function*

The application of BART to causal inference was first considered in [142]. BART can be immediately applied to estimate the average treatment effect by setting $\mu_0 \sim \text{BART}(\pi_{\mathscr{T}}, \pi_{\mathscr{M}})$ and $\mu_1 \sim \text{BART}(\pi_{\mathscr{T}}, \pi_{\mathscr{M}})$ independently. Alternatively, we can treat the exposure $A_i = a$ in the same fashion as the confounders, with $\mu_a(x) = \mu(a,x)$ and $\mu \sim \text{BART}(\pi_{\mathscr{T}}, \mathscr{M}_{\mathscr{T}})$. In either case, the posterior distribution of the average causal effect can then be obtained by computing

$$\Delta = \int \{\mu_1(x) - \mu_0(x)\}\, F_X(dx). \tag{5.12}$$

As the distribution of the confounders is typically unknown, we also need a model for F_X. A simple option is the *Bayesian bootstrap*; the posterior distribution of F_X becomes

$$F_X = \sum_{i=1}^{N} \omega_i\, \delta_{X_i}, \qquad \omega \sim \text{Dirichlet}(1, \ldots, 1)$$

where δ_x denotes a point-mass distribution at x; see Section 3.2.3 for details on the Bayesian bootstrap. An algorithm for posterior sampling of Δ is given in Algorithm 5.4.

Algorithm 5.4 Algorithm for obtaining samples of Δ from a BART model

1. Run the MCMC algorithm for sampling from the BART posterior, obtaining T samples $\mu_a^{(1)}(x), \dots, \mu_a^{(T)}(x)$.
2. For $t = 1, \dots, T$:
   (a) Sample $\omega^{(t)} \sim \text{Dirichlet}(1, \dots, 1)$.
   (b) Set $\Delta^{(t)} = \sum_{i=1}^{N} \omega_i^{(t)} \left\{\mu_1^{(t)}(X_i) - \mu_0^{(t)}(X_i)\right\}$.
3. Return the samples $\Delta^{(1)}, \dots, \Delta^{(T)}$.

Example 5.4.1 (*MEPS example*)**.** We use BART to estimate the causal effect of smoking on log medical expenditures for the MEPS dataset, using expression (5.12) with $\mu_1(x)$ and $\mu_0(x)$ estimated using Bayesian additive regression trees, controlling for possible confounding variables. As potential confounders, we consider age, sex, marital status, education level, income level, race, and how often an individual uses a seat belt when riding in a car. Seat belt usage is used to control for the personality trait of being a "risk taker," as individuals who are risk takers are likely to both smoke and engage in non-smoking activities that put their health at risk.

Samples from the posterior distribution of Δ are given in Figure 5.9. Based on these results, a naive interpretation is that neither approach supports the hypothesis that smoking increases medical expenditures. Indeed, even ignoring confounding, a standard t-test for a difference between smoking and non-smoking groups does not give evidence that these two groups are different. As we will see, the lesson here is not that smoking and medical expenditures are unrelated; rather, the effect of smoking on medical expenditures is heterogeneous and is small for most of the population. Additionally, much of the effect is *mediated* by the effect of smoking on health. In Chapter 13 we show that, after accounting for health status as a mediator, there is evidence of a positive indirect effect of smoking on medical expenditures.

5.4.2 Regularization-Induced Confounding and Bayesian Causal Forests

A benefit of Bayesian approaches is their ability to perform *shrinkage* estimation in a principled fashion. This is necessary in nonparametric problems, where some amount of regularization is required to perform estimation. As pointed out by [143], however, in causal inference, care must be taken in how regularization is performed.

One possible consequence of naive regularization is *regularization-induced confounding*. Regularization-induced confounding occurs when the prior used induces an inductive bias toward models in which selection bias is minimal. Roughly speaking, regularization-induced confounding occurs when the "simplest" explanation of differences in $\mu_1(x)$ and $\mu_0(x)$ is to attribute them to the treatment $A = a$ rather than

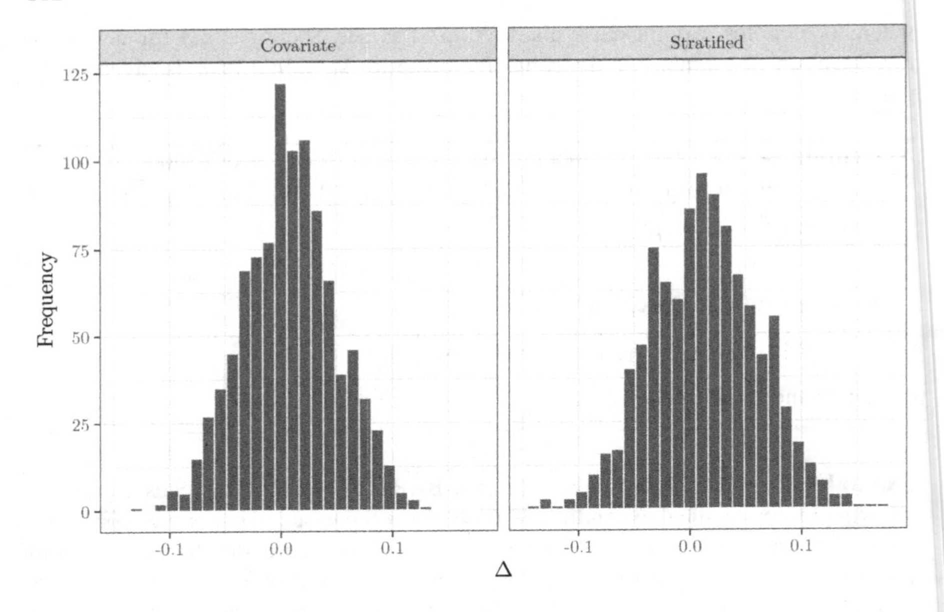

Figure 5.9 *Samples of* Δ *obtained from the BART fit to the MEPS data. The left panel treats the treatment exposure* A_i *as a covariate in the same fashion as the confounders, while the right panel fits separate BART models for* $A_i = 1$ *and* $A_i = 0$.

potential confounders $X = x$. This is easiest to see in the context of linear regression. Consider the model

$$Y_i(a) = \alpha + a\Delta + X_i\beta + \varepsilon_i(a), \tag{5.13}$$

where X_i is a univariate confounder. Suppose that we used a prior where Δ has high variance but β is shrunk heavily toward zero. If X_i is highly correlated with A_i, then the model has the choice of using either the X_i or A_i to explain $E\{Y_i(a) \mid A_i = a, X_i = x\}$. The purpose of including the confounder X_i is to control for this possibility, with our preference being to attribute the effect to X_i rather than A_i. Instead, the opposite will occur univariate by virtue of penalizing β, the model will be encouraged to attribute the effect of X_i to A_i instead. Hence, even though we have controlled for the relevant confounder X_i, we have *regularized* its effect out and we have *effective* confounding.

In the univariate case, the regularization-induced confounding problem of (5.13) is unlikely to occur because we have no reason to use an informative prior for either Δ or β. The problem is more acute in high-dimensional and nonparametric problems where some amount of regularization of $\mu_a(x)$ is required. It is shown by [143] that regularization-induced confounding occurs for the BART models described in Section 5.4.1, and they introduce *Bayesian causal forests* partially to address this problem.

The *Bayesian causal forest* (BCF) modifies the BART models described thus far in two ways. First, an estimate of the propensity score $\widehat{e}(x)$ is included as a predictor in the model, where $\widehat{e}(x)$ is estimated *prior* to fitting the response. Second, a BART model is specified separately for the prognostic effects of the predictors and the interaction between the covariates and the treatment. The Bayesian causal forest is written as

$$\mu_a(x) = \mu_0\{x, \widehat{e}(x)\} + a\,\Delta(x). \tag{5.14}$$

The function $\mu_0(\cdot)$ captures the effect of the covariates on the response not attributable to their effect on the treatment, while $\Delta(\cdot)$ captures the effect of the treatment (which is allowed to interact with the covariates).

The decomposition (5.14) serves several purposes. By modeling $\Delta(x)$ separately from $\mu_0\{x, \widehat{e}(x)\}$, we allow for more natural prior elicitation. For example, in observational studies, we often expect a priori that the treatment effect will be small relative to the effect of the confounders, and by using this approach, we can directly choose the prior to control the magnitude of $\Delta(x)$ relative to $\mu_0(x)$. By contrast, if we specify independent priors for $\mu_1(x)$ and $\mu_0(x)$, we necessarily impose that $\text{Var}\{\Delta(x)\} = 2\,\text{Var}\{\mu_1(x)\}$, implying that we believe a priori the treatment effect will be large. By using (5.14), conversely, we can directly specify the prior variance $\text{Var}\{\Delta(x)\}$.

On the surface, incorporating $\widehat{e}(x)$ into $\mu_0\{x, \widehat{e}(x)\}$ seems redundant: we already have the confounders X_i included, so there does not appear to be anything gained by including $\widehat{e}(X_i)$ as a predictor in a non linear model, as the model is flexible enough to capture any relationship between $Y_i(a)$ and X_i. Incorporating $\widehat{e}(x)$ into the model, however, is the key to avoiding regularization-induced confounding. Regularization-induced confounding essentially occurs when it requires less model complexity to attribute the effect of confounders to the treatment; the idea behind including $\widehat{e}(x)$ directly is to allow the model to attribute the effect of the confounders to $\widehat{e}(x)$, rather than to the treatment, in a parsimonious way.

Given this setup, the BCF prior is specified in essentially the same fashion as the usual BART prior, with some minor modifications. First, we can shrink toward a homogeneous treatment effect by having the prior for $\Delta(x)$ encourage the trees of the ensemble to have depth 0; this can be accomplished by decreasing γ in the branching process. We can also directly control $\text{Var}\{\Delta(x)\}$ by choosing its associated σ_μ to be small. An extension of Bayesian causal forests to mediation is described in the case study in Chapter 14.

5.5 BART Models for Other Data Types

BART is not limited to the regression model $Y_i \sim N\{g(X_i), \sigma^2\}$. The prior $\mu_a(x) \sim \text{BART}(\pi_{\mathscr{T}}, \mathscr{M}_{\mathscr{T}})$ can be applied to any unknown function. We have already seen how the Gaussian response model can be used to fit a probit regression, which can be used for both binary response models and as a model for the propensity score $e(x) = \Phi\{g(x)\}$ [27].

Similar extensions of BART can be obtained for many exponential family models used in practice. For example, [144] introduced an extension of BART to the setting of Poisson log-linear models with

$$Y_i(a) \sim \text{Poisson}\{\lambda_i(a)\}, \qquad \log \lambda_i(a) = g_a(X_i).$$

Interestingly, a Bayesian backfitting approach can be used for this Poisson model as well: this is possible because for the model $Y_i \sim \text{Poisson}\{\exp(\mu + \xi_i)\}$ the prior $\mu \sim \log\text{Gam}(\alpha_\mu, \beta_\mu)$ is conditionally conjugate. The quantities $\xi_i^{(m)} = \sum_{j \neq m} \text{Tree}(X_i; \mathcal{T}_j, \mathcal{M}_j)$ then play the role of the backfit residuals in a Gibbs sampler. Similar extensions are also possible for the Gamma distribution, with $Y_i(a) \sim \text{Gam}\{\alpha, \alpha\lambda_i(a)\}$ using the same model as above or the Weibull distribution for survival data. These models are developed in [122] and [145]; other important extensions to the setting of survival analysis are given by [121, 146, 147].

5.6 Summary

BART is a powerful tool for both estimating the outcome regression and propensity score. It is also fairly easy to use for practitioners, due to the availability of packages, several of which focus on causal inference in particular. We propose an alternative to estimate functions in Chapter 7, Gaussian processes.

BART is less useful as a tool for solving missing data problems, particularly when components of X_i are missing, or when Y_i is a longitudinal vector subject to non-monotone missingness. In these cases, we require a method that allows for imputation of missing data in a relatively straight forward and flexible way. While there are BART methods that accomplish this goal [148], there are more tools available within the mixture modeling framework we describe in Chapter 6.

Chapter 6

Dirichlet process mixtures and extensions

DOI: 10.1201/9780429324222-6

6.1 Motivation for Dirichlet Process Mixtures

In the previous chapter, we described a flexible Bayesian semiparametric model (BART) for the mean of a continuous or binary response, which can be used to estimate average causal effects. The applications of BART we discussed model the conditional mean using an ensemble of regression trees with an (additive) normal residual (for continuous responses). For inference on non-mean causal effects, such as quantile effects or any other functional of the causal effect distribution, it is essential to model the entire conditional distribution flexibly rather than just the conditional mean. As we will see, *Dirichlet process mixtures* (DPMs) can be used for this purpose by modeling the joint distribution of the outcome and covariates.

Another drawback of BART is that it does not immediately handle missing covariates, as the distribution of the covariates is not modeled. The DPM handles missing covariates "automatically" via data augmentation (see Chapter 3) under the assumption that the covariates are subject to ignorable missingness (see Section 2.5) and can be extended for non-ignorable missing covariates (Lindberg and Daniels, working paper).

In this chapter, we introduce several Dirichlet process mixture models that we recommend to use for the observed data distribution in causal inference and missing data problems. We start by introducing Dirichlet process priors.

6.2 Dirichlet Process Priors

The *Dirichlet process* (DP) describes a model for a random *probability distribution* F on some space Θ. It is most easily explained in terms of the *stick-breaking* construction of [149].

Definition 6.2.1. A random probability distribution F on a space Θ is said to be a *Dirichlet process* if it can be represented as a countably supported discrete distribution

$$F = \sum_{j=1}^{\infty} w_j \, \delta_{\theta_j}, \tag{6.1}$$

where the point mass distributions are such that $\theta_j \overset{\text{ind}}{\sim} H$ for some *base distribution* H and the weights w_j are independent of the θ_j's and have the *stick-breaking* form

$$w_j = \beta_j \prod_{k=1}^{j-1}(1-\beta_k), \qquad \beta_k \overset{\text{ind}}{\sim} \text{Beta}(1,\alpha).$$

We write $F \sim \text{DP}(\alpha, H)$ to denote the fact that F is a Dirichlet process with base distribution H and *concentration parameter* α.

The distribution of the weights $\{w_j\}_{j=1}^{\infty}$ is called the *Griffiths-Engen-McCloskey*, or GEM, distribution, and we write $w \sim \text{GEM}(\alpha)$ [150].

The term "stick-breaking" for the weights comes from the following visualization: we start with a "stick" of length 1 and remove $100\beta_1\%$ of the stick and assign this to w_1; then, from the remaining stick of length $(1-\beta_1)$, we break of $100\beta_2\%$ of it and assign this piece to w_2, and so forth. If $Y \sim F$, then the weights w_k are such that

$$\Pr(Y=\theta_k) = \underbrace{\Pr(Y \neq \theta_j \text{ for all } j<k)}_{(1-\beta_1)\cdots(1-\beta_{k-1})} \times \underbrace{\Pr(Y=\theta_k \mid Y \neq \theta_j \text{ for all } j<k)}_{\beta_k}.$$

An alternate definition of the Dirichlet process – which is less direct, but explains our interest in it and helps us interpret α — is as follows.

Definition 6.2.2. A random probability distribution F on a space Θ is said to be a *Dirichlet process* with *base distribution* H and *concentration parameter* α if, for any finite partition $(A_1,\ldots,A_p)$ of Θ, we have

$$(F(A_1),\ldots,F(A_p)) \sim \text{Dirichlet}(\alpha H(A_1),\ldots,\alpha H(A_p)).$$

It follows from the above definition and properties of the Dirichlet distribution that, for any set A, the first two moments of $F(A)$ are given by

$$E\{F(A)\} = H(A) \qquad \text{and} \qquad \text{Var}\{F(A)\} = \frac{H(A)\{1-H(A)\}}{\alpha+1}.$$

It is now clear that, as $\alpha \to \infty$, $F(A) \to H(A)$. For this reason, α is referred to as the concentration (or *precision*) parameter, as it captures how closely the distribution F is to the base distribution H.

Just as the Dirichlet distribution is conjugate to multinomial sampling on a finite set $\{1,\ldots,J\}$, it turns out that the Dirichlet process is conjugate to *iid sampling* from a distribution F. Specifically, suppose that $[Y_1,\ldots,Y_n \mid F] \overset{\text{iid}}{\sim} F$. Then the posterior distribution of F conditional on $[Y_1,\ldots,Y_n]$ is also a Dirichlet process, but with updated parameters

$$H^\star = \frac{\alpha}{\alpha+n}H + \frac{n}{\alpha+n}\mathbb{F}_n \qquad \text{and} \qquad \alpha^\star = \alpha+n,$$

where $\mathbb{F}_n = \frac{1}{n}\sum_{i=1}^{n} \delta_{Y_n}$ denotes the empirical distribution of the data. Note that the updated base distribution is a weighted combination of our prior "guess" for F (H) and the information contained in the data $(\mathbb{F}_n)$. The updated precision parameter $\alpha^\star = \alpha + n$ is incremented by our sample size n and reveals that we can regard α as a "prior" sample size for our guess for F. For fixed α, as n grows, the posterior for F is mostly driven by the observed Y's. For small α, the posterior is mostly driven by the empirical distribution; as $\alpha \to 0$, in fact, the posterior distribution for F converges to the Bayesian bootstrap (see Chapter 3, Section 3.2.3).

A problem with Dirichlet process priors $F \sim \text{DP}(\alpha, H)$ is that the realizations from the prior are, with probability 1, discrete (even if H is continuous). While they can still be useful when the Y_i's are actually continuous, there are some purposes for which the discrete nature of the Dirichlet process creates problems. This motivates the use of Dirichlet process *mixtures* of continuous distributions. We introduce these mixtures next.

6.3 Dirichlet Process Mixtures (DPMs)

DPMs extend the discrete DP to continuous distributions by having realizations from $F \sim \text{DP}(\alpha, H)$ correspond to *latent variables* rather than to the data itself. The DPM model is written as

$$\begin{aligned}[Y_i \mid \theta_i, F] &\overset{\text{ind}}{\sim} p(y_i; \theta_i), \\ [\theta_i \mid F] &\overset{\text{iid}}{\sim} F, \\ F &\sim DP(\alpha, H).\end{aligned} \tag{6.2}$$

DPMs get their name from the fact that the parameters of the distribution in the first level of (6.2) follow a distribution F which is itself given a DP prior.

The DPM can be rewritten as an infinite mixture of distributions using the stick-breaking construction in (6.1) as

$$p(y;\theta) = \sum_{j=1}^{\infty} w_j p(y;\theta_j), \tag{6.3}$$

where $w_j = \beta_j \prod_{k=1}^{j-1}(1-\beta_k)$, $\beta_j \sim \text{Beta}(1,\alpha)$, and $\theta_j \sim H$.

Computation with Dirichlet process mixtures is complicated by the fact that (6.3) depends on an infinite number of parameters. There are several ways to bypass this issue. The simplest is to truncate the sum in (6.3) at some "large" number K. Fortunately, the stick-breaking weights decay exponentially quickly, so typically we can take K fairly small. On average, the first K weights add up to

$$E\{1-(1-\beta_1)\cdots(1-\beta_K)\} = 1 - \left(\frac{\alpha}{\alpha+1}\right)^K.$$

For example, when $\alpha = 1$, the first 20 weights add up to (on average) $1-(1/2)^{20} \approx 1$, while for $\alpha = 10$ the weights only add up to $1-(10/11)^{20} \approx 0.85$ (see Figure 6.1 and Figure 6.2).

Ishwaran and James [151] approximate (6.3) with a finite mixture and show the error in this approximation is bounded by a function of α, which facilitates the choice of the number of mixture components for the approximation. The error in the approximation can be quantified as

$$\|f_K(Y_1,\ldots,Y_n) - f_\infty(Y_1,\ldots,Y_n)\|_1 \approx 4n\exp\left\{-\frac{K-1}{\alpha}\right\}$$

where $f_K(y_1,\ldots,y_n)$ denotes the joint distribution (after marginalizing out the parameters) of $(Y_1,\ldots,Y_n)$ in (6.2) when using a truncation level K, and $\|g\|_1 \equiv \int |g|\, dy_1 \cdots dy_n$. Clearly as $K \to \infty$ the difference between the exact and approximate marginals goes to 0, with larger values of α requiring a larger K. A preliminary run with many components can be used to estimate α and determine an acceptable error for the given data (i.e., how large K should be). In practice, K is often chosen such that $E\{(1-\beta_1)\cdots(1-\beta_K)\} \approx 0.01$; after finding K, we set $\beta_K = 1$ (i.e., we assign the entire remainder of the stick to w_K).

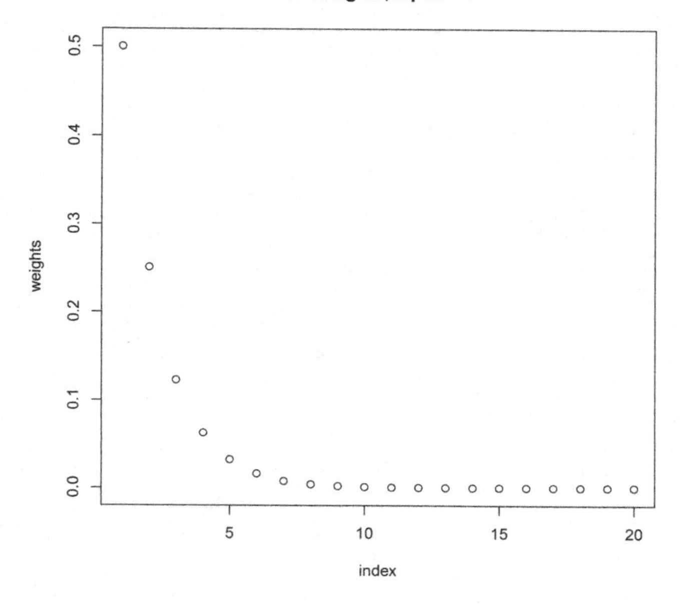

Figure 6.1 *Average DP weights for $\alpha = 1$.*

Given this truncation approximation, DPMs can also be implemented as a (finite) latent class model

$$\begin{aligned} [Y_i \mid Z_i, \{\theta_k\}^\star] &\sim p(y \mid \theta^\star_{Z_i}) \\ [\theta^\star_k \mid H] &\sim H \\ [Z_i \mid w] &\sim \text{Categorical}(w) \\ [w \mid \alpha] &\sim \text{GEM}(\alpha), \end{aligned} \tag{6.4}$$

where $\theta^\star_1, \ldots, \theta^\star_K$ denote the *unique* values of the θ_i's from (6.2). This approximation and formulation allows DPMs to be easily fit in software like `rjags` (see Chapter 11).

Example 6.3.1 *(Dirichlet process mixture (DPM) of normal distributions)***.** The DPM of normals model sets $p(y;\theta)$ to be a normal distribution with mean θ and variance σ^2. One possible, conditionally conjugate, specification takes

$$\begin{aligned} [Y_i \mid Z_i, \{\theta_k\}^\star] &\sim N(y \mid \theta^\star_{Z_i}, \sigma^2) \\ [\theta^\star_k \mid H] &\sim H = N(\mu, \tau^2) \\ [Z_i \mid w] &\sim \text{Categorical}(w) \\ [w \mid \alpha] &\sim \text{GEM}(\alpha) \\ \sigma^{-2} &\sim \text{Gam}(a, b). \end{aligned}$$

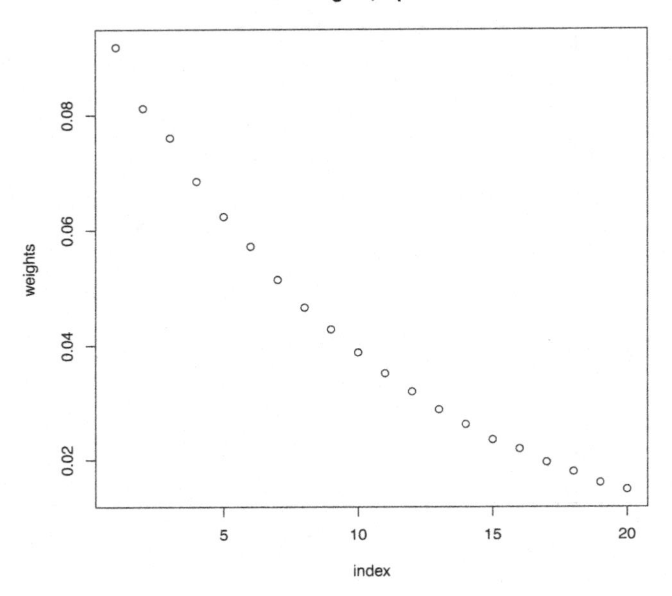

Figure 6.2 *Average DP weights for $\alpha = 10$.*

Effectively, after truncation, this is just a finite mixture of normals with weights following a GEM distribution.

Example 6.3.2 *(Dirichlet process mixture (DPM) of bivariate normal distributions).* Now consider a joint model for (Y,L) (say, the outcome and a confounder in a causal inference problem). We use a DPM of bivariate normal distributions for this joint distribution. Once again we can write this as an infinite mixture

$$p(y,\ell;\theta) = \sum_{j=1}^{\infty} w_j N(y,\ell;\mu_j,\Sigma_j),$$

where $w \sim \text{GEM}(\alpha)$ and $\theta_j \overset{\text{iid}}{\sim} H$; here, $p(\cdot)$ is a bivariate normal distribution with $\theta_j = (\mu_j, \Sigma_j)$, $\mu_j = (\mu_{jY}, \mu_{jL})$, and

$$\Sigma_j = \begin{pmatrix} \sigma_{jY}^2 & \rho_j \sigma_{jY} \sigma_{jL} \\ \rho_j \sigma_{jY} \sigma_{jL} & \sigma_{jL}^2 \end{pmatrix}.$$

A normal-inverse Wishart distribution is often specified for H to facilitate computations, though more flexible alternatives are possible [152].

The DPM of bivariate normals induces the following conditional distribution for $[Y_i \mid L_i]$ [153]:

$$[Y \mid L = \ell] \sim \sum_{j=1}^{\infty} w_j(\ell) N(y \mid \gamma_{0j} + \gamma_{1j}\ell, \sigma_j^2) \tag{6.5}$$

where

$$w_j(\ell) = \frac{w_j N(\ell \mid \mu_{j\ell}, \sigma^2_{j\ell})}{\sum_{j'=1}^{\infty} w_{j'} N(\ell \mid \mu_{j'\ell}, \sigma^2_{j'\ell})} \tag{6.6}$$

and $\gamma_{0j} = \mu_{jY} - \frac{\sigma_{jY}}{\sigma_{j\ell}} \rho_j \mu_{j\ell}$, $\gamma_{1j} = \frac{\sigma_{jY}}{\sigma_{j\ell}} \rho_j$, $\sigma_j^2 = (1-\rho_j^2)\sigma^2_{jY}$. The resulting distribution has mean

$$E(Y \mid L = \ell, \mu_j, \Sigma_j) = \sum_{j=1}^{\infty} w_j(\ell)(\gamma_{0j} + \gamma_{1j}\ell),$$

which is nonlinear and non-additive in L due to the weights $w_j(\ell)$ being a function of L. This conditional distribution itself is a mixture of normal distributions with *covariate dependent* weights $w_j(\ell)$.

Example 6.3.3 *(Dirichlet process mixture (DPM) of product multinomials)*. To model the joint distribution of binary variables $Y_i = (Y_{i1}, \ldots, Y_{iJ})$, we can use a DPM of multinomial distributions given as

$$f(y \mid \omega) = \sum_{k=1}^{\infty} w_k \left\{ \prod_{j=1}^{J} \gamma_{kj}^{y_j} (1 - \gamma_{kj})^{1-y_j} \right\} \tag{6.7}$$

with $w \sim \text{GEM}(\alpha)$ and $\gamma_{kj} \sim \text{Beta}\{\rho_{\gamma_j} a_{\gamma_j}, (1-\rho_{\gamma_j}) a_{\gamma_j}\}$. This model assumes the binary variables are locally independent, similar to the specification of covariates in Section 6.3.3. This type of model has also been called an infinite product-multinomial model [154]. Models of this form have been used to perform multiple imputations for large-scale surveys under MAR [155]. We use this model in the case study in Chapter 12.

DPMs provide a (i) flexible joint model, e.g., (Y, L), and (ii) a flexible regression model, $[Y \mid L]$ (flexible mean, residual, and conditional distribution). We will provide further details in the context of causal inference and missing data later in this chapter.

6.3.1 *Posterior Computations*

There are various algorithms available for sampling from the posterior distribution of a DPM, including the *blocked Gibbs sampler* [156], the *slice sampler* [157], and the *Pólya urn sampler* [79].

The blocked Gibbs sampler

The blocked Gibbs sampler relies on the truncation approximation to the Dirichlet process mixture to design a simple (albeit approximate) Gibbs sampling algorithm for DPMs. Suppose that we have adopted a finite truncation level K for the DPM; then, the posterior distribution associated with the latent class representation (6.4) is (proportional to)

$$\left[\prod_{k=1}^{K} \left\{ \prod_{i: z_i = k} w_k \, p(y_i \mid \theta_k^\star) \right\} H(\theta_k) \times \prod_{k=1}^{K-1} \text{Beta}(\beta_k \mid 1, \alpha) \right] \text{Gam}(\alpha \mid a, b). \tag{6.8}$$

From this, it is straight forward to derive the full conditionals for the *blocks* of parameters $\{\theta_k\}_{k=1}^K$, $\{\beta_k\}_{k=1}^{K-1}$, and $\{Z_i\}_{i=1}^n$. The blocked Gibbs sampler is given in Algorithm 6.1, which simply applies Gibbs sampling as described in Section 3.3.2 to the posterior (6.8).

Algorithm 6.1 Blocked Gibbs sampler for DPMs truncated to K components

1. *Update parameters within each mixture component.* This step updates the parameters in each mixture component. For an empty cluster (say, cluster k), the parameter $\theta_k^\star$, is sampled from H. For the non-empty clusters (say cluster j), the parameters $\theta_j^\star$ are sampled from

$$\pi(\theta_j^\star \mid z, \alpha, y) \propto H(d\theta_j^\star) \prod_{i:z_i=j} p(y_i \mid \theta_j^\star).$$

   Note if H is conjugate for $\theta^\star$, this is a conjugate update. We provide specific details on this for a different sampler in Section 6.4.2.1.

2. *Update cluster membership Z_i.* For each unit, Z_i is sampled from a multinomial distribution with the kth (of the K) multinomial probability parameters proportional to

$$p_{ik} \propto w_k p(y_i \mid \theta_k^\star).$$

3. *Update the weights.* The full conditional of $(\beta_1, \ldots, \beta_K)$ is given by

$$\beta_j \overset{\text{ind}}{\sim} \text{Beta}\left(1 + M_j, \alpha + \sum_{k=j+1}^{N} M_k\right), j = 1, \ldots, K-1,$$

   where M_k is the number of units with $Z_i = k$.

4. *Update the concentration parameter.* For a $\text{Gam}(a, b)$ prior on the mass parameter α, the full conditional for α is a $\text{Gam}(K + a - 1, b - \log p_K)$ where $p_K = \exp\{\sum_{k=1}^{K-1} \log(1 - \beta_k)\}$ [158].

Pólya urn samplers

A fundamental challenge with performing MCMC with DPMs is that the distribution F is infinite dimensional; practically speaking, it is impossible for us to store the entirety of F. To bypass this issue, the blocked Gibbs sampler effectively truncates the F so that $F \approx \sum_{k=1}^K w_k \delta_{\theta_k}$ for some large (but finite) K.

An alternative approach is to instead *integrate out* F and perform MCMC directly on $\{\theta_1, \ldots, \theta_n\}$ from (6.2). Remarkably, it can be shown [159, 79] that the conditional distribution of θ_i can be written as

$$[\theta_i \mid \theta_{-i}] \sim \frac{1}{n-1+\alpha} \sum_{j \neq i} \delta_{\theta_j} + \frac{\alpha}{n-1+\alpha} H,$$

where $\theta_{-i} = (\theta_1, \ldots, \theta_{i-1}, \theta_{i+1}, \ldots, \theta_n)$. This is referred to as the *Pólya urn representation* of the distribution of $\{\theta_1, \ldots, \theta_n\}$. When combined with the likelihood $p(y_i \mid \theta_i)$, following [79], this yields the full conditional for θ_i given by the mixture distribution

$$[\theta_i \mid \theta_{-i}, y_i] \sim \sum_{j \neq i} q_{i,j}\, \delta_{\theta_j} + r_i H_i \tag{6.9}$$

where $q_{i,j} \propto p(y_i \mid \theta_j)$ and $r_i \propto \alpha \int p(y_i \mid \theta)\, H(d\theta)$, normalized so that $\sum_{j \neq i} q_{i,j} + r_i = 1$. Here, the distribution H_i corresponds to the posterior distribution we would obtain if we set $\theta_i \sim H$ and observed only y_i.

The above observations suggest a Gibbs sampler that operates by iteratively sampling from $[\theta_i \mid \theta_{-i}, y_i]$ for $i = 1, \ldots, n$ many times. In practice this strategy has some shortcomings. In terms of bookkeeping, many of the θ_j's will be identical, so we do not need to explicitly keep track of $n-1$ mixture components, but can instead reduce to cluster indicator variables Z_i that keep track of the *unique* values of the θ_i's. Additionally, sampling from (6.9) is only convenient when H_i is available in closed form, which practically occurs only when $H(d\theta)$ is conjugate to $p(y \mid \theta)$.

To address these shortcomings, [79] introduced a large number of alternative algorithms that are based on the Pólya urn representation but (a) are more streamlined in terms of keeping track of the unique values of θ_j and (b) do not require H to be conjugate to $p(y \mid \theta)$. An example of such an algorithm is given in Algorithm 6.2. It also includes an additional update of the unique values of the θ_j's from their full conditionals to improve the mixing of the algorithm (see Step 1 of Algorithm 6.2), as well as an update for the concentration parameter α under a gamma prior (the validity of this update is not obvious, but is justified by [159]).

Other approaches

There are several other classes of samplers that can be used to fit DPMs, which may be more convenient depending on the particular problem one is dealing with. One approach is to work with a truncation of the DPM using (6.4) but with the latent classes Z_i integrated out; in this case, we target the posterior distribution

$$\pi(\theta, w \mid y) \propto \left\{ \prod_{i=1}^{n} \sum_{k=1}^{K} w_k\, p(y_i \mid \theta_k) \right\} \times \prod_{k=1}^{K} H(d\theta_k) \times \prod_{k=1}^{K-1} \text{Beta}(\beta_k \mid 1, \alpha). \tag{6.10}$$

This approach is required to fit DPMs in the software package `Stan` because `Stan` does not (as of this writing) allow for discrete parameters like latent classes. The parameters (θ, w) can then be sampled via HMC using (6.10), which can be done automatically in `Stan`.

A more advanced version of the blocked Gibbs sampler is the *slice sampler* of [157], which uses an adaptive truncation level K that changes over the course of the sampling algorithm. We defer details to [157]; the downside of this approach is that it is somewhat more complicated to implement than the blocked Gibbs sampler.

Algorithm 6.2 Polya urn sampler for DPMs

1. *Update the mixture component parameters.* This step updates the parameters in each mixture component. Let $\theta_1^\star, \ldots, \theta_K^\star$ denote the *unique* values of the θ_i's, and let $Z_i = j$ if Y_i is associated with $\theta_j^\star$. We update $\theta_j^\star$ by sampling from
$$\pi(\theta_j^* \mid z, y, \theta_{-j}^*) \propto H(d\theta_j^*) \prod_{i:z_i=j} p(y_i \mid \theta_j^*).$$
Note if H is conjugate for θ, this is a conjugate update.
2. *Update cluster membership.* Let K be the current number of clusters (i.e., the number of unique values of Z_i). We use $-i$ notation to indicate that the ith value is removed (e.g., y^{-i} is the vector of outcomes excluding subject i). We sample Z_i from a multinomial distribution with probabilities $\Pr(Z_i = j \mid Z^{-i}, \theta^*, y)$ defined as follows:
$$\Pr(Z_i = j \mid Z^{-i}, \theta^\star, y) \propto \begin{cases} n_j^{-i} p(y_i; \theta_j^*), & \text{for } j \in \{1, \ldots, K^{-i}\} \\ \alpha p^\star(y_i), & \text{for } j = K^{-i} + 1 \end{cases}$$
where n_j^{-i} is the number of subjects in cluster j (excluding subject i) and $p^\star(y_i) = \int p(y_i \mid \theta) H(d\theta)$. This step is done for each $i = 1, \cdots, n$. If conjugate priors are specified for H, then $p^\star(y_i)$ will be available in closed form. We provide details if this is not the case later in Section 6.4.2.1.
3. *Update the concentration parameter.* For a $\text{Gam}(a, b)$ prior on the mass parameter α, to sample α first draw $\eta \sim \text{Beta}(\alpha + 1, n)$ and then draw α from
$$\pi \, \text{Gam}(a + K, b - \log(\eta)) + (1 - \pi) \, \text{Gam}(a + K, b - \log(\eta)),$$
where $\pi = \frac{\frac{K}{n(1-\log(\eta))}}{1 + \frac{K}{n(1-\log(\eta))}}$ and K is the number of unique values for Z_i [159].

Comparison of samplers

In practice, there are several important considerations in choosing between the blocked Gibbs sampler and the Pólya urn scheme.

1. The blocked Gibbs sampler can be implemented in general purpose Gibbs sampling software packages such as `jags`, whereas the Pólya urn scheme cannot. This makes it especially convenient when one needs to go beyond DPMs of multivariate normals or incorporate categorical variables.
2. The blocked Gibbs sampler preserves draws of the distribution F so that the density of the data can be estimated; the Pólya urn scheme, but contrast, integrates out F and hence requires further processing to get (approximate) samples of F.
3. Pólya urn schemes do not rely on specifying a truncation level.
4. While it is difficult to say anything universal about the mixing of the blocked Gibbs sampler relative to Pólya urn sampler, [156] show that the one-at-a-time

updates of the Pólya urn sampler can result in slow mixing; this issue can be exacerbated in large datasets.

Fitting DPMs in practice

There are various specialized software options for fitting DPMs. The DPM of normals model can be fit using the `R` package `dirichletprocess` (which uses the Pólya urn sampler) or the `bnpqte` package (which can use either the blocked Gibbs or Pólya urn samplers). If a truncation approximation is used [158], then the posterior distribution can be sampled easily using `rjags` (using the hierarchical model (6.4) to implement the blocked Gibbs sampler in Algorithm 6.1 though it does not exploit all the conjugacy in the algorithm) or `rstan` (using HMC with the model representation (6.10)).

6.3.2 *DPMs for Causal Inference and Missing Data*

We now introduce some examples to illustrate the use of DPMs in missingness and causal inference problems.

Example 6.3.4 (*Ignorable missingness in a bivariate longitudinal response*)**.** As introduced in Chapter 2, under ignorable missingness only a model for $f(y \mid \omega)$ needs to be specified. For a continuous response, y, a DPM of normals can be specified as in Example 6.3.2, and data augmentation can be used to impute the missing outcomes as part of a Gibbs sampler. These models might be appropriate for a clinical trial with a baseline and single response after treatment, assuming that missingness in the response after treatment is missing at random.

Example 6.3.5 (*Causal inference using the propensity score*)**.** DPMs allow for the estimation of any (i.e., not just average) causal effects that can be expressed in terms of the marginal distributions of the potential outcomes $f\{Y(a) = y\}$; all functionals of this distribution are flexibly modeled. Using the propensity score in a regression setting (as in Chapter 1), we can specify the joint distribution of the outcome and propensity score using a DPM of bivariate normals. We can use the DPM of bivariate normals introduced in Example 6.3.2 with L replaced by an estimated propensity score $\widehat{e}(L)$. Using the distribution of the potential outcomes, causal effects can be computed using g-computation under ignorability as

$$\Pr\{Y(a) < y\} = \int \int_{-\infty}^{y} f\{t \mid A = a, \widehat{e}(\ell)\}\, dt\, F_L(d\ell),$$

where the conditional distribution on the right hand side has the form given in (6.5), with L replaced by $\hat{e}(L)$, and is fit separately for each value of A. The distribution of the confounders can be estimated using the Bayesian bootstrap (cf. Chapter 3), and the propensity score can be estimated nonparametrically using BART (see Chapter 5).

This model was used to evaluate the effect of two drug therapies on serum creatinine change using EHRs from two affiliated tertiary care institutions. The continuous outcome (Y) was the maximum difference in serum creatinine from baseline during the five-day follow-up. The propensity score was estimated using BART and included 33 baseline confounding variables (L). The study objective was to estimate quantile causal effects between two treatment groups: piperacillin-tazobactam ($A = 1$) and cefepime ($A = 0$). We provide a full analysis of this data in Chapter 8.

Example 6.3.6 (*Causal inference using g-computation with confounders*). Causal estimands can also be obtained from a DPM of multivariate normals in the case of a continuous response and continuous covariates (separately for each value of A) using the g-formula,

$$\Pr\{Y(a) < y\} = \int \Pr(Y < y \mid A = a, L = \ell)\ F_L(d\ell),$$

where the marginal distribution of L takes the form,

$$f(L = \ell) = \sum_{a=0}^{1} \left[\sum_{j=1}^{\infty} w_j^a N(\ell | a; \mu_j, \Sigma_j) \right] \pi^a (1-\pi)^{1-a}$$

where the term in the brackets is from fitting the model separately for each value of A and the conditional is computed as in Example 6.3.2. To sample F_L, we first model $A \sim \text{Bernoulli}(\pi)$ with a Beta prior on π. Then we (i) sample π from its posterior; (ii) sample A from a Bernoulli distribution given π; and (iii) sample from $[L \mid A]$ and use the sampled L to compute the integral (ignoring the sampled A). Further details on sampling can be found in Section 6.3.1.

This model could be used for the serum creatinine study by jointly modeling the response with the (continuous) confounders using a DPM (cf. Chapter 8).

Example 6.3.7 (*Non-ignorable missingness*). For non-ignorable missingness, a joint model for the outcome $Y = (Y_1, \ldots, Y_J)$ and missingness indicators $R = (R_1, \ldots, R_J)$ is needed. For simplicity, we assume monotone missingness which allows R to be replaced by $S = \sum_{j=1}^{J} R_j$ as discussed in Chapter 2. A simple DPM to consider would be

$$\begin{aligned}
[Y_i \mid \theta_i] &\sim p_y(y \mid \theta_i) \\
[S_i \mid \omega_i] &\sim p_s(s \mid \omega_i) \\
[(\theta_i, \omega_i) \mid F] &\sim F \\
F &\sim DP(\alpha, H_{0\theta} \times H_{0\omega}).
\end{aligned}$$

where p_y is a J-dimensional multivariate normal distribution, $\theta_i = (\mu_i, \Sigma_i)$, and p_s is a multinomial distribution with probability vector, ω_i. The conditionally conjugate choices for the base distribution would be a normal-inverse Wishart distribution for θ and an independent Dirichlet distribution for ω. From this, we can extract the observed data distribution. Let $\bar{y}_s = (y_1, \ldots, y_s)$ and $\bar{y}_{-s} = (y_{s+1}, \ldots, y_J)$. Then the observed data distribution is

$$\sum_{k=1}^{\infty} w_k p_s(s; \omega_k) p_y(\bar{y}_s, \ell; \theta_k),$$

where the w_k's are the stick-breaking weights given under (6.3). Identification of the extrapolation distributions, $p(\bar{y}_{-s}|\bar{y}_s, S=s) : s = 1, \ldots, J-1$ was discussed in Chapter 4. As with the previous examples, the DPM can be fit separately for each value of A.

A similar model was used to analyze data from a multicenter randomized double-blind clinical trial aimed to assess the safety and efficacy of a test drug relative to a placebo and an active control group. The response was an assessment of schizophrenia symptoms (PANSS) which was scheduled to be collected at baseline, day 4, and then weeks 1, 2, 3, and 4. Dropout on the arms ranged from 19% to 33%. The outcome of interest was the expected change from baseline to week 6. We provide a full analysis of this data in Chapter 11 which also includes an approach to incorporate information about the reasons for dropout.

Example 6.3.6 suggested several complications. First, how best to address non continuous covariates (including a binary treatment)? Second, within the components of the mixture, some explicit dependence between Y and L would likely lead to better small sample properties; note this is easily addressed without computational difficulties with a continuous response and covariates (just a multivariate normal). To address these issues, Shahbaba and Neal [160] introduced a modified DPM of models described in the next subsection.

6.3.3 *Shahbaba and Neal DPM*

DPMs of (multivariate) normals do not easily handle correlated continuous and categorical predictors. As an alternative, Shahbaba and Neal [160] introduced the following DPM:

$$
\begin{aligned}
[Y_i \mid L_i, \theta_i] &\sim p(y \mid \ell, \theta_i) \\
[L_{i,j} \mid \omega_{ij}] &\sim p_j(\ell_j \mid \omega_{ij}), j = 1, \ldots, p \\
[(\theta_i, \omega_i) \mid F] &\sim F \\
F &\sim DP(\alpha, H_{0\theta} \times H_{0\omega}),
\end{aligned}
$$

where $\omega_i = (\omega_{i1}, \ldots, \omega_{ip})$, $p(y \mid \ell, \theta_i)$ is a generalized linear model as opposed to $p(y \mid \theta_i)$, and here the covariates, L_j, are now assumed locally (within component) independent. This DPM can be written using the stick-breaking construction,

$$
f(y, \ell \mid \theta) = \sum_{j=1}^{\infty} w_j p(y \mid \ell, \theta_j) \prod_{k=1}^{p} p_k(\ell_k \mid \omega_{jk}). \tag{6.11}
$$

Shahbaba and Neal then use this model to derive a flexible conditional distribution model for y,

$$
f(y \mid \ell, \theta) = \sum_{j=1}^{\infty} w_j(\ell) p(y \mid \ell, \theta_j) \qquad \text{where} \qquad w_j(\ell) = \frac{w_j \prod_k p_k(\ell_k \mid \omega_{jk})}{\sum_m w_m \prod_k p_k(\ell_k \mid \omega_{mk})}.
$$

This joint DP mixture model is quite flexible, allowing both the outcome and the covariates to be continuous or discrete. Additionally, the local independence of covariates makes it easy to specify conjugate priors.

Example 6.3.8 (*Binary regression with a continuous and binary covariate*). Consider a binary regression with two covariates, one continuous and one binary,

$$
\begin{aligned}
[Y_i \mid L_i, \theta_i] &\sim p_y(y \mid \ell, \theta_i) \\
[L_{i,1} \mid \omega_{1i}] &\sim p_1(\ell_1 \mid \omega_{1i}) \\
[L_{i,2} \mid \omega_{2i}] &\sim p_2(\ell_2 \mid \omega_{2i}) \\
[(\theta_i, \omega_i) \mid F] &\sim F \\
F &\sim DP(\alpha, H_{0\theta} \times H_{0\omega})
\end{aligned}
$$

where $\omega_i = (\omega_{1i}, \omega_{2i})$, p_y is a Bernoulli distribution with success probability $\Phi(\theta_{1i} + \theta_{2i}\ell_1 + \theta_{3i}\ell_2)$, p_1 is a normal distribution, p_2 is a Bernoulli distribution, $\omega_{1i} = (\mu_i, \sigma_i^2)$, $H_{0\theta}$ is a normal distribution, and $H_{0\omega}$ is the product of a normal-inverse gamma distribution and a Beta distribution. Note H is conjugate for all components of ω and for θ with data augmentation (cf. Chapter 3, Section 3.5). Of course, this model does not require local independence for all covariates, L. For example, one could still use a multivariate normal for the continuous covariates and/or a large multinomial for the binary covariates or more complex specifications [161].

6.3.4 *Priors on Parameters of the Base Measure*

For computational reasons (cf. Section 6.3.1), conjugate priors for the base measure are preferred. We provide some suggestions here based on specifications in [162, 99, 163]. For a q-dimensional multivariate normal mean and covariance matrix, following [162], we assume $m_Z \sim N(m_Z^\star, S_Z^\star)$, $S_Z \sim \mathscr{W}^{-1}(\nu_s, \Omega_Z^\star)$, and $\Psi_Z \sim \mathscr{W}(\nu_\Psi, \Psi_Z^\star)$ where $\mathscr{W}^{-1}$ is the inverse Wishart parameterized such that $E(S_Z) = S_Z^\star/(\nu_s - q - 1)$ and where $\mathscr{W}$ is the Wishart parameterized such that $E(\Psi_Z) = \nu_\Psi \Psi_Z^\star$. The hyperparameters (ν_Z, ν_s, ν_Ψ) are given the (default) value $2(q+1)$ and $m_Z^\star$ is set to the mean of the data. Both $S_Z^\star$ and $(\Omega_Z^\star)/(\nu_s - q - 1)$ are set to $\hat{\Sigma}/2$ where $\hat{\Sigma}$ is the MLE of the covariance matrix. Finally, $\Psi_Z^\star$ is specified as $\nu_\Psi \Psi_Z^\star/(\nu_z - q - 1) = \hat{\Sigma}/2$. An alternative specification for the mean and variance of a univariate normal (as well as for specific longitudinal normal models) is given in the supplementary materials of [99].

For regression parameters, we suggest normal priors with mean the MLE from fitting the parametric regression model and variance a constant (proportional to n) times the asymptotic covariance matrix; in [163] the constant was specified as $n/5$ (based on simulations). Note these are conjugate for linear regressions and probit and logit regressions (with data augmentation). For parameters of Bernoulli distributions, we recommend Beta priors with default parameters equal to one.

Note that the priors introduced above are just some possible specifications for DPMs, and all have some weak data dependence. We provide details on variations in the context of the case studies in Chapters 8-15.

6.4 Enriched Dirichlet Process Mixtures (EDPMs)

Unfortunately, the DPM models discussed above have issues when incorporating large numbers of covariates (or interest, auxiliary, and/or confounders) due to the fact that the "likelihood" for cluster k is $p(y \mid \ell, \theta_k)\prod_{j=1}^{p} p(\ell_j \mid \omega_k)$. This implies that, in determining the cluster membership, the outcome model only gets $(1/p)^{\text{th}}$ of the weight of the covariates. As such, if p is large, the model may dedicate most of its resources to modeling the distributions of the confounders rather than the outcome of interest [164].

An *enriched Dirichlet process mixture* (EDPM), introduced in [165, 164], is a more flexible DPM model that overcomes this problem. We still assume that we have an outcome Y and continuous and/or categorical covariates L. An EDPM is specified as

$$\begin{aligned}
[Y_i \mid L_i, \theta_i] &\sim p(y \mid \ell, \theta_i) \\
[L_{i,j} \mid \omega_i] &\sim p(\ell_j \mid \omega_i), \\
[(\theta_i, \omega_i) \mid F] &\sim F \\
F &\sim \text{EDP}(\alpha_\theta, \alpha_\omega, H).
\end{aligned}$$

The expression $F \sim \text{EDP}(\alpha_\theta, \alpha_\omega, H)$ means that $F(d\theta, d\omega) = F_\theta(d\theta) \times F_{\omega|\theta}(d\omega \mid \theta)$ where $F_\theta \sim \text{DP}(\alpha_\theta, H_\theta)$ and $F_{\omega|\theta} \sim \text{DP}(\alpha_\omega, H_{\omega|\theta})$, with base measures $H = H_\omega \times H_{\omega|\theta}$. Note that the precision parameter for the L clusters can depend on θ, i.e., $\alpha_{\omega|\theta}$. This specification has several advantages: (i) it allows many ℓ-clusters (important for local dependence) without having to create additional y-clusters (via nested clustering of L clusters with in Y clusters); (ii) simple (independent) models for the components of ℓ make it easy to include many covariates (as with the Shahbaba and Neal specification); and (iii) it directly addresses ignorable missingness in the covariates (similarly to a regular DPM) since it jointly models Y and L. The first of these advantages is unique to EDPMs.

The joint distribution of (Y, L) has the following *square-breaking* construction:

$$p(y; \theta) = \sum_{j=1}^{\infty} \left\{ \gamma_j p(y \mid \ell; \theta_j) \sum_{k}^{\infty} \gamma_{k|j} p(\ell; \omega_{k|j}) \right\},$$

where

$$\begin{aligned}
\gamma_j &= \gamma'_j \prod_{\ell<j} (1 - \gamma'_\ell), & \gamma'_\ell &\sim \text{Beta}(1, \alpha_\theta) & \theta_j &\overset{\text{iid}}{\sim} H_\theta \\
\gamma_{k|j} &= \gamma'_{k|j} \prod_{\ell<k} (1 - \gamma'_{\ell|j}), & \gamma'_{\ell|j} &\sim \text{Beta}(1, \alpha_\omega) & \omega_{k|j} &\overset{\text{iid}}{\sim} H_{\omega|\theta}.
\end{aligned}$$

We see that the usual stick-breaking weights appear for the y-clusters, but, within each y-cluster, there are additional stick-breaking weights for the ℓ-clusters. The EDPM induces the following conditional distribution for the outcome ℓ:

$$p(y \mid \ell) = \sum_{j=1}^{\infty} w_j(\ell) p(y \mid \ell, \theta_j), \tag{6.12}$$

where

$$w_j(\ell) = \frac{\gamma_j \sum_{l=1}^{\infty} \gamma_{l|j} p(\ell \mid \omega_{l|j})}{\sum_{h=1}^{\infty} \gamma_h \sum_{l=1}^{\infty} \gamma_{l|h} p(\ell \mid \omega_{l|h})}.$$

which has a similar level of flexibility as the DPM model for the conditional. The weights are similar to the DPM weights in (6.6). Priors on the parameters of the base measure can be specified as in the DPM (see Section 6.3.4).

6.4.1 *EDPM for Causal Inference and Missing Data*

The EDPM has many advantages for causal inference and ignorable missingness with auxiliary covariates. We illustrate via some examples next.

Example 6.4.1 (*Ignorable missingness with auxiliary covariates*)**.** Assume a bivariate longitudinal continuous outcome, (Y_1, Y_2), with missingness in Y_2, p auxiliary covariates V (q_c continuous and $p - q_c$ binary), and a binary treatment A. We then specify the following EDPM model:

$$\begin{aligned}
[Y_i \mid V_i, \theta_i] &\sim p_y(y \mid v, \theta_i) \\
[V_{i,j} \mid \omega_i] &\sim p_c(v_j \mid \omega_{ci}) \quad (j = 1, \ldots, q_c), \\
[V_{i,j} \mid \omega_i] &\sim p_b(v_k \mid \omega_{bi}) \quad (j = q_c + 1, \ldots, p) \\
[A_i | \omega_i] &\sim p_b(a | \omega_{ai}) \\
[(\theta_i, \omega_i) \mid F] &\sim F \\
F &\sim \text{EDP}(\alpha_\theta, \alpha_\omega, H),
\end{aligned}$$

where p_y is a bivariate normal distribution with mean $(V_i^\top \beta_{1i}, V_i^\top \beta_{2i})$ and covariance matrix Σ_i, p_c is a normal distribution with mean μ_{ij} and variance τ_{ij}^2, p_b is a Bernoulli distribution with mean π_{ij} (for $V_{i,j}$) or π_i (for A_i), and the EDPM parameters are given by $\theta_i = (\beta_i, \Sigma_i)$ and $\omega_i = (\{\mu_{ij}, \tau_{ij}^2 : j = 1, \ldots, q_c\}, \{\pi_{ij} : j = q_c + 1, \ldots, p\}, \pi_i)$. The conditional distribution of $Y_j \mid V$ takes the form given in (6.12). For computations, it is easy to use data augmentation to fill in the missing Y_2, conditional on cluster assignment as in Chapter 3, Section 3.5.

For inference, it is typically of interest to integrate out the auxiliary covariates

$$f(y \mid a) = \int f(y \mid a, v) \, F(dv \mid a),$$

where if A is randomized we often assume $F(v \mid a) = F(v)$. The distribution $p(y \mid a, v)$ takes the form as in (6.12) with $L = (A, V)$. MC integration as described in Chapter 3 can be used to compute the integral above; in particular, we sample L (as will be described in Section 6.4.2.2) and keep V if $A = a$ and discard it otherwise.

Example 6.4.2 (*Causal inference with point treatment and many confounders*)**.** As discussed above, for causal settings with many confounders, the EDPM should provide improved inference based on $Y \mid A, L$. We now consider a binary response and p

confounders, q_c continuous and $p - q_c$ binary, and a binary treatment, A. The EDPM takes the following form:

$$
\begin{aligned}
[Y_i \mid L_i, \theta_i] &\sim p_y(y \mid \ell, \theta_i) \\
[L_{i,j} \mid \omega_i] &\sim p_c(\ell_j \mid \omega_{ci}) : j = 1, \ldots, q_c, \\
[L_{i,k} \mid \omega_i] &\sim p_b(\ell_k \mid \omega_{bi}), k = q_c + 1, \ldots, q_c + q_b \\
[A_i \mid \omega_i] &\sim p_b(a|\omega_{ai}) \\
[(\theta_i, \omega_i) \mid F] &\sim F \\
F &\sim \text{EDP}(\alpha_\theta, \alpha_\omega, H).
\end{aligned}
$$

where p_y is a Bernoulli distribution with mean $g(L_i^\top \theta_i)$, p_c is a normal distribution with mean μ_{ij} and variance τ_{ij}^2, and p_b is a Bernoulli distribution with mean π_{ij}, $\omega_i = (\{\mu_{ij}, \tau_{ij}^2 : j = 1, \ldots, q_c\}, \{\pi_{ij} : j = q_c + 1, \ldots, p\}, \pi_i)$.

This model was used to analyze data from the Veterans Aging Cohort Study (VACS) to compare two types of antiretroviral therapy (ART) regimens for HIV/HCV coinfected patients on death within two years after treatment initiation. Numerous confounders were available and included in the observed data model. We provide a full analysis of this data in Chapter 9.

Example 6.4.3 (*Causal mediation with a single mediator)***.** In this case, we can implement an EDPM with $[Y \mid L]$ replaced with $[Y \mid M, L]$ and L replaced by $[M \mid L]$ and L. An EDPM in this setting might take the form

$$
\begin{aligned}
[Y_i \mid M_i, L_i, \theta_i] &\sim p_y(y \mid m, \ell, \theta_i) \\
[M_i \mid L_i, \omega_i] &\sim p_m(m \mid \ell, \omega_{mi}) \\
[L_{i,j} \mid \omega_i] &\sim p_c(\ell_j \mid \omega_{ci}) : j = 1, \ldots, q_c, \\
[L_{i,k} \mid \omega_i] &\sim p_b(\ell_k \mid \omega_{bi}), k = q_c + 1, \ldots, p \\
[A_i|\omega_i] &\sim p_b(a|\omega_{ai}) \\
[(\theta_i, \omega_i) \mid F] &\sim F \\
F &\sim \text{EDP}(\alpha_\theta, \alpha_\omega, H).
\end{aligned}
$$

where, assuming a continuous outcome and mediator, the first two conditional distributions are normal linear regressions. Under the sequential ignorability assumptions in Chapter 1, (1.5)-(1.6), we can identify the mean potential outcome needed to compute natural direct and indirect effects using the "mediational" g-formula:

$$E[Y\{1, M(0)\}] = \int E(Y \mid M = m, A = 1, L = \ell)\, F_{M|A=0,L=\ell}(dm)\, F_L(d\ell).$$

Notice that, on the right hand side, the conditional mean of the outcome is for treatment $A = 1$, but we average the mediator distribution under what it would be if treatment was set $A = 0$. To compute the NIE, we compute $\text{NIE} = E[Y\{1, M(1)\}] - E[Y\{1, M(0)\}]$ where the former would be computed as

$$E[Y\{1, M(1)\}] = \int E(Y \mid M = m, A = 1, L = \ell)\, F_{M|A=1,L=\ell}(dm)\, F_L(d\ell).$$

With a flexible model specification as here, this should be approximately the sample mean among those randomized to $A = 1$. We introduced a general g-computation algorithm for this setting in Chapter 3 and provide further details on computations for EDPMs in Section 6.4.2.

This model could be used in randomized trials to assess mediation. A simpler version of this model was used to evaluate mediators in STRIDE, a randomized clinical trial to assess the effectiveness of interventions delivering individually tailored messages in increasing physical activity in sedentary adults. We provide a full analysis of this study in Chapter 13.

We could extend this approach to multiple mediators and a three-level extension of the EDPM [166] with M clusters within Y clusters and L clusters within (M,Y) clusters. This would provide similar advantages to the two-level EDPM where now the covariates L do not dominate the clustering of the mediators (M) or outcome (Y), and the mediators do not dominate the outcome.

6.4.2 Posterior Computations

We first describe MCMC computations for the EDPM (i.e., obtaining the posterior distribution of the parameters of the observed data) using an approach similar to the Pólya urn sampler for the DPM in Section 6.3.1 (an alternative algorithm would be a blocked Gibbs sampler with a truncation approximation, as described in [167].). We then describe how to post-process the output, perform g-computation, and compute the posterior distribution of the causal effects under certain specifications of the observed data model and identifying restrictions.

6.4.2.1 MCMC

We describe here an algorithm based on a generalized Pólya urn scheme that is an extension of "Algorithm 8" of [79] to accommodate the nested clustering. We expand the cluster membership index Z_i from the DPM to $Z_i = (Z_{i,y}, Z_{i,x})$. Note that the value of $Z_{i,x}$ is only meaningful in conjunction with $Z_{i,y}$, as it describes the subcluster assignment within the cluster $Z_{i,y}$. The basic steps in the Gibbs sampler are as follows. We sample Z_i for each subject, and then, given Z, we sample the parameters θ and ω from their full conditional distributions. Denote by θ_j^* the θ that is associated with the jth of the currently non-empty clusters; $\omega_{l|j}^*$ is defined similarly.

Denote by K the current number of y-clusters (i.e., the number of unique values of Z_y) and by K_j the number of x-subclusters of the jth y-cluster. We again use $-i$ notation to indicate that the ith value is removed (e.g., y^{-i} is the vector of outcomes excluding subject i).

For the step involving drawing cluster membership, each subject can be assigned to one of the non-empty clusters, to one of m new y-clusters, or to one of m new x-clusters within each existing y-cluster. Note that the m new clusters are used when non-conjugate priors are used; otherwise there is just "one" new cluster. In the below, we provide details for updating cluster membership for both the non-conjugate (m clusters) and conjugate (one cluster) cases.

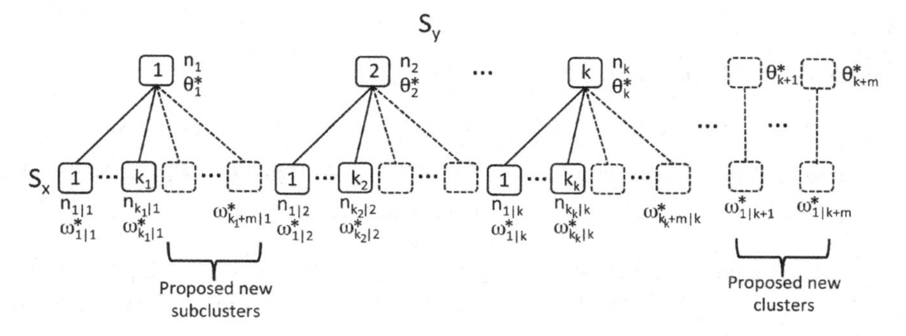

Figure 6.3 *Graphical depiction of the cluster membership update.*

Step 1 *Update cluster membership* Z_i. First, we need to do some relabeling and generate some auxiliary parameters for potential new clusters, before we can update Z_i. For subject i, denote by K^{-i} the number of unique y-clusters currently non-empty if you exclude subject i, and K_j^{-i} the number of unique x-subclusters of the jth y-cluster that are non-empty if you exclude subject i. Label these clusters that are currently occupied by the subjects other than i $\{1,\cdots,K^{-i}\}$ for the y-clusters and $\{1,\cdots,K_j^{-i}\}$ for the x-subcluster of the jth y-cluster. Denote by n_j^{-i} and $n_{l|j}^{-i}$ the number of subjects in the corresponding subclusters, excluding subject i. If the current value of Z_i is in one of these clusters, then draw values of θ^* and ω^* from prior distributions $H_{0\theta}$ and $H_{0\omega}$ for y-clusters $\{K^{-i}+1,\cdots,K^{-i}+m\}$. If the current value of Z_i is in an existing y-cluster (say, y-cluster j) but not an existing x-subcluster, then set its current cluster to $Z_i=(j,K_j^{-i}+1)$ and draw $m-1$ values of ω^* from its prior distribution and assign them to $\{K_j^{-i}+2,\cdots,K_j^{-i}+m\}$. In addition, draw m sets of parameters from the priors for the other clusters and subclusters. Finally, if the current value of Z_{yi} does not correspond with any of the K^{-i} y-clusters, then $Z_{yi}=K^{-i}+1$ and then draw θ^* and ω^* for y-clusters $\{K^{-i}+2,\cdots,K^{-i}+m\}$ and x-subclusters $\{K_j^{-i}+1,\cdots,K_j^{-i}+m\}$ for $j=1,\cdots,K^{-i}$. At this point, all of the occupied and extra clusters have θ^* and ω^* parameters associated with it. We can now draw a new value of Z_i.
Draw a new value Z_i as follows.

$$
\Pr(Z_i=(j,l)\mid Z_{-i},\theta^*,\omega^*,x,y)
$$
$$
=\begin{cases}
b\frac{n_j^{-i}n_{l|j}^{-i}}{n_j^{-i}+\alpha_\omega}p(y_i\mid x_i,\theta_j^*)p(x_i\mid\omega_{l|j}^*), & \text{for } j\in\{1,\ldots,K^{-i}\} \text{ and } \ell\in\{1,\ldots,K_j^{-i}\}\\
b\frac{n_j^{-i}\alpha_\omega/m}{n_j^{-i}+\alpha_\omega}p(y_i\mid x_i,\theta_j^*)p(x_i\mid\omega_{l|j}^*), & \text{for } j\in\{1,\ldots,K^{-i}\} \text{ and } \ell\in\{K_j^{-i},\ldots,K_j^{-i}+m\}\\
b\frac{\alpha_\theta}{m}p(y_i\mid x_i,\theta_j^*)p(x_i\mid\omega_{l|j}^*), & \text{for } j\in\{K^{-i},\ldots,K^{-i}+m\}
\end{cases}
$$

where b is a constant such that $1=\sum_{j=1}^{K^{-i}+m}\sum_{l=1}^{K_j^{-i}+m}\Pr(Z_i=(j,l)\mid Z_{-i},\theta^*,\omega^*,x,y)$. This step is done for each $i=1,\cdots,n$. A graphical depiction of this update can be found in Figure 6.3. Note that if conjugate priors are specified as $H_{0\omega}$ and $H_{0\theta}$, then there is no need to generate m new clusters. The probability for drawing a new value

Z_i is then

$$\Pr(Z_i = (j,l) \mid Z_{-i}, \theta^*, \omega^*, x, y)$$
$$= \begin{cases} b\frac{n_j^{-i} n_{l|j}^{-i}}{n_j^{-i}+\alpha_\omega} p(y_i \mid x_i, \theta_j^*) p(x_i \mid \omega_{l|j}^*), & \text{for } j \in \{1,\ldots,K^{-i}\} \text{ and } \ell \in \{1,\ldots,K_j^{-i}\} \\ b\frac{n_j^{-i}\alpha_\omega}{n_j^{-i}+\alpha_\omega} p(y_i \mid x_i, \theta_j^*) p^\star(x_i), & \text{for } j \in \{1,\ldots,K^{-i}\} \text{ and } \ell \in \{K_j^{-i},\ldots,K_j^{-i}+1\} \\ b\alpha_\theta p^\star(y_i \mid x_i) p^\star(x_i), & \text{for } j \in \{K^{-i},\ldots,K^{-i}+1\} \end{cases}$$

where b is a constant such that $1 = \sum_{j=1}^{K^{-i}+1} \sum_{l=1}^{K_j^{-i}+1} \Pr(Z_i = (j,l) \mid Z_{-i}, \theta^*, \omega^*, x, y)$ and $p^\star(y_i \mid x_i) = \int p(y_i \mid x_i, \theta) H_{0\theta}(d\theta)$ and $p^\star(x_i) = \int p(x_i \mid \omega) H_{0\omega}(d\omega)$. This step is done for each $i = 1, \cdots, n$.

Step 2 *Update parameters θ^* and ω^* within each mixture component.* For each unique j in $Z_y = \{Z_{y1}, \cdots, Z_{yn}\}$, update θ_j^* from

$$\pi(\theta_j^* \mid z, y, x; \theta_{-j}^*, \omega^*) \propto h_{0\theta}(\theta_j^*) \prod_{i:Z_{yi}=j} p(y_i \mid x_i; \theta_j^*).$$

This update will be a standard update from a Bayesian regression and will be available in closed form if a conjugate prior is specified. For each unique (j,l) in $Z = \{Z_1, \cdots, Z_n\}$, update $\omega_{l|j}^*$ from

$$\pi(\omega_{l|j}^* \mid s, y, x, \theta^*, \omega_{-l|j}^*) \propto h_{0\omega}(\omega_{l|j}^*) \prod_{i:Z_i=(j,l)} p(x_i \mid \omega_{l|j}^*).$$

Consider the situation where the first p_1 variables in L are binary, and the remaining p_2 variables are continuous. We assume

$$p(x_{i,r}; \omega_i) = \text{Bernoulli}(\pi_i^r)$$

with

$$h_{0\omega}(\pi_i^r) = \text{Beta}(a_x, b_x),\ r = 1, \cdots, q-1+p_1.$$

Thus, we update $\pi_{l|j}^{r*}$ from $\text{Beta}\left(a_x + \sum_{i:Z_i=(j,l)} x_{i,r}, b_x + n_{l|j} - \sum_{i:Z_i=(j,l)} x_{i,r}\right)$. For the last p_2 x-variables, we assume

$$p(x_{i,r} \mid \omega_i) = N(x_{i,r} \mid \mu_i^r, \tau_i^{2,r})$$

with

$$h_{0\omega}(\mu_i^r \mid \tau_i^{2,r}) = N(\mu_i^r \mid \mu_0, \tau_i^{2,r}/c_0),$$
$$h_{0\omega}(\tau_i^{2,r}) = \text{Inv-}\chi^2(\tau_i^{2,r} \mid \nu_0, \tau_0^2).$$

We can then update $\mu_{l|j}^{r*}$ and $\tau_{l|j}^{2,r*}$ from normal and scale inv-χ^2 distributions.

$$\tau_{l|j}^{2,r*} \sim \text{Inv-}\chi^2\left(\nu_0 + n_{l|j}, \frac{\nu_0\tau_0^2 + (n_{l|j}-1)s_{l|j}^{2,r} + \frac{c_0 n_{l|j}}{c_0+n_{l|j}}(\bar{x}_{l|j}^r - \mu_0)^2}{\nu_0 + n_{l|j}}\right)$$

$$\mu_{l|j}^{r*} \sim N\left(\frac{\frac{c_0}{\tau_{l|j}^{2,*}}\mu_0 + \frac{n_{l|j}}{\tau_{l|j}^{2,*}}\overline{x}_{l|j}^r}{\frac{c_0}{\tau_{l|j}^{2,*}} + \frac{n_{l|j}}{\tau_{l|j}^{2,*}}}, \frac{1}{\frac{c_0}{\tau_{l|j}^{2,*}} + \frac{n_{l|j}}{\tau_{l|j}^{2,*}}}\right),$$

where $\overline{x}_{l|j}^r$ and $s_{l|j}^{2,r}$ are the sample mean and sample standard deviation, respectively, of the rth covariate among subjects with $s = (j,l)$.

Step 3 *Update mass parameters.* Under a $\text{Gam}(a_0, b_0)$, to update α_θ, first draw $\eta \sim \text{Beta}(\alpha_\theta + 1, n)$. Next, set draw α_θ from

$$\pi\,\text{Gam}(a_0 + K, b_0 - \log(\eta)) + (1-\pi)\,\text{Gam}(a_0 + K - 1, b_0 - \log(\eta)),$$

where $\pi = \frac{\frac{K}{n(1-\log(\eta))}}{1+\frac{K}{n(1-\log(\eta))}}$ [159]. To update α_ω, we use Metropolis-Hastings, where the full conditional of α_ω is proportional to

$$\pi(\alpha_\omega)\alpha^{\sum_{j=1}^{K}(K_j-1)}\prod_{j=1}^{K}(\alpha_\omega + n_j)\beta(\alpha_\omega + 1, n_j)$$

where $\beta(a,b) = \Gamma(a)\Gamma(b)/\Gamma(a+b)$ is the beta function.

Note if α_ω depends on the θ cluster, i.e., $\alpha_{\omega|\theta}$, then the update in Step 3 can now be sampled exactly using a similar update to that used for α_θ in Step 3.

Missing covariates Within the above algorithm, missing covariates (L's) can be dealt with using data augmentation under an assumption of ignorable missingness. Because we have already specified a full model for (Y, A, L), we simply need to obtain draws of missing L's from the appropriate conditional posterior distribution at each iteration of the Gibbs sampler. Suppose L_{ir} is a binary covariate that is missing for subject i. At each step in the Gibbs sampler, we do the following. Denote by ω_i^r the current value of binomial probability parameter for the rth covariate. Note that this value of ω is based on the cluster assigned to subject i. Denote by $X_i^{[k]}$ the original vector X_i in which covariate L_{ir} is set to a value of k. We draw L_{ir} from a binomial distribution with probability

$$\frac{\omega_i^r p(y_i \mid x_i^{[1]}, \theta_i)}{\omega_i^r p(y_i \mid x_i^{[1]}, \theta_i) + (1-\omega_i^r)p(y_i \mid x_i^{[0]}, \theta_i)}.$$

Otherwise, to draw values for missing continuous covariates, we can use a Metropolis-Hastings algorithm in general. The posterior for a missing continuous covariate L_{ir} is proportional to $p(\ell_{ir} \mid \omega_i)p(y_i \mid x_i; \theta_i)$. Note that with a normal linear regression for the response and a normal distribution for ℓ_{ir}, we can do a closed form conjugate normal update (and for a binary response with a probit link, a closed form conjugate normal update can also be used with data augmentation as discussed in Chapter 3). We also note that data augmenting missing covariates for the DPM follows similarly.

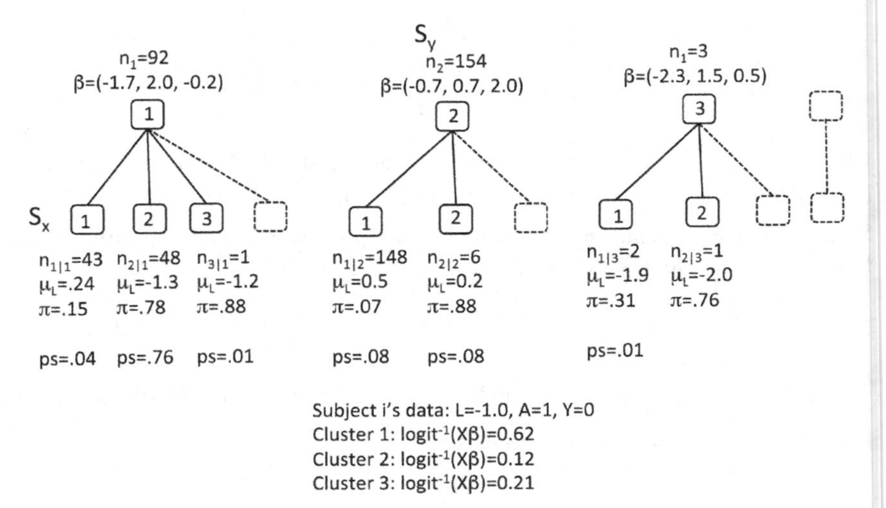

Figure 6.4 *Example of a cluster membership update for an individual subject.*

Example 6.4.4 (*Linear regression with mixed covariates*). Assume Y_i is a continuous outcome and we have two covariates $X_i = (L_i, A_i)$ with L_i continuous and A_i binary:

$$\begin{aligned}
[Y_i \mid X_i, \theta_i] &\sim p(y \mid x, \theta_i) \\
[L_i \mid \omega_i] &\sim p(\ell \mid \omega_{i1}), \\
[A_i \mid \omega_i] &\sim p(a \mid \omega_{i2}), \\
[(\theta_i, \omega_i)] \mid F &\sim F \\
F &\sim \text{EDP}(\alpha_\theta, \alpha_\omega, H).
\end{aligned}$$

where $p(y \mid x, \theta_i)$ is a normal density with mean $x^\top \beta_i$ and variance σ_i^2 (so $\theta_i = (\beta_i, \sigma_i^2)$), $p(\ell \mid \omega_{i1})$ is a normal density with mean μ_{i1} and variance τ_i^2, and $p(a \mid \omega_{i2})$ is a Bernoulli distribution with parameter π_i (so $\omega_{i1} = (\mu_{i1}, \tau_i^2)$ and $\omega_{i2} = \pi_i$). If we specify H to be the product of a normal-inverse gamma distribution for θ_i, a normal inverse-gamma distribution for ω_{i1}, and a Beta distribution for ω_{i2}, then we will have conjugate updates in the MCMC algorithm for all these parameters conditional on the cluster assignments Z_i. Figure 6.4 shows the options for cluster assignment for an individual with $L_i = -1$ and $A_i = 1$ (the binary covariate here) and $Y_i = 0$ for a binary outcome. The p's are the probabilities of assignment to each existing cluster. Note the computations for this example could be simplified by using a probit link (instead of a logit link) and using data augmentation.

For causal inference (and certain non-ignorable missing data settings), we need to post-process the MCMC draws using g-computation. Details are provided in the following section.

6.4.2.2 Post-processing Steps (after MCMC): g-Computation

Once we have obtained draws of the parameters from the posterior distribution, we can compute any functional of the marginal distributions of the potential outcomes. Here, we illustrate on expectations of the potential outcomes $E\{Y(a)\}$, $E\{Y(a) \mid V\}$, or $E\{Y(a) \mid A = a'\}$. Suppose, for example, that we would like to obtain draws from the posterior of $E\{Y(a)\}$. Assume a GLM within clusters with $E(Y \mid A = a, L = \ell, \theta_i) = g^{-1}(x_i^\top \beta_i)$ where $x_i^\top = (1, a^\top, \ell^\top)$. As in Chapter 3, we use Monte Carlo integration to integrate over L; in particular, given the current values of the parameters $\{\theta^*, \omega^*, z\}$ obtained from the Gibbs sampler as

$$E(Y \mid A = a, L = \ell, \theta^*, \omega^*, z) = \frac{w_{k+1}(a,\ell)E_0(Y \mid a,\ell) + \sum_{j=1}^k w_j(a,\ell)E(Y \mid a,\ell,\theta_j^*)}{w_{k+1}(a,\ell) + \sum_{j=1}^k w_j(a,\ell)} \tag{6.13}$$

where $w_{k+1}(a,\ell) = [\alpha_\theta/(\alpha_\theta + n)]p^\star(x)$ and

$$w_j(a,\ell) = \frac{n_j}{\alpha_\theta + n}\left\{\frac{\alpha_\omega}{\alpha_\omega + n_j}p^\star(x) + \sum_{\ell=1}^{K_j}\frac{n_{\ell|j}}{\alpha_\omega + n_j}p(x \mid \omega_{\ell|j}^*)\right\}$$

The terms $p^\star(x)$ and $E_0(Y \mid a,\ell)$ are the distribution and mean, respectively, after integrating the parameters over the prior; that is $p^\star(x) = \int p(x;\omega)\, H_{0\omega}(d\omega)$ and $E_0(Y \mid a,\ell) = \int E(Y \mid a,\ell,\theta)\, H_{0\theta}(d\theta)$. For non-conjugate distributions, $p^\star(x)$ needs to be computed using Monte Carlo integration or approximation. Note that the expectation in (6.13) is the expected value of the predictive distribution of Y with F integrated out. The actual causal parameters will be functionals of F. However, this posterior predictive is a good, simple-to-use approximation.

We can then obtain a draw from the marginal distribution of $E\{Y(a)\}$ by integrating over the marginal distribution of L using MC integration. In particular, for each posterior sample of the observed data parameters, we do MC integration of the current estimate of the marginal distribution of L; this can be viewed as sampling from the posterior predictive distribution of L. For this, we must first draw N_L samples of (L,Z) as described in Algorithm 6.3.

Once we have obtained M values (L^m, Z^m), we can approximate the integral as

$$E\{Y(a)\} \approx \frac{1}{M}\sum_{m=1}^{M} E(Y \mid A = a, L = \ell^m, \theta_{z_y^m}^*, \omega_{z_x^m|z_y^m}^*, z^m),$$

as discussed in Chapter 3. Computing this separately for $a = 1$ and $a = 0$, for example, would allow us to obtain a draw of a causal effect, such as $E\{Y(1)\} - E\{Y(0)\}$. For causal effect conditional on $V = v$ (V is a subset of L) or $A = a$, we essentially repeat the above steps, but integrate over the conditional distribution of $L_{-v} \mid V = v$ (where $L = (L_{-v}, V)$) or $L \mid A = a$, respectively, rather than the marginal distribution of L. To sample from $L_{-v} \mid V = v$, see Algorithm 6.4.

Algorithm 6.3 Bayesian g-computation for ACE

For $m = 1, \ldots, N_L$

1. Draw Z_y^m from a $(K+1)$-dimensional multinomial distribution with probabilities $\left(\frac{n_1}{\alpha_\theta+n}, \cdots, \frac{n_K}{\alpha_\theta+n}, \frac{\alpha_\theta}{\alpha_\theta+n}\right)$.
2. If $Z_y^m < (K+1)$, draw Z_x^m from a $(K_j+1) =$ dimensional multinomial distribution with probabilities
$$\left(\frac{n_{1|j}}{\alpha_\omega+n_j}, \cdots, \frac{n_{K_j|j}}{\alpha_\omega+n_j}, \frac{\alpha_\omega}{\alpha_\omega+n_j}\right)$$
otherwise, set $Z_x^m = 1$.
3. Draw L^m from $p(x \mid \omega^*_{z_x^m|z_y^m})$; if $Z_y^m = (K+1)$ or if $Z_x^m = (K_j+1)$ (i.e., if a new cluster is opened up), $\omega^*_{z_x^m|z_y^m}$ is drawn from the prior distribution.

Algorithm 6.4 Bayesian g-computation for CACE

1. Draw Z_y^m from a multinomial $\{1, \cdots, K+1\}$ with probabilities $\left(\frac{n_1 \sum_j p(v|\omega_{j|1})}{\alpha_\theta \int p(v|\omega)dF(\omega)+\sum_k n_k \sum_\ell p(v|\omega_{\ell|k})}, \cdots, \frac{n_K \sum_j p(v|\omega_{j|k})}{\alpha_\theta \int p(v|\omega)dF(\omega)+\sum_k n_k \sum_\ell p(v|\omega_{\ell|k})}, \frac{\alpha_\theta \int p(v|\omega)dF(\omega)}{\alpha_\theta \int p(v|\omega)dF(\omega)+\sum_k n_k \sum_\ell p(v|\omega_{\ell|k})}\right)$.
2. If $Z_y^m < (K+1)$, draw Z_x^m from a multinomial $\{1, \cdots, K_j+1\}$ with probabilities $\left(\frac{n_{1|j} f(v|\omega_{1|j})}{\alpha_\omega \int p(v|\omega)dF(\omega)+\sum_k n_{k|j} f(v|\omega_{k|j})}, \cdots, \frac{n_{K_j|j} f(v|\omega_{K_j|j})}{\alpha_\omega \int p(v|\omega)dF(\omega)+\sum_k n_{k|j} f(v|\omega_{k|j})}, \frac{\alpha_\omega \int p(v|\omega)dF(\omega)}{\alpha_\omega \int p(v|\omega)dF(\omega)+\sum_k n_{k|j} f(v|\omega_{k|j})}\right)$;
otherwise, set $Z_x^m = 1$.
3. Draw L_{-v}^m from $p(\ell_{-v} \mid \omega^*_{z_x^m|z_y^m})$, where, if $Z_y^m = (K+1)$ or if $Z_x^m = (K_j+1)$ (i.e., if a new cluster is opened up), $\omega^*_{z_x^m|z_y^m}$ is drawn from the prior distribution.

Computing quantile causal effects follows similarly to the means above. Given current values of the parameters, $\{\theta^*, \omega^*, z\}$, from the Gibbs sampler, we can compute the desired quantile in closed form from $\Pr\{(Y < y \mid \theta^*, \omega^*)\}$ (ignoring z) and then average over (θ^*, ω^*).

Because this is a post-processing step, its computation is not needed for the Gibbs sampler to sample the posterior of the observed data model parameters. Therefore, to improve computational efficiency, draws of the causal effect parameters do not need to be obtained for every draw of the Gibbs sampler (typical thinning in inference from MCMC; see Section 3.3.6 of Chapter 3) and could be done in parallel (i.e., the M MC integrations for g-computation); note, with M "machines," the time to do the g-computation would be the time to do one Monte Carlo integration (in addition, the AGC approach in Example 3.6.4 can further increase efficiency).

Causal Mediation

Recall that from Chapter 1 we identify the mean potential outcome needed to compute natural direct and indirect effects using a "mediational" g-formula:

$$E\{Y(1,M(0)\} = \int E(Y \mid M=m, A=1, L=\ell)\ F_{M|A=0,L=\ell}(dm)\ F_L(d\ell).$$

We can use the general algorithm from Example 3.6.2 applied to the EDPM (see Example 6.4.3). We obtain a posterior draw from the marginal distribution of $E\{Y(1,M(0))\}$ by first integrating over the marginal distribution of L using MC integration as we did for computing $E\{Y(a)\}$; from this, we obtain a sample $(z_x^k, \omega^*_{c,z_x^k}, \omega^*_{b,z_x^k})$. We then sample m^k from $p_m(m \mid A=0, L=\ell^k, \omega^*_{m,z_x^k})$.

Once we have obtained N_L values (m^k, ℓ^k, s^k), we can approximate the integral as follows:

$$E[Y\{1,M(0)\}] \approx \frac{1}{N_L}\sum_{k=1}^{N_L} E(Y \mid M=m^k, A=1, L=\ell^k, \theta^*_{z_y^k}, \omega^*_{z_x^k|z_y^k}, z^k),$$

where $E(Y \mid M=m^k, A=1, L=\ell^k, \theta^*_{z_y^k}, \omega^*_{z_x^k|z_y^k}, z^k)$ is computed as in (6.13). Similarly, we compute $E[Y\{a,M(a)\}]$,

$$E[Y\{a,M(a)\}] = \int E(Y \mid M=m, A=a, L=\ell)\ F_{M|A=a,L=\ell}(dm)\ F_L(d\ell).$$

We illustrate this in the case study in Chapter 13.

6.5 Summary

We have introduced various DPMs of distributions that can be used in missing data and causal inference problems and illustrated their use for various specific settings. We have also reviewed computations, some of which can be done in standard softwares (`rstan` or `rjags`). In Chapters 8-13, we will provide further details about their use in a series of case studies. In the next chapter, we introduce how to use a DPM with a Gaussian process for modeling a conditional distribution directly.

Chapter 7

Gaussian process priors and dependent Dirichlet processes

DOI: 10.1201/9780429324222-7

7.1 Motivation: Alternate Priors for Functions and Nonparametric Modeling of Conditional Distributions

In Chapter 5, we discussed priors based on decision trees for functions such as the outcome regression function $\mu_a(x)$ or the propensity score $e(x)$. In this chapter we focus on an alternative prior for functions, Gaussian process (GP) priors. GP priors are popular in the machine learning literature. From a Bayesian perspective, GPs are advantageous because they result in smooth function estimates and, taking advantage of the properties of the multivariate normal distribution, have posterior distributions that are relatively easy to sample from.

In Chapter 6 we introduced DPMs, which are very useful for modeling marginal and joint distributions. While one can obtain a conditional distribution indirectly from a joint distribution, it may be preferable to model a conditional distribution directly and not also model the distribution of the random variables being conditioned on, in particular, if there is no missingness in these random variables. In this chapter we introduce dependent Dirichlet process priors for this purpose. As noted in Chapters 1 and 2, conditional distributions (e.g., of the outcome given the treatment and confounders) are often needed in causal inference and missing data problems.

GPs can also be used in conjunction with *dependent Dirichlet process* (DDP) priors as a nonparametric prior for the functional mean part of a conditional distribution. We introduce this approach in Section 7.3. First, we begin with a review of GP priors.

7.2 Gaussian Process Priors

Recall from Chapter 5 that there are many situations in which we would like to nonparametrically model a function, $g(x)$. For a binary outcome model (or the propensity score)

$$[Y_i \mid X_i = x] \sim \text{Bernoulli}[\Phi\{g(x)\}],$$

or for a normal outcome model

$$[Y_i \mid X_i = x] \sim N\{g(x), \sigma^2\},$$

we need to model the unknown function $g(x)$. One option (cf. Chapter 5) is tree-based approach such as BART: $g(x_i) = \sum_{m=1}^{M} \text{Tree}(X_i; \mathscr{T}_m, \mathscr{M}_m)$. A nonparametric alternative (with pros and cons discussed later) is to specify a GP prior for g.

GPs are frequently used as a prior distribution for an unknown function [168]; a formal definition follows.

Definition 7.2.1. Let $g(\cdot)$ be a real-valued stochastic process (i.e., a function on a set $\mathscr{X}$ such that $g(x)$ is a random variable for each $x \in \mathscr{X}$). We say that $g(x)$ is a *Gaussian process* (GP) on $\mathscr{X}$ with mean function $m(\cdot)$ and covariance function $\kappa(\cdot,\cdot)$ if, for any finite set of points $\{x_1, \ldots, x_n\} \subseteq \mathscr{X}$, the vector $(g(x_1), \ldots, g(x_n))^\top$ has a multivariate normal distribution such that $E\{g(x_i)\} = m(x_i)$ and $\text{Cov}\{g(x_i), g(x_j)\} = \kappa(x_i, x_j)$, and we write $g \sim \text{GP}\{m(\cdot), \kappa(\cdot,\cdot)\}$.

7.2.1 Normal Outcomes

Suppose we assume the conditional distribution of Y_i given X_i can be written as

$$[Y_i \mid X_i] \sim N\{g(X_i), \sigma^2\}.$$

A parametric Bayesian approach would assume $g(\cdot)$ has a known form, e.g., $g(x) = x^\top \beta$, and then specify priors for the parameters β and σ^2. However, this implies that *a priori* there is no uncertainty about the functional form of $g(x)$.

Alternatively, we could specify a prior distribution for the function to reflect the fact that we are uncertain of its form. To motivate the GP approach, suppose our best guess is that $g(x)$ is approximately an additive linear function in x. We therefore might wish to center the prior on that belief. Further, we might believe that $g(x_j)$ should be very close to $g(x_k)$ if x_j is close to x_k. In this case, we could specify a prior such that the correlation between $g(x_j)$ and $g(x_k)$ is higher the closer x_j and x_k are to each other. Suitably specified GP priors allow for us to accomplish both of these goals.

To encode the above information, we might decide to shrink to a linear relationship where $E(Y_i \mid X_i = x) \approx x^\top \beta$. To do this, we can choose the prior mean to be $m(x) = x^\top \beta$ with a hyperprior placed on β. In areas of x-space where we have a substantial amount of data, the prior mean $m(x)$ may not matter much; however, in areas of the x-space where we do not have much data, shrinking toward linearity may improve predictions. We note that there is a duality between modeling this in the mean function $m(x)$ or the covariance $\kappa(x,x')$; specifying $\beta \sim N(0, \sigma_\beta^2)$, for example, is equivalent to using the kernel $\kappa(x,x') + \sigma_\beta^2 x^\top x'$. For this reason, it is common in the machine learning literature to set $m(x) = 0$ and model all of the desired structures in the covariance function [169].

The Matérn kernel is a popular choice for κ and has considerable flexibility [168]. If two points x and x' are Euclidean distance $d = d(x,x')$ apart, then the covariance between $g(x)$ and $g(x')$ under the Matérn kernel is

$$\kappa(x,x') = \eta \frac{2^{1-\nu}}{\Gamma(\nu)} \left(\rho\, d \sqrt{2\nu}\right)^\nu K_\nu \left(\rho\, d \sqrt{2\nu}\right)$$

with positive parameters (ν, η, ρ), where K_ν denotes the modified Bessel function of the second kind; roughly speaking, ρ controls how quickly the correlation between $g(x)$ and $g(x')$ decays as d increases, η controls the overall magnitude of $g(x)$, and ν controls the "roughness" of the function, with $g(x)$ being k-times mean-square differentiable only if $\nu > k$.

Another general form for a kernel is

$$\kappa(x,x') = \eta \exp\left(- \sum_{k=1}^{p} \rho_k^\gamma |x_k - x_k'|^\gamma \right), \tag{7.1}$$

where $0 < \gamma \leq 2$ and p is the number of covariates; [168] refer to this as the γ-exponential family. If covariates are standardized, it will often be sensible in practice to specify $\rho_1 = \cdots = \rho_p$ (i.e., same distance between any two covariates counts equally). For $R = 1$, we have an exponential kernel, and for $R = 2$, we have a squared-exponential kernel; this pair of choices also happens to correspond to special cases of the Matérn kernel with $\nu = 1/2$ and $\nu \to \infty$, respectively.

As a particular example, suppose $m(x) = x^\top \beta$ and the $(j,\ell)^{\text{th}}$ element of $\kappa(x,x')$ is

$$\kappa(x,x') = \eta \exp(-\rho \|x - x'\|^2) + b\delta_{j\ell}, \tag{7.2}$$

where ρ and η are parameters, b is typically set to a small value (e.g., 0.01) and $\delta_{j\ell} = I(j = \ell)$; the term $b\,\delta_{j\ell}$ is commonly referred to as a *nugget* or *jitter* term, and it serves the practical purpose of avoiding numerically non-singular covariance matrices. With this specification, we have centered our prior for the function at $x^\top \beta$, while our uncertainty about the functional form is reflected in $\kappa(x,x')$. If η is close to 0, then this prior implies that there is little variation of the function from the mean, $m(x)$. Large η suggests the opposite. The parameter ρ represents how rapidly correlation between $g(x_j)$ and $g(x_\ell)$ decays as the distance between x_j and x_ℓ increases. With this specification of κ, covariates would typically first be standardized, so that distance is all in standard deviation units.

Example 7.2.1 (*Simulated data example).* We consider simulated data with one covariate X_i and an outcome Y_i, where the true function is $g(x) = 0.3x^3$. Figure 7.1 displays posterior estimates and 95% credible intervals for the function $g(x)$. We specified two different GP priors, one with $m(x) = 0$ (left hand side plots) and one with $m(x) = \beta_0 + \beta_1 x$. In both cases we used the specification (7.2) for $\kappa(x,x')$. Consider first the estimate of $g(x)$ when $n = 20$ (first row of plots). In this small sample size setting, we can see the curve shrinking toward the prior mean (plotted in gray). We also can see that there is more uncertainty about the curve in in the middle of the plot where there is no observed data. This is in contrast to linear regression, where there would be more certainty about the mean of y for values near the middle of the x distribution. For the lower two plots with larger sample size, the curve fits the data well and there is very little influence of the prior mean on the estimate of the function.

In the simple example above, we considered only a single covariate. Now consider the situation where there are p covariates and we specify $\kappa(\cdot,\cdot)$ as in (7.1). If all of the X_i's are binary, then $|x_{i\ell} - x_{j\ell}|$ records whether subjects i and j differ on the lth covariate. If some covariates are a priori viewed as stronger confounders, then ρ_ℓ could have a prior centered on larger values for those covariates. Otherwise one might set $\rho = \rho_1 = \cdots = \rho_\ell$. In that case, $\sum_{\ell=1}^{p} |x_{i\ell} - x_{j\ell}|$ is a count of the number of variables where subjects i and j differ, and ρ represents how fast the correlation between points decays as disagreement increases. If the lth covariate is continuous and has been standardized, then a 1 unit (1 SD) difference between $x_{i\ell}$ and $x_{j\ell}$ is treated the same as if two binary covariates were not equal.

The Posterior Distribution

Gaussian processes are conjugate to the normal sampling model in the following sense: if $Y_i \sim N\{g(X_i), \sigma^2\}$ and $g(\cdot) \sim \text{GP}(m,\kappa)$, then the posterior for $g(\cdot)$ is also

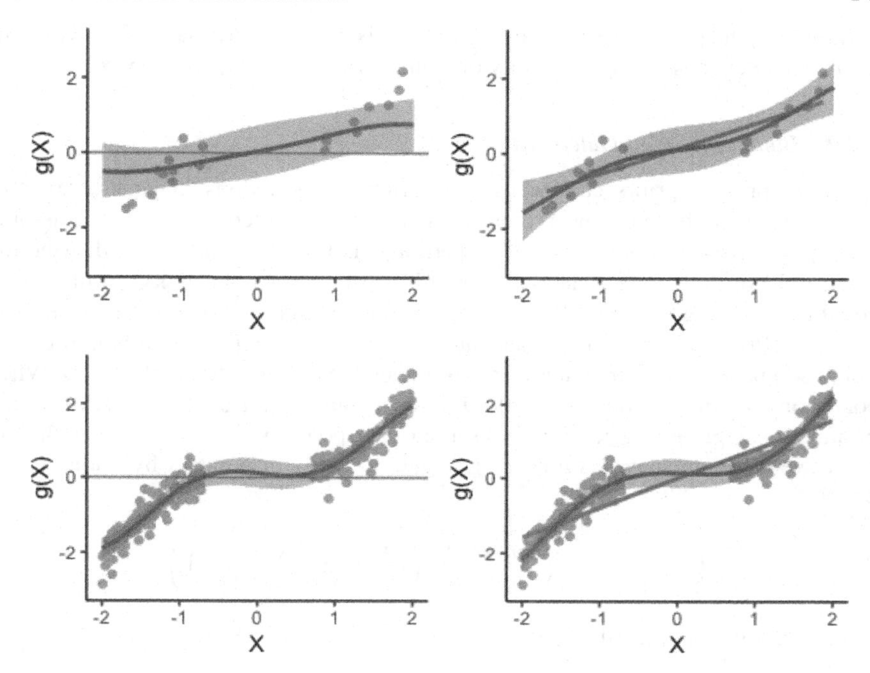

Figure 7.1 *Illustration of the posterior distribution of $g(x)$ from Example 7.2.1. Each plot displays data points (gray dots), the prior mean of $g(x)$ (gray line), the posterior mean of $g(x)$ (dark curve), and a 95% credible band. The first row takes $n = 20$, while the second row takes $n = 200$. The prior mean functions were $m(x)$ (left column) and $m(x) = \beta_0 + \beta_1$ (right column).*

a GP. Let $\mathbf{g} = g(\mathbf{x}) = (g(x_1), \ldots, g(x_n))^\top$ and suppose that we are interested in the posterior of $\widetilde{\mathbf{g}} = g(\widetilde{\mathbf{x}})$ where $\mathbf{x} = (\widetilde{x}_1, \ldots, \widetilde{x}_m)$ is a collection of covariate values of interest. Because both $f(y \mid \mathbf{g})$ and $\pi(\mathbf{g})$ are normal, the joint distribution $f(y, \widetilde{\mathbf{g}})$ is also normal and is given by

$$\begin{pmatrix} y \\ \widetilde{\mathbf{g}} \end{pmatrix} \sim N\left(\begin{pmatrix} m(\mathbf{x}) \\ m(\widetilde{\mathbf{x}}) \end{pmatrix}, \begin{pmatrix} \kappa(\mathbf{x},\mathbf{x}) + \sigma^2 \mathrm{I} & \kappa(\widetilde{\mathbf{x}},\mathbf{x}) \\ \kappa(\mathbf{x},\widetilde{\mathbf{x}}) & \kappa(\widetilde{\mathbf{x}},\widetilde{\mathbf{x}}) \end{pmatrix} \right),$$

where $\kappa(\mathbf{x}, \widetilde{\mathbf{x}})$ is a matrix with $(j,k)^{\text{th}}$ entry $\kappa(x_j, \widetilde{x}_k)$. Therefore, from well-known properties of the multivariate normal distribution, the distribution of $[\widetilde{\mathbf{g}} \mid X, Y, \eta, \rho, \sigma]$ is normal with mean

$$m(\widetilde{\mathbf{x}}) + \kappa(\widetilde{\mathbf{x}},\mathbf{x})[\kappa(\mathbf{x},\mathbf{x}) + \sigma^2 \mathrm{I}]^{-1}(y - m(\mathbf{x})) \tag{7.3}$$

and variance

$$\kappa(\widetilde{\mathbf{x}},\widetilde{\mathbf{x}}) - \kappa(\widetilde{\mathbf{x}},\mathbf{x})[\kappa(\mathbf{x},\mathbf{x}) + \sigma^2 \mathrm{I}]^{-1}\kappa(\mathbf{x},\widetilde{\mathbf{x}}). \tag{7.4}$$

The posterior mean of $\widetilde{\mathbf{g}}$ in (7.3) is a weighted average of the prior mean $m(\widetilde{\mathbf{x}})$ and the training data (Y, X), where the training data will get more weight if $\widetilde{x}$ is close to

x. Similarly, the posterior variance of $\widetilde{\mathbf{g}}$ in (7.4) is the prior covariance $\kappa(\widetilde{\mathbf{x}},\widetilde{\mathbf{x}})$ minus a term that depends on the proximity of $\mathbf{x}$ and $\widetilde{\mathbf{x}}$ via the covariance, $\kappa(\widetilde{\mathbf{x}},\mathbf{x})$.

7.2.2 Binary or Count Outcomes

Gaussian process priors can be used as priors for functions in general, regardless of whether the outcome is normal. For a binary outcome, we might assume $[Y_i \mid X_i = x, g] \sim \text{Bernoulli}[h\{g(x)\}]$ where $h(\cdot)$ is the cdf of a logistic distribution (logit link) or the cdf of a standard normal distribution (probit link). Similarly, for count data we might set $[Y_i \mid X_i = x, g] \sim \text{Poisson}[\exp\{g(x)\}]$. In either case, we can set $g \sim \text{GP}(m,\kappa)$. For non-normal outcomes, we lose conjugacy, although conditional conjugacy is possible in the case of the probit link (see Example 7.2.3). When conditional conjugacy does not hold, inference can be carried out via HMC using `rstan` to sample $\mathbf{g} = g(\mathbf{x})$ at the observed data points $\mathbf{x} = (x_1,\ldots,x_n)$; with $\mathbf{g}$ in hand, we can then sample values of $\widetilde{\mathbf{g}} = g(\widetilde{\mathbf{x}})$ at held-out points $\widetilde{\mathbf{x}}$ by noting that a priori

$$\begin{pmatrix} \mathbf{g} \\ \widetilde{\mathbf{g}} \end{pmatrix} \sim N\left(\begin{pmatrix} m(\mathbf{x}) \\ m(\widetilde{\mathbf{x}}) \end{pmatrix}, \begin{pmatrix} \kappa(\mathbf{x},\mathbf{x}) & \kappa(\widetilde{\mathbf{x}},\mathbf{x}) \\ \kappa(\mathbf{x},\widetilde{\mathbf{x}}) & \kappa(\widetilde{\mathbf{x}},\widetilde{\mathbf{x}}) \end{pmatrix} \right),$$

so that the full conditional of $\widetilde{\mathbf{g}}$ is

$$\widetilde{\mathbf{g}} \sim N\left(m(\widetilde{\mathbf{x}}) + \kappa(\widetilde{\mathbf{x}},\mathbf{x})[\kappa(\mathbf{x},\mathbf{x})]^{-1}(\mathbf{g}-m(\mathbf{x})), \kappa(\widetilde{\mathbf{x}},\widetilde{\mathbf{x}}) - \kappa(\widetilde{\mathbf{x}},\mathbf{x})[\kappa(\mathbf{x},\mathbf{x})]^{-1}\kappa(\mathbf{x},\widetilde{\mathbf{x}})\right).$$

7.2.3 Priors on GP Parameters

Often, m or κ will involve parameters that will also require prior distributions. For example, if $m(x) = x^\top\beta$, we might specify $\pi(\beta) \propto 1$ or $\pi(\beta) = N(\beta \mid 0, \tau_0^2 \mathrm{I})$, where τ_0^2 is a large number. Essentially, here decisions about how to specify $\pi(\beta)$ are similar to decisions in parametric regression models.

Suppose $\kappa(\cdot,\cdot)$ is from (7.2), and we therefore need priors for ρ and η. In practice, sometimes η and ρ are fixed and treated as known. The main reason to do this is it will speed up computations by only requiring us to invert $\kappa(\mathbf{x},\mathbf{x})$ once. If, instead, priors are selected, a goal is to avoid overfitting. We might therefore prefer priors that penalize straying too far away from our prior guess at the function, $m(x)$. This would involve a prior for η with more weight toward values close to 0. Gamma or inverse gamma priors are common choices for η and ρ.

7.2.4 Posterior Computations

We now show how to perform computations for several GP models. For the purpose of this section, we let $\mathbf{x} = (x_1,\ldots,x_n)^\top$ denote an $n \times p$ design matrix, with $m(\mathbf{x}) = (m(x_1),\ldots,m(x_n))^\top$ and $\kappa(\mathbf{x},\mathbf{x})$ an $n \times n$ matrix with $(i,j)^{\text{th}}$ entry $\kappa(x_i,x_j)$.

Example 7.2.2 (*Gaussian process model for normal outcomes*). Consider the following hierarchical model:

$$\begin{aligned} Y_i &\sim N\{g(X_i), \sigma^2\}, \\ g &\sim \mathrm{GP}(m, \kappa), \\ m(x) &= x^\top \beta, \\ \kappa(x, x') &= \eta \exp(-\rho \|x - x'\|^2), \end{aligned}$$

with hyperpriors $\beta \sim N(\beta_0, \Sigma_\beta^0)$, $\sigma^2 \sim \mathrm{InvGam}(a, b)$, $\rho \sim \mathrm{InvGam}(a_\rho, b_\rho)$, and $\eta \sim \mathrm{InvGam}(a_\eta, b_\eta)$. Inference can proceed via a Gibbs sampler with some embedded Metropolis-Hastings (MH) steps. Specifically, given the current value of $\mathbf{g} = (g(x_1), \ldots, g(x_n))^\top$, the observed outcome $\mathbf{y} = (y_1, \ldots, y_n)^\top$, and the current values of β, η, and ρ, we can update parameters as in Algorithm 7.1.

Algorithm 7.1 Gibbs sampler for GP with normal outcomes

1. Update σ^2 from $\mathrm{InvGam}\left\{\sigma^2 \mid \frac{n}{2}, \frac{\|\mathbf{y}-\mathbf{g}\|_2^2}{2}\right\}$.
2. Update β from $N(\beta \mid \mu_\beta^{\mathrm{post}}, \Sigma_\beta^{\mathrm{post}})$, where

$$\begin{aligned} \Sigma^{\mathrm{post}} &= \left\{\Sigma_{\beta_0}^{-1} + \mathbf{x}^\top \kappa(\mathbf{x}, \mathbf{x})^{-1} \mathbf{x}\right\}^{-1}, \\ \mu_\beta^{\mathrm{post}} &= \Sigma_\beta^{\mathrm{post}} \left\{\Sigma_{\beta_0}^{-1} \beta_0 + \mathbf{x}^\top \kappa(\mathbf{x}, \mathbf{x})^{-1} \mathbf{g}\right\}. \end{aligned}$$

3. Update $\mathbf{g}$ from $N(\mathbf{g} \mid \mu_g^{\mathrm{post}}, \Sigma_g^{\mathrm{post}})$, where

$$\begin{aligned} \Sigma_g^{\mathrm{post}} &= \kappa(\mathbf{x}, \mathbf{x}) - \kappa(\mathbf{x}, \mathbf{x}) \{\kappa(\mathbf{x}, \mathbf{x}) + \sigma^2 I\}^{-1} \kappa(\mathbf{x}, \mathbf{x}) \\ \mu_g^{\mathrm{post}} &= \mathbf{x}\beta + \kappa(\mathbf{x}, \mathbf{x}) \{\kappa(\mathbf{x}, \mathbf{x}) + \sigma^2 I\}^{-1} (\mathbf{y} - \mathbf{x}\beta). \end{aligned}$$

4. Update the parameters ρ and η (within $\kappa(x, x')$) via MH steps.

A computational challenge here is that Steps 2 and 3 involve inverting an $n \times n$, matrix when ρ and η are not fixed. For large n, this becomes infeasible; however, this can be minimized if ρ and η are fixed (so just one inversion) or if given a discrete prior (so just the number of support points of the prior) as the inversions can be done and stored before running the algorithm. Generally, computing the posterior in a GP model has computational costs which scale like n^3. Approaches for faster GP algorithms are an active area of research [170, 171, 172].

MCMC yields a set of M draws, $\{\sigma^{2(m)}, \beta^{(m)}, \eta^{(m)}, \rho^{(m)}, \mathbf{g}^{(m)}\}_{1:M}$, approximately from the posterior $\pi(\mathbf{g}, \sigma^2, \beta, \eta, \rho \mid \mathbf{y}, \mathbf{x})$. We can then obtain draws of $g(\widetilde{\mathbf{x}})$ for any $\widetilde{\mathbf{x}}$ of interest (see (7.3) and (7.4)).

Example 7.2.3 (*Gaussian process model for probit regression*). A computationally simple approach to modeling binary data with a Gaussian process is to use a probit regression model and use the data augmentation approach of Chapter 3 to sample from the posterior; we discuss this in detail now, although we note that similar approaches can be made to work for the logistic link function [71].

Recall the binary outcome $Y_i \sim \text{Bernoulli}[\Phi\{g(X_i)\}]$ can be thought of as a manifestation of whether an underlying (latent) continuous variable Z_i is greater than 0, i.e., $Y_i = I(Z_i > 0)$ and $f(Z_i = z \mid X_i = x, g) = N\{z \mid g(x), 1\}$, where the prior for $g(\cdot)$ and its parameters are the same as in Example 7.2.2. We can then use the Gibbs sampling algorithm in Algorithm 7.2 to sample from the posterior.

Algorithm 7.2 Gibbs sampler for GP with probit regression

1. Draw Z_i from a truncated normal with mean parameter $\mathbf{g}_i$, variance parameter 1, truncated to be $(-\infty, 0)$ if $y_i = 0$ or $(0, \infty)$ if $y_i = 1$.
2. Update β from $N(\beta \mid \mu_\beta^{\text{post}}, \Sigma_\beta^{\text{post}})$, where

$$\Sigma^{\text{post}} = \left\{\Sigma_{\beta_0}^{-1} + \mathbf{x}^\top \kappa(\mathbf{x},\mathbf{x})^{-1}\mathbf{x}\right\}^{-1},$$

$$\mu_\beta^{\text{post}} = \Sigma_\beta^{\text{post}}\left\{\Sigma_{\beta_0}^{-1}\beta_0 + \mathbf{x}^\top \kappa(\mathbf{x},\mathbf{x})^{-1}\mathbf{g}\right\}.$$

3. Update $\mathbf{g}$ from $N(\mathbf{g} \mid \mu_g^{\text{post}}, \Sigma_g^{\text{post}})$, where

$$\Sigma_g^{\text{post}} = \kappa(\mathbf{x},\mathbf{x}) - \kappa(\mathbf{x},\mathbf{x})\{\kappa(\mathbf{x},\mathbf{x}) + I\}^{-1}\kappa(\mathbf{x},\mathbf{x})$$

$$\mu_g^{\text{post}} = \mathbf{x}\beta + \kappa(\mathbf{x},\mathbf{x})\{\kappa(\mathbf{x},\mathbf{x}) + I\}^{-1}(\mathbf{Z} - \mathbf{x}\beta).$$

4. Update the parameters ρ and η (within $\kappa(x,x')$) via MH steps.

7.2.5 *GP for Causal Inference*

We now illustrate the use of GPs on two causal inference problems.

Example 7.2.4 (*Causal inference with a point treatment and a continuous outcome*). Consider the standard observational setting of a point treatment A_i and outcome $Y_i(a)$, with a vector of confounders L_i. Assuming consistency, SUTVA, positivity, and ignorability, we can define the regression function

$$\mu_a(\ell) = E\{Y_i(a) \mid L_i = \ell\} = E(Y_i \mid A_i = a, L_i = \ell),$$

as in Chapter 5. Assuming that the outcome is continuous, it is then straight forward to specify either separate GP priors for $\mu_0(\cdot)$ and $\mu_1(\cdot)$ or specify a single GP prior $\mu_a(\ell)$ with distance arguments (a,ℓ). Alternatively, as with Bayesian Causal Forests, we can specify $\mu_a(\ell) = \mu_0(\ell) + a\Delta(\ell)$ and place independent GP priors on $\mu_0(\cdot)$ and $\Delta(\cdot)$.

The conditional causal effect, given a particular ℓ, is $\Delta(\ell) = \mu_1(\ell) - \mu_0(\ell)$. Posterior draws for $\Delta(\ell)$ can be computed as $\Delta^{(m)}(\ell) = \mu_1^{(m)}(\ell) - \mu_0^{(m)}(\ell)$, for $m = 1, \cdots, M$, where $\mu_a^{(m)}(\ell)$ is the mth sample from the posterior of $\mu_a(\cdot)$.

Similarly, the average causal effect is given by

$$\Delta = \int \{\mu_1(\ell) - \mu_0(\ell)\} F_L(d\ell).$$

As discussed in Section 3.2.3, we need a model for F_L in order to integrate over the marginal distribution of the confounders. Using the Bayesian bootstrap (see Section 3.2.3), at each MCMC step, we sample $\varpi^{(m)} \sim \text{Dirichlet}(1,\ldots,1)$ and then compute $\Delta^{(m)}$ as

$$\Delta^{(m)} = \sum_{i=1}^{n} \varpi_i^{(m)} \{\mu_1^{(m)}(\ell_i) - \mu_0^{(m)}(\ell_i)\}.$$

An attractive feature of the GP approach for causal inference is that there is automatically more uncertainty about the function $\mu_a(\ell)$ for values of ℓ that are not close to the values observed in the original dataset (see Figure 7.1); this is useful because the distribution $f(L_i = \ell \mid A_i = 1, \theta)$ might be quite different from $f(L_i = \ell \mid A_i = 0, \theta)$. In addition, in any finite sample of observed values of x_i in each subgroup, $A_i = 0$ and $A_i = 1$ might have areas without overlap (i.e., practical violations of the positivity assumption). Without loss of generality, suppose we have sorted the observations so that the first n^* observations are untreated subjects and the next $n - n^*$ are treated. If $\mu_0(\ell)$ and $\mu_1(\ell)$ are given independent priors, then $\mu_0(\ell)$ will be trained on $\ell_1,\ldots,\ell_{n^*}$, but each posterior draw of $\Delta^{(m)}$ requires draws of $\mu_0(\ell)$ for $\ell \in \{\ell_{n^*+1},\ldots,\ell_n\}$ as well. For untreated subjects whose ℓ_i's are far away from the ℓ_i's of the treated subjects, we will have a large amount of uncertainty about $\mu_1(\ell)$, leading to more uncertainty in Δ. This is desirable in the sense that we have less information about the causal effect in areas where there is no substantial overlap [173].

Example 7.2.5 (*Causal inference with point treatment, continuous outcome, and distance on the propensity score*)**.** In Example 7.2.4 the covariance matrix is comprised of distances between the whole vector of p covariates between any two subjects. Rather than directly condition on the p covariates in the outcome model, we could instead condition on the propensity score $e(\ell)$. We now define

$$\mu_a(e) = E\{Y_i(a) \mid e(L_i) = e\} = E(Y_i \mid A_i = a, e(L_i) = e).$$

We can then specify a GP prior for $\mu_a(e)$, essentially treating $e(L_i)$ as a one-dimensional covariate. Everything else would proceed as in Example 7.2.4. This is similar to Example 6.3.5, except here there is a focus on the average causal effect, and we use a GP (and assume normal residuals) instead of a fully nonparametric DPM.

7.3 Dependent Dirichlet Process Priors

So far in this chapter we have considered conditional distributions for $[Y_i \mid X_i, \theta]$ of the form $Y_i = g(X_i) + \varepsilon$ where we have assumed that $g(\cdot)$ is unknown but that the distribution for ε has a known parametric form. Essentially, we have been taking a nonparametric approach for $g(\cdot)$ but a parametric approach for the error distribution (similar to BART). In this section, we describe a nonparametric approach to the conditional distribution itself. We will now directly model the conditional distribution of $[Y_i \mid X_i, \theta]$. This is distinct from Chapter 6, where the conditional distributions were "derived" from a joint distribution $[Y_i, X_i \mid \theta]$. Thus, this approach avoids the need to model the joint distribution, which can be both computationally expensive and statistically difficult when there are many covariates.

In Chapter 6 we showed how DP priors can be used to model density functions and joint distributions using DPMs. Recall from Section 6.3 that we can write a DPM as

$$\begin{aligned}
[Y_i \mid \theta_i] &\overset{\text{ind}}{\sim} p(y_i \mid \theta_i), \\
[\theta_i \mid F] &\overset{\text{iid}}{\sim} F, \\
F &\sim DP(\alpha, H),
\end{aligned}$$

or, equivalently, using the stick-breaking formulation,

$$f(y \mid \theta) = \sum_{j=1}^{\infty} w_j p(y \mid \theta_j),$$

where $w_j = \beta_j \prod_{k=1}^{j-1}(1-\beta_k)$, $\beta_j \overset{\text{iid}}{\sim} \text{Beta}(1, \alpha)$, and $\theta_j \overset{\text{iid}}{\sim} H$.

Now, suppose we would like to directly estimate the conditional distribution $f(y \mid x, \theta)$. In other words, we are interested in the collection of distributions $\{P_x : x \in \mathscr{X}\}$ where $\mathscr{X}$ is the covariate space. We describe extensions of DPM models to conditional distributions. These are known as *dependent Dirichlet process* (DDP) mixture models [174, 175].

We can write a DDP mixture model as an infinite mixture

$$p\{y \mid x, \theta(x)\} = \sum_{j=1}^{\infty} w_j(x) p\{y \mid \theta_j(x)\}, \tag{7.5}$$

where $\theta(x) = (\theta_1(x), \theta_2(x), \ldots)$ and $w_j(\cdot)$ are random functions with the property that $w_j(x) = w_j'(x) \prod_{k=1}^{j-1}\{1 - w_k'(x)\}$ with $w_j'(x) \sim \text{Beta}(1, \alpha)$ for all x, and with $\theta_j(\cdot) \overset{\text{iid}}{\sim} H$. This ensures that, for any fixed $x \in \mathscr{X}$, $p\{y \mid x, \theta(x)\}$ is a DPM.

We will primarily consider *fixed weight* DDPs, which take $w_j'(x) \equiv w_j'$ for all x, i.e., the weights $w_j(x)$ do not depend on x. For continuous data, we will consider the model $p\{y \mid \theta_j(x)\} = N\{y \mid \theta_j(x), \sigma_j^2\}$ with $\theta_j(x)$ given a Gaussian process prior. Thus, the conditional distribution $p\{y \mid x, \theta(x)\}$ can be specified with a DDP (for the outcome around the mean) and a GP (for the mean), which we denote DDP+GP.

A specific example of a DDP+GP is

$$\begin{aligned}
p\{y \mid x, \theta(x)\} &= \sum_{j=1}^{\infty} w_j N\{y \mid \theta_j(x), \sigma_j^2\}, \\
\theta_j(x) &\sim \text{GP}\{m_j, \kappa(x, x)\}, \\
m_j(x) &= x^\top \beta_j, \\
\kappa(x, x') &= \eta \exp\left(-\rho \|x - x'\|^2\right) + 0.01 I(x = x'),
\end{aligned}$$

where $w_j = w'_j \prod_{k=1}^{j-1}(1-w'_k)$, $w'_j \stackrel{\text{iid}}{\sim} \text{Beta}(1,\alpha)$, $\sigma_j^2 \stackrel{\text{iid}}{\sim} \text{InvGam}(a,b)$, and $\beta_j \stackrel{\text{iid}}{\sim} N(0,\Sigma_\beta)$. Typically, we would also specify priors for (η,ρ,α).

Essentially, we now have a mixture model where, for the jth component of the mixture, we have a standard GP model as described earlier. Thus, these models can be thought of mixtures of local GP regression models.

7.3.1 *Sampling Algorithms*

One approach to inference for DDP+GP models is to use the blocked Gibbs sampler, which we recall truncates the infinite mixture to only K components for some suitably large K (see Section 6.3.1). Denote by Z_i the random variable that denotes cluster membership for subject i. A blocked Gibbs sampler for fitting a DDP+GP model is given in Algorithm 7.3.

Algorithm 7.3 Gibbs sampler for DDP+GP

Given the current values of the parameters, we do the following steps.

1. Update Z_i from a multinomial distribution with probability that Z_i is category j equal to $\frac{w_j N\{y_i|g_j(x_i),\sigma_j^2\}}{\sum_{k=1}^K w_k N\{y_i|g_k(x_i),\sigma_k^2\}}$ for $j=1,\ldots,K$.
2. Update $w'_j \sim \text{Beta}(1+n_j, \alpha+\sum_{k>j} n_k)$, where $n_j = \sum_i I(Z_i = j)$.
3. For each $k=1,\ldots,K$, update $g_k(\cdot)$ and σ_k^2 using the data for which $Z_i = k$ using the steps described in Algorithm 7.1.

7.3.2 *DDP+GP for Causal Inference*

We provide an example of DDP+GP for causal inference next.

Example 7.3.1 (*Causal inference using g-computation with confounders*)**.** Suppose we have a continuous outcome Y_i, treatment A_i, and confounders L_i and are interested in the average causal effect $\Delta = \int\{\mu_1(\ell)-\mu_0(\ell)\}\,F_L(d\ell)$. We specify a DDP+GP for the observed data. Specifically, given the values of $\mathbf{g} = (g_j(a_i,\ell_i): i=1,\ldots,n)$, we can sample $\widetilde{\mathbf{g}} = (g_j(1-a_i,\ell_i): i=1,\ldots,n)$ from its conditional distribution given $\mathbf{g}$ as described in Section 7.2.1.

Denote the mth draw of $\mu_a(\ell_i)$ by $\mu_a^{(m)}(\ell_i)$. Using the Bayesian bootstrap, at each MCMC step, we sample $\varpi^{(m)} \sim \text{Dirichlet}(1,\ldots,1)$. We can then compute $\Delta^{(m)}$ as

$$\Delta^{(m)} = \sum_{i=1}^{n} \varpi^{(m)}\{\mu_1^{(m)}(\ell_i) - \mu_0^{(m)}(\ell_i)\}.$$

If we have M post-burn-in MCMC draws from the posterior, we then have a collection of draws of causal effect parameters $\{\Delta^{(1)},\ldots,\Delta^{(M)}\}$.

A modified version of this model and algorithm was proposed in [35] to directly parameterize a causal model and do the marginalization within the MCMC algorithm. We provide details in the case study in Chapter 10.

The DDP+GP model has been used in other causal settings including to model conditional event time distributions in a semi competing risk setting, where principal stratum causal effects were computed in post-processing steps [176]; see the case study in Chapter 15. In a dynamic treatment strategy setting (see Chapter 1), Xu et al. [177] modeled longitudinal conditional distributions using DDP+GP priors.

7.3.3 Considerations for Choosing between Various DP Mixture Models

As discussed in Chapter 6, DPMs or EDPMs also can be used to estimate conditional distributions. For example, the implied conditional distribution $f(y \mid \ell)$ from a DPM in the simple case of a single ℓ is given in (6.5). A key difference between that derived conditional distribution and the one from the DDP+GP in (7.5) is the flexible form of the local mean in the latter. Therefore, fewer mixture components might be needed to have an accurate approximation of the distribution.

An important consideration in deciding between a DPM or EDPM model on the one hand or a DDP+GP model on the other is whether there is missingness in the covariates. When there is missingness, then we will need to explicitly model the distribution of the covariates, and so a generative model (such as an EDPM) may be more convenient, as we can use a single model for the purpose of both flexibly modeling the outcome and modeling the covariates. A generative model allows imputation of the missing data while guaranteeing the imputation and outcome models are compatible [163].

7.4 Summary

We have introduced Gaussian process (GP) priors and dependent Dirichlet process (DDP) mixture models that can be used for various causal inference problems. We also showed how DDPs combined with GPs can be used to model conditional distributions, particularly for continuous outcomes. The utility of these models is further illustrated in the case studies in Chapters 10 and 15.

Part III

Case studies

Chapter 8

Causal inference on quantiles using propensity scores

DOI: 10.1201/9780429324222-8

8.1 EHR Data and Questions of Interest

In this chapter, we use electronic health records (EHR) data to understand the effect of two drug therapies on serum creatinine (SCr) from two affiliated tertiary care institutions at the University of Florida. Elevated SCr can result in acute kidney injury (AKI), and understanding the causal effect of the two drug therapies on the SCr change can provide evidence-based guidelines for choosing appropriate drug therapies and reduce the risk of AKI. The distribution of SCr change is right-skewed as shown in Figure 8.1, and the greater the increase in SCr level, the more severe AKI the patients can develop; therefore, causal effects on the upper quantiles are of particular interest to clinicians.

Specifically, we will compare the effect of piperacillin-tazobactam and cefepime to an antibiotic regime of vancomycin on SCr change. Large SCr increases can result in AKI; for example, an increase of 0.3 mg/dl or larger can be used to define first-stage AKI. The population consists of inpatient adults admitted to UF Health Shands Hospital from January 1, 2012 to December 31, 2013 and UF Health Jacksonville Hospital from January 3, 2013 to December 31, 2013, with the demographic, diagnostic, and procedural information, lab results, medication administration, and other clinical variables obtained from EHRs. Patients on vancomycin therapy can enter the cohort and are divided into two groups based on the addition of either piperacillin-tazobactam or cefepime during vancomycin therapy or up to 3 days after the last vancomycin administration. The first day on piperacillin-tazobactam or cefepime is the index day. Patients are required to have at least 5 days of follow-up after the index date. Patients with severe blood loss or end-stage renal disease at the time of index date are excluded. The outcome is difference between the maximum SCr during the 5-day follow-up and baseline SCr.

All covariates are evaluated during the baseline period defined as within one day of the index date. A total of 33 baseline confounding covariates are used for

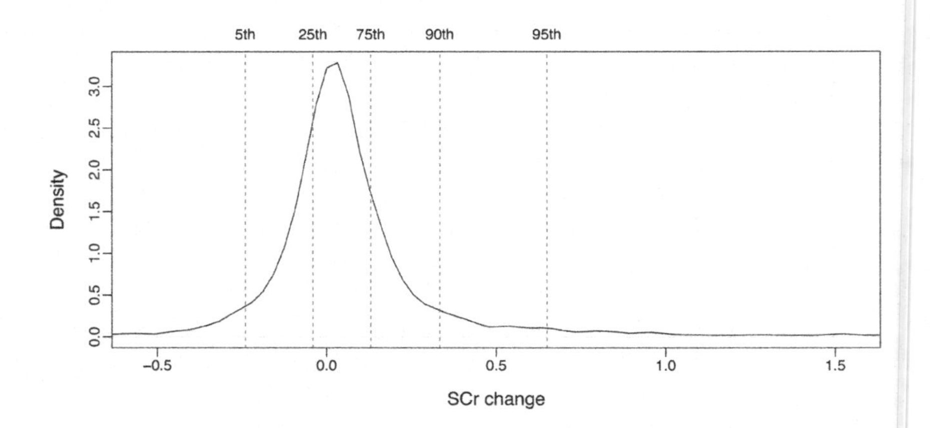

Figure 8.1 *The estimated density for the observed serum creatinine change for the EHR data with 5th, 25th, 75th, 90th, and 95 quantiles marked.*

this study: gender, race, age on admission, systolic blood pressure, mean arterial pressure, body temperature, serum albumin level, baseline SCr, creatinine clearance, blood urea nitrogen to SCr ratio, in intensive care unit, use of contrast, received cardiac surgery, comorbidities including sepsis, heart failure, pneumonia, diabetes, acute myocardial infarction, anemia, hypertension, liver disease, chronic kidney disease, skin tissue infection, bacteremia, urinary tract infection, infectious arthritis, and medication administrations including use of diuretics, use of angiotensin-converting enzyme inhibitors or angiotensin II receptor blockers, use of cephalosporins, use of albumin, use of non-steroidal anti-inflammatory drugs, use of other high nephrotoxic medications, and vancomycin total daily dose.

The study objective is to estimate quantile causal effects between two groups: piperacillin-tazobactam ($A_i = 1$) and cefepime ($A_i = 0$). Since clinicians are more interested in larger increases in SCr (upper quantiles), we focus on the 50th, 75th, and 90th quantile effects.

The final study population has 4087 patients with 1481 in the piperacillin-tazobactam group and 2606 in the cefepime group. Of the 33 covariates, 7 have missing values impacting 1894 patients.

Causal Estimand of Interest Let the qth quantile for the distribution of the potential outcomes for treatment $A_i = a$ be $F_a^{-1}(q)$. For this application, we focus on the quantile causal effects

$$F_1^{-1}(p) - F_0^{-1}(p) \qquad \text{for} \qquad p \in \{0.5, 0.75, 0.90\}$$

comparing piperacillin-tazobactam.

8.2 Methods

We combine the models/methods introduced in Chapters 3, 5, and 6 to estimate quantile causal effects. DPMs allow for estimation of any functional of the distribution of the potential outcomes, rather than just the mean (see Chapter 6). Using the propensity score in a regression setting (as introduced in Chapter 1), we can specify the joint distribution of the outcome and the propensity score using a DPM of bivariate normal distributions. In particular, we can use the DPM of bivariate normals introduced in Example 6.3.2 with L_i replaced by the estimated propensity score $\widehat{e}(L_i)$ and then fit the DPM separately for each value of the treatment $A_i = 0, 1$.

As described in Chapter 1, causal effects can be computed using the distribution of the potential outcomes, which are computed using g-computation under SUTVA, consistency, and ignorability,

$$F_a(y) = \Pr\{Y_i(a) < y \mid \theta\} = \int \int_{-\infty}^{y} f\{Y_i = t \mid A_i = a, \widehat{e}(\ell), \theta\}\, dt\, F_L(d\ell). \qquad (8.1)$$

From the DPM of bivariate normals, we obtain the conditional distribution in (8.1) (see Example 6.3.2). The two remaining pieces are estimating the propensity score, $e(\ell)$, and the distribution of the propensity score. We estimate the propensity score nonparametrically using probit BART (see Chapter 5). The distribution of the confounders (which induces the distribution of the propensity score) can be estimated using the Bayesian bootstrap (see Chapter 3); this latter avoids the need for sampling $e(\ell)$ from the DPM (and subsequent Monte Carlo Integration). We refer to this approach as *BART-DPM* [27, 178].

Computational Details

We provide details on the steps to compute the quantile causal effects for our setting in Algorithm 8.1. The above approach accounts for the uncertainty in the propensity score, the conditional distribution of outcome given the propensity score, and the distribution of the confounders (through the propensity score). It can be implemented using the R package `BNPqte` [178].

Priors

We use the default BART-probit prior as described in Chapter 5. The prior for the baseline distribution of the DPM is a conjugate normal-inverse Wishart distribution, $H(d\mu, d\Sigma) = N(d\mu \mid m_1, \Sigma/k_0)\mathscr{W}^{-1}(d\Sigma \mid \nu_1, \Psi_1)$, with hyperpriors $m_1 \sim N(m_2, S_2)$, $k_0 \sim \text{Gam}(\tau_1, \tau_2)$, and $\Psi_1 \sim \mathscr{W}^{-1}(\nu_2, \Psi_2)$. Let r_y and r_e be the *range* of the outcome and propensity score in the sample, and c_y and c_e be the center of the outcome and propensity score in the sample. The default hyperprior parameters for the baseline distribution [162] are specified as $m_2 = (c_y, c_2)$, $S_2 = \text{diag}(r_y^2/4, r_e^2/4)$, and $\Psi_2 = \nu_2 S_2^{-1}$. The degrees of freedom are set to $\nu_1 = 4$ and $\nu_2 = 4$ to allow the largest possible dispersion. We specify $E(1/k_0) = \tau_2/(\tau_1 - 1) = 2$. Such a specification ensures that the marginal prior for μ has $E(\mu) = m_2$ and $\text{Cov}(\mu) = 2S_2$, and the marginal prior for Σ has $E(\Sigma) = S_2$, which scales to the data appropriately. This prior specification for the DPM is similar to the default suggestions in Chapter 6.

8.3 Analysis

There was missingness in some of the covariates as noted in Section 8.1. We used a sequential BART approach to impute the missing covariates [148]; this approach orders the covariates by the amount of missingness and then creates the joint distribution of the missing covariates using a sequence of BART models (e.g., $[x_2|x_1]$, $[x_3|x_1, x_2]$, etc.). The approach assumes ignorable missingness in the covariates, conditional on treatment assignment, but unconditional on the outcome. We discuss this further in Section 8.4.

We plot the estimated probability densities and the cdfs of potential outcomes with 95% CIs in Figure 8.2 and calculate the quantile causal effect with 95% CIs in Table 8.1. The 90th quantile causal effect comparing piperacillin-tazobactam with cefepime is 0.2 mg/dL (95% CI: 0.11–0.29) for the proposed approach with the 90th quantile on the SCr change for piperacillin-tazobactam therapy 0.46 mg/dL (95% CI: 0.39 and 0.55), and for cefepime therapy 0.27 mg/dL (95% CI: 0.24 and 0.3). All three quantile causal effects are positive, indicating that patients on piperacillin-tazobactam therapy have larger SCr increases compared to cefepime therapy; the quantile causal effects become larger as the quantile is increased as well. The

Algorithm 8.1 Algorithm to compute the posterior distribution of quantile causal effects

1. Sample from the posterior distribution of the parameters of the propensity score model, which is specified as a BART probit model (cf. Chapter 5, Section 5.3.1).
2. For each posterior sample from the propensity score model, $m = 1,\dots,M$:
   (a) Calculate the propensity score $e^{(m)}(\ell_i)$ for $i = 1,\dots,n$.
   (b) Sample the parameters of the DPM model (cf. Example 6.3.2 for each treatment group ($A_i = a$) using the blocked Gibbs sampler (cf. Section 6.3.1)).
   (c) For each posterior sample from the DPM model $m' = 1,\dots,M'$:
      i. Calculate the conditional cdf of the outcome given the propensity score and $A_i = a$ on a set of grid points of y values $(g_1,\dots,g_S)$ using the truncation approximation to the DPM (Section 6.3.1) and the notation from Example 6.5,

$$F^{(mm')}(g_s \mid e^{(m)}(\ell_i), A_i = a) = \sum_{k=1}^{K} w_k^{(mm')}\{e^{(m)}(\ell_i)\}\,\Phi(g_s \mid \gamma_{0k}^{(mm')} + \gamma_{1k}\,e^{(m)}(\ell_i), \sigma^{2(mm')}),$$

      for $s = 1,\dots,S$ and $i = 1,\dots,n$, where $\Phi(\cdot \mid \mu,\sigma^2)$ is the cdf of a $N(\mu,\sigma^2)$ distribution.

      ii. Sample $(\omega_1^{(mm')},\dots,\omega_n^{(mm')}) \sim$ Dirichlet$(1,\dots,1)$ as the weights for the Bayesian bootstrap (cf. Section 3.2.3).

      iii. Calculate the cdf of $Y_i(a)$ by marginalizing the conditional distribution

$$F_a^{(mm')}(g_s) = \int F_a^{(mm')}(g_s \mid e^{(m)}(\ell))\, F_L^{(mm')}(d\ell) = \sum_{i=1}^{n} \omega_i^{(mm')}\, F^{(mm')}(g_s \mid e^{(m)}(\ell_i), A_i = a),$$

      where g_s ($s = 1,\dots,S$) is the same set of grid points as in Step 2(c)i, $F^{(mm')}$ is from Step 2(c)i, and $\omega^{(mm')}$ is from Step 2(c)ii.

      iv. For each treatment group, calculate the quantile of interest $F_a^{-1(mm')}(p)$ by finding the grid point such that $F_a^{(mm')}(g_s) \approx p$. The quantile causal effect is computed as $F_1^{-1(mm')}(p) - F_0^{-1(mm')}(p)$.
3. After obtaining the samples of the quantile causal effect $F_1^{-1(mm')}(p) - F_0^{-1(mm')}(p)$ for $m = 1,\dots,M$ and $m' = 1,\dots,M'$, calculate the estimated causal quantile effect by averaging the samples and compute credible intervals (CI) using the relevant percentiles.

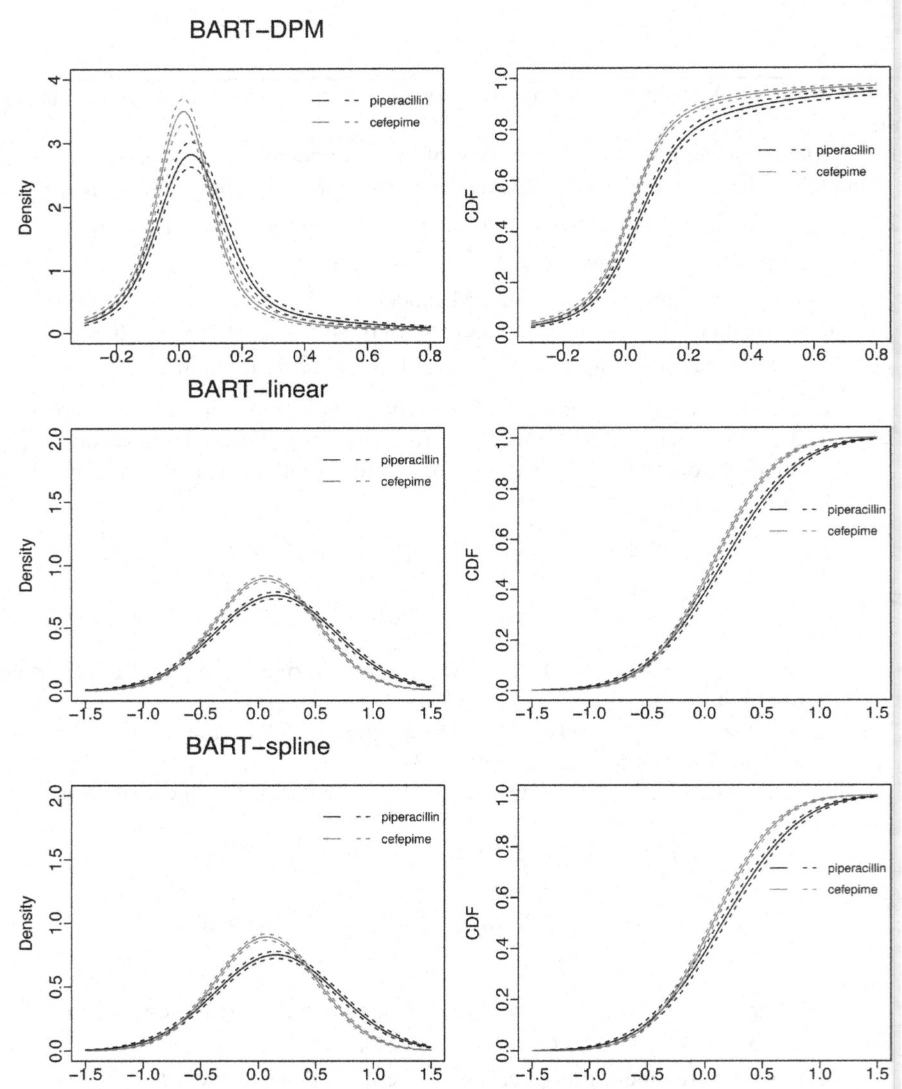

Figure 8.2 *Estimated densities and cdfs with 95% CIs for potential outcomes for the BART-DPM approach, the approach using a linear normal model on the propensity score as the conditional distribution of outcome (BART-linear), and the approach using a linear normal model on B-splines of the propensity score as the conditional distribution of outcome (BART-spline).*

Table 8.1 *Estimated 50th, 75th, and 90th quantile causal effects on serum creatinine change comparing piperacillin-tazobactam with cefepime for the BART-DPM approach, the approach using a linear normal model on the propensity score as the conditional distribution of outcome (BART-linear), the approach using a linear normal model on B-splines of the propensity score as the conditional distribution of outcome (BART-spline), and three frequentist approaches: a doubly robust (DR) estimator, a hybrid estimator, and TMLE.*

	Quantile causal effect estimates and 95% CI		
	50th	75th	90th
BART-DPM	0.04(0.03,0.05)	0.07(0.05,0.09)	0.20(0.11,0.29)
BART-linear	0.08(0.05,0.12)	0.14(0.10,0.18)	0.19(0.14,0.24)
BART-spline	0.08(0.05,0.12)	0.14(0.10,0.18)	0.19(0.15,0.24)
DR	0.03(0.01,0.06)	0.06(0.04,0.09)	0.18(0.11,0.29)
Hybrid	0.03(0.01,0.05)	0.06(0.04,0.09)	0.19(0.12,0.29)
TMLE	0.03(0.02,0.05)	0.06(0.04,0.08)	0.17(0.11,0.26)

results show that cefepime therapy is preferred to piperacillin-tazobactam in preventing AKI.

Model comparisons

For comparison, we also estimate the quantile causal effects using two Bayesian semiparametric approaches and three frequentist estimators. In terms of the two Bayesian approaches, the first assumes that the conditional distribution of the outcome is a linear normal model on the propensity score, i.e., $[Y_i \mid e(L_i) = e, \theta] \sim N(\beta_0 + \beta_1 e, \sigma^2)$, and the second is a semiparametric model that uses B-splines to model the relationship between the outcome and the propensity score, i.e., $[Y_i \mid e(L_i) = e, \theta] \sim N(\beta_0 + \sum_{b=1}^{B} \beta_b \psi_b(e), \sigma^2)$, where $\psi_b(e)$ is the bth B-spline basis function evaluated at e for $1 \leq b \leq B$ with $B = 6$; we refer to these as BART-linear and BART-spline. The three frequentist approaches considered are (i) a doubly robust (DR) estimator proposed by [179], (ii) a hybrid estimator proposed by [179], and (iii) a targeted maximum likelihood estimator (TMLE) proposed by [180]. The DR estimator combines the outcome regression (OR) model with an inverse-probability-weighted (IPW) approach, while the hybrid estimator combines the OR model with a propensity score (PS) stratification approach. The OR model for $[Y_i \mid A_i, L_i, \theta]$ is assumed to be a normal linear model after a Box-Cox transformation of the outcome for each treatment, and the PS model is a simple logistic regression model. The TMLE is estimated in three steps. First the PS and the cdf of the outcome are estimated; second, the quantiles are estimated based on the current cdf of the outcome; and third, the cdf of the outcome is updated based on an exponential submodel for the density of the outcome. The last two steps are iterated until convergence.

For the frequentist approaches, we combine bootstrapping with multiple imputation; specifically, we generate 100 bootstrapped samples from the original dataset with missing data, and for each bootstrapped sample, we generate ten imputed datasets using the sequential BART approach. Finally, we calculate the frequentist estimators using the 1000 datasets.

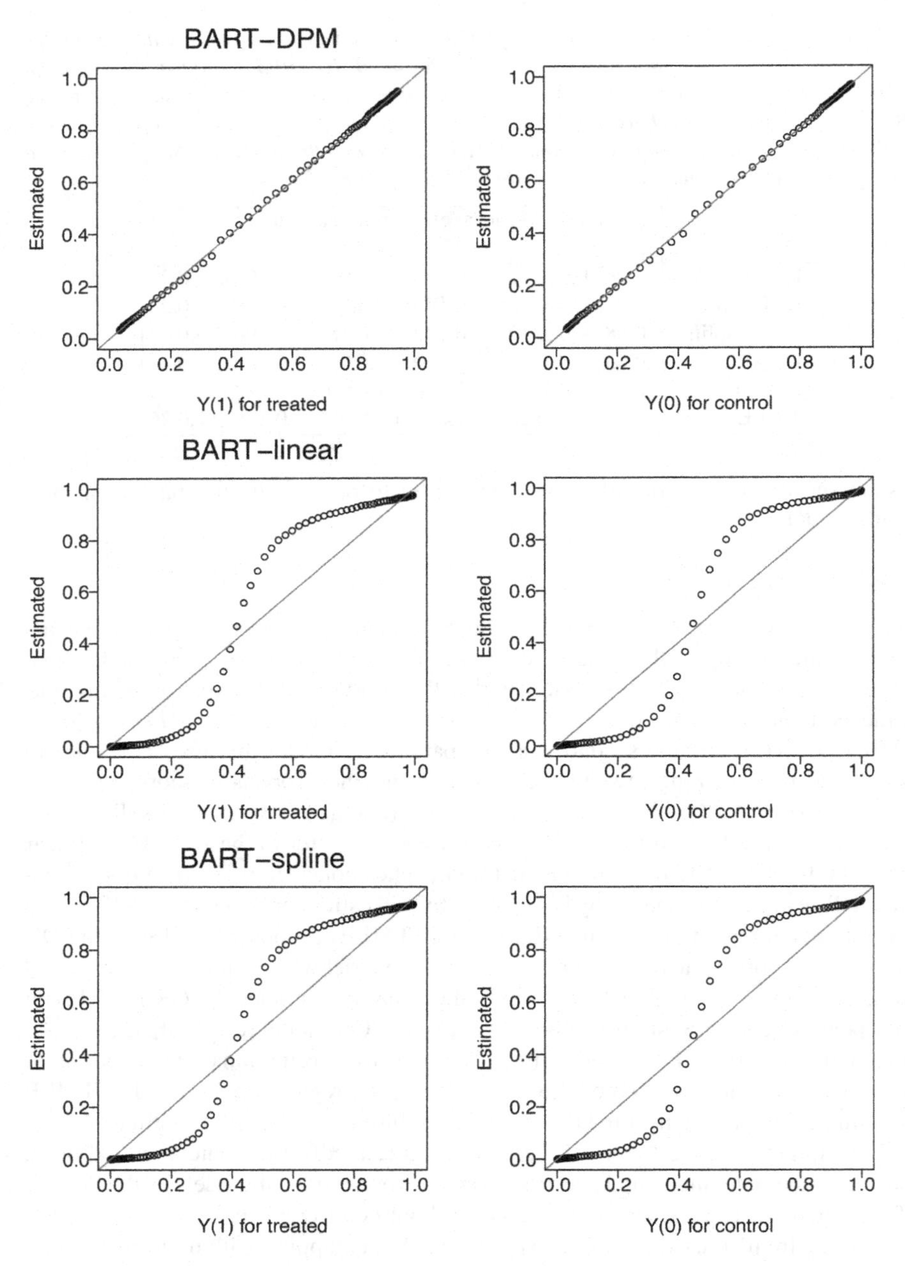

Figure 8.3 *QQ plots comparing the empirical cdf (x-axis) of $Y(1)$ for treated and $Y(0)$ for control with estimated cdf (y-axis) using the BART-DPM approach, the approach using a linear normal model on the propensity score as the conditional distribution of outcome (BART-linear), and the approach using a linear normal model on B-splines of the propensity score as the conditional distribution of outcome (BART-spline).*

The results for the proposed approach and the three frequentist approaches are quite similar for this dataset, though we note small efficiency gains (in terms of CI width) in the median for the BART-DPM over the other approaches and for the 75th quantile over the double robust and hybrid approaches (see Figure 8.2 and Table 8.1).

Model fit

To examine the fit of the three Bayesian approaches, we calculate the marginal (over the propensity score distribution for each treatment) cdf of the outcome for each treatment (which is just the cdfs of the potential outcomes $Y_i(1)$ (piperacillin-tazobactam) and $Y_i(0)$ (cefepime)) using the three approaches. Specifically, we compute the marginal cdfs

$$\Pr(Y_i \leq y \mid A_i = a) = \int_{-\infty}^{y} \int f(Y_i = t \mid e(L_i) = e, A_i = a) f\{e(L_i) = e \mid A_i = a\} \, de \, dt$$

for $a \in \{0,1\}$ and compare the estimated cdf using the Bayesian approaches with the empirical cdf of the observed outcome in each treatment group using QQ plots as shown in Figure 8.3. We see that the estimated cdf of the proposed approach matches well with the observed data, but the two other simpler Bayesian approaches do not.

8.4 Conclusions

We have shown how to implement the flexible BART-DPM approach to compute estimates and quantify uncertainty in the quantile causal effects. For the EHR example, the posteriors supported cefepime therapy over piperacillin-tazobactam in preventing AKI. Due to the outcome regression only including the propensity score (a single covariate as opposed to the entire vector of confounders), inference should be more accurate for "extreme" quantiles since, conditional on the propensity score, we only need to estimate a two-dimensional distribution (as opposed to a $p+1$-dimensional distribution, where p is the number of confounders).

We note that the approach here is not fully Bayesian [181] since the outcome regression would also need to condition on L (the confounders) in addition to the propensity score. Also, missing confounders in the propensity score were imputed conditional on other confounders and treatment, but not the outcome. In Chapter 9, we use an EDPM to compute posteriors for causal effects with implicit imputation of missing confounders using both treatment and the outcome. We also note that both BART (Chapter 5) and the DDP-GP approach (Chapter 7) model conditional distributions but do not easily deal with missing confounders. Large sample properties of the proposed approach are explored in [182]; empirically, the BART-DPM approach performs well in simulations relative to the competitors in Table 8.1, and we refer the reader to [27] for more details.

Chapter 9

Causal inference with a point treatment using an EDPM model

DOI: 10.1201/9780429324222-9

9.1 Hepatic Safety of Therapies for HIV/HCV Coinfection

Standard therapy for human immunodeficiency virus (HIV)-infected patients involves combinations of antiretroviral drugs (ART). There is concern that long-term use of these drugs for individuals who are coinfected with chronic hepatitis C virus (HCV) are at greater risk for drug-related liver injury. One class of drugs that are commonly included in ART regimens are nucleoside reverse transcriptase inhibitors (NRTIs). Some drugs in the NRTI class (didanosine, stavudine, zidovudine, and zalcitabine) might cause depletion of mitochondrial DNA, leading to liver injury. We label these drugs as mtNRTIs. Our interest is in comparing outcomes for HIV/HCV coinfected individuals who are taking ARTs that include mtNRTIs versus other NRTIs. In particular, for this analysis, we will focus on risk of death within 2 years of treatment initiation. Data for the study come from the Veterans Aging Cohort Study (VACS; [183]). The study population was HIV/HCVcoinfected patients from 2002 to 2009 who newly initiated a ART regimen that included NRTIs. A total of $n = 1747$ patients were included in the study. Information on potential confounding variables was obtained during the year prior to the ART initiation date. Follow-up was for 2 years.

Baseline confounding variables (L_i) included age at baseline (years), year of ART initiation, race/ethnicity (White, Black, Hispanic, and other), body mass index, diabetes mellitus, alcohol dependence/abuse, injection/non-injection drug abuse, exposure to other antiretrovirals associated with hepatotoxicity (i.e., abacavir, nevirapine, saquinavir and tipranavir), CD4 count, HIV RNA, alanine aminotransferase (ALT), aspartate aminotransferase (AST), and fibrosis-4 (FIB-4) score.

Although data were well captured for demographic and clinical variables, laboratory data was subject to missingness. In particular, about 5% of patients had at least one laboratory value that was missing. The percentage of missing data for each variable was as follows: ALT 1.3%, AST 2.5%, CD4 1.8%, and FIB-4 3.1%.

We adopt the following notation. We let Y_i denote the outcome, with $Y_i = 1$ if the subject died within two years after initiation of ART ($Y_i = 0$ otherwise). We let the treatment $A_i = 1$ if the ART regimen included mtNRTIs and 0 otherwise. We let L_i denote the confounding variables (nine binary and seven continuous).

Estimand of Interest Our goal is to estimate the causal relative risk

$$\psi_{rr} = \frac{E\{Y_i(1)\}}{E\{Y_i(0)\}},$$

which is the ratio of the probability of death if a subject received the ART regimen that included an mtNRTI to the probability of death if the subject received an ART regimen that included other NRTIs.

9.2 Methods

We apply enriched Dirichlet process mixtures (EDPMs), as described in Chapter 6, to the data in the VACS study [163]. Part of the motivation for this modeling choice is that there are missing confounding variables. The use of a generative model enables multiple imputation of these covariates within the algorithm unlike the approach in Chapter 8. This avoids the need to specify separate, possibly incompatible models for multiple imputation prior to data analysis [184]. We choose an EDPM approach because it can handle many covariates, while not weighing them too heavily in the overall likelihood. Further, it enables simple (local) covariate models that make the imputation step straightforward.

Our EDPM model is as follows. We first sorted the 16 covariates so that the first 9 are the binary ones and the last 7 are continuous. We then standardized the continuous covariates to have mean 0 and variance 1 so that regression coefficients for the covariates share a common scale (one unit corresponds to a standard deviation change in the covariate). We then let $X_i = (A_i, L_i)$ denote the combination of the treatment and confounders and set

$$
\begin{aligned}
[Y_i \mid X_i, \beta_i] &\sim \text{Bernoulli}\{\text{logit}^{-1}(\beta_{0i} + X_i^\top \beta_{1i})\}, && (9.1)\\
[X_{ij} \mid \pi_{ij}] &\sim \text{Bernoulli}(\pi_{ij}), \quad (j = 1, \ldots, 10), && (9.2)\\
[X_{ij} \mid \mu_{ij}, \tau_{ij}^2] &\sim N(\mu_{ij}, \tau_{ij}^2), \quad (j = 11, \ldots, 17), && (9.3)\\
[\beta_i, \pi_i, \mu_i, \tau_i^2 \mid P] &\sim F, && (9.4)\\
F &\sim \text{EDP}(\alpha_\beta, \alpha_\omega, H), && (9.5)
\end{aligned}
$$

where $\omega_i = (\pi_i, \mu_i, \tau_i^2)$. We specify (local) logistic regression models for the outcome and simple independent Bernoulli or normal models for the covariates. The EDPM prior on the parameters induces nested clustering. As a result, the proposed outcome model is a mixture of logistic regression with weights dependent on covariates (cf. (6.12) from Chapter 6). The specification of H is given next.

Priors

We now describe the base measure H for $(\beta_i, \pi_i, \mu_i, \tau_i^2)$. For the outcome model, we specify $\beta_i \sim N(\beta_0, c\Sigma_0^\beta)$. As in Chapter 6, Section 6.3.4, we set β_0 and Σ_β^0 to the maximum likelihood estimate and its asymptotic variance from an ordinary logistic regression of Y_i on X_i. In the data analyses, we used $c = n/5$, as this value worked well in a variety of simulation studies. We assume conjugate priors $\pi_{ij} \sim \text{Beta}(1,1)$ for the binary covariates. For the continuous covariate parameters (which have been standardized), we specified $\tau_{ij}^2 \sim \text{InvGam}(1,1)$ and $[\mu_{ij} \mid \tau_{ij}^2] \sim N(0, 2^2)$. Finally, we specified $\text{Gam}(1,1)$ priors for the concentration parameters α_θ and α_ω.

Computations

For general details on computations, see the description of the generalized Pólya urn sampler in Section 6.3.1. In brief, the MCMC algorithm alternates between (i) updating the cluster membership, (ii) updating the parameters of the mixture components, and (iii) imputing (via data augmentation) the missing data. These three steps are repeated until a sufficient number of draws have been obtained for accurate posterior inference.

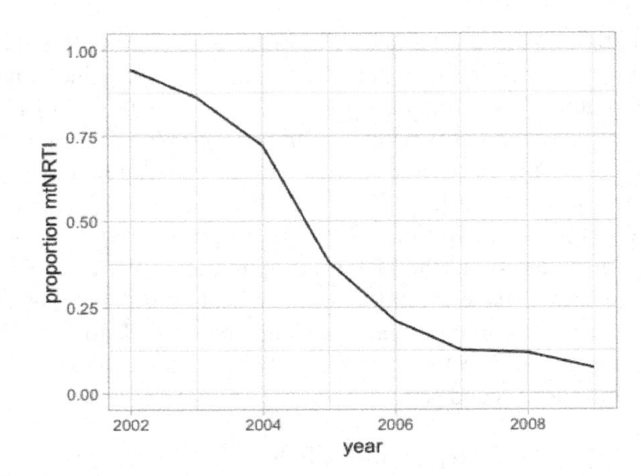

Figure 9.1 *Proportion of subjects initiating ART therapy with an mtNRTI (rather than other NRTIs) over time.*

After collecting samples of the parameters from the Gibbs sampler, we then post-process the output to compute the causal effect using g-computation; see Algorithm 6.3.

9.3 Analysis

We begin with some exploratory analyses. We first look at treatment over time. As can be seen from Figure 9.1, initiating an ART regiment with an mtNRTI decreased over time, going from a large majority of cases in 2002 to a small minority of cases in 2009. Other relevant things may have changed during the period from 2002 to 2009, such as other aspects of patient care and the baseline risk (underlying health) of patients at treatment initiation. Year of ART initiation will therefore be an important variable to control for.

We next computed standardized differences between the treatment groups for each covariate. Standardized differences are commonly used to assess balance in covariates between the treatment arms in causal inference analyses. As a rule-of-thumb, absolute standardized differences of < 0.1 are often considered consistent with what we would expect in a randomized trial (cf. Section 1.3.3). Figure 9.2 displays the standardized differences in this study, sorted from largest to smallest. The largest standardized difference was for the year of ART initiation (yr), which is not unexpected given Figure 9.1. The other standardized differences that were > 0.1 were age, exposure to other antiretrovirals associated with hepatotoxicity (`tox_hep`), alcohol dependence (`alcohol`), and BMI< 25 (`bmi_lt25`). It is possible that confounding is not too severe in this example due to treatment being driven by the adoption of the newer therapy over time rather than by clinical factors. However, because this is an observational study, we still want to adjust for these variables in a flexible way.

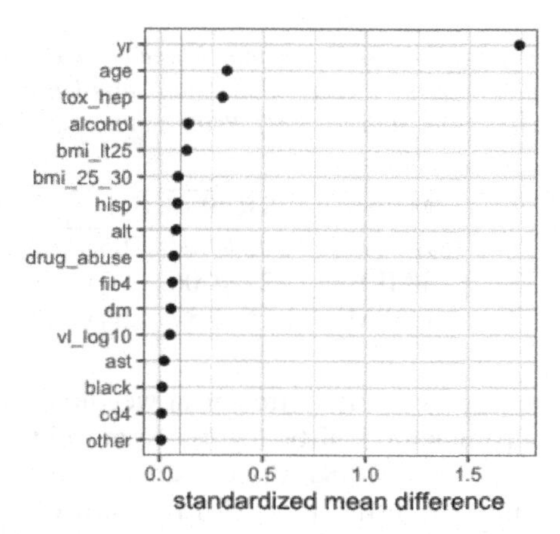

Figure 9.2 *Standardized mean differences for the covariates in the HIV/HCV coinfection study.*

There were 76 deaths out of 836 patients in the mtNRTI group, and 89 deaths out of 911 patients in the other NRTI group. Therefore, the unadjusted estimate of the relative risk was 0.93.

We ran 20,500 iterations each for three chains of the Gibbs sampler, each with different starting values. The Gelman–Rubin convergence diagnostic (cf. Section 6.3.1) was 1.04, providing evidence of convergence. Convergence appeared to be reached by iteration 500. We used every 100th draw (post-burn-in) for post-processing calculation for the causal relative risk. Across iterations, the number of y-clusters tended to be about 4, while the number of x-subclusters typically ranged from 1 to 7. The posterior median and 95% credible intervals for the EDPM concentration parameters were 0.63 (0.21, 1.43) for α_θ and 0.74 (0.43, 1.16) for α_ω.

As can be seen in Table 9.1, the EDPM approach estimated an RR of 1.16. If the true value of the RR is above 1, then starting an ART regimen that includes mtNRTIs has a high risk of mortality in the first 2 years. The 95% credible interval is fairly wide and includes 1. Therefore, after carefully adjusting for confounding using this flexible approach, we did not find conclusive evidence of a difference in risk between the treatment regimens.

Model comparisons

We implemented two frequentist estimators as comparisons to the EDPM approach. To address missing covariates for these approaches, we implemented a multiple imputation approach [185]. Twenty imputed datasets were created using chained equations (MICE) [186]. MICE is a common approach to impute missing data that sequentially samples from the full conditional distributions of the missing data; it has been criticized given the property that these conditionals do not typically correspond to a valid joint distribution [187], and like most imputation procedures, it is likely incompatible with the observed data model [184]. The first method for

Table 9.1 *Results from analysis of the VACS data. Three methods were used: inverse probability of treatment weighting (IPTW), targeted minimum loss-based estimation (TMLE), and an enriched Dirichlet process mixture (EDPM). Point estimates and 95% confidence/credible intervals are reported.*

Method	RR (95% CI)
IPTW	1.02 (0.97, 1.08)
TMLE	1.22 (1.06, 1.47)
EDPM	1.16 (0.87, 1.54)

comparison used inverse probability of treatment weighting (IPTW) to estimate the causal relative risk; a parametric logistic regression model was used to estimate the propensity score. The second method for comparison is a doubly robust estimator using a targeted minimum loss estimation (TMLE) approach; we used super learning to estimate the mean functions and included the following learners: mean (`mean`), generalized linear models (`glm`), generalized linear model with stepwise selection (`step`), generalized additive models (`gam`), random forest (`randomForest`), and generalized linear models with elastic net regularization `glmnet`. These algorithms were chosen to provide a reasonable range of types of models, including parametric and semi-parametric regression, as well as a tree-based approach (random forest). TMLE was implemented via the `R` package `tmle` [188]. The results are given in the first two row of Table 9.1. IPTW had much narrower CI and had a point estimate closer to 1. However, IPTW is the only one of the three methods that relied on parametric models. TMLE and EDPM had results that were more similar. The CI from TMLE was narrower and did not include 1.

In previous work we compared IPTW, TMLE, and EDPM in a variety of simulation studies, including for binary outcomes [163]. We considered some scenarios with missing covariate data and with complex relationships between the outcome and covariates. We generally found that EDPM performed well, with close to nominal coverage and small bias. We found that the EDPM approach that accounts for missing data within the algorithm generally performed well (low bias and accurate coverage), even compared with IPTW and TMLE that were applied to the full dataset with no missing data.

9.4 Conclusions

We have demonstrated the use of an EDPM approach for the point treatment setting when there are missing covariates. An advantage of this approach is that missing data imputation can be included within the same algorithm as posterior inference (under an assumption of ignorable missingness (cf. Section 2.5)) and does not require specifying a separate, likely incompatible imputation model. In previous work, we found the EDPM approach to causal inference had good frequentist properties (bias and coverage) in a variety of settings [163] as noted above.

Chapter 10

DDP+GP for causal inference using marginal structural models

DOI: 10.1201/9780429324222-10

10.1 Changes in Neurocognitive Function among Individuals with HIV

In the early 2000s it was observed that HIV-infected individuals experienced declining neurocognitive performance [189]. Further, in these populations recent heavy alcohol was found to be associated with decreased neurocognitive function [190]. At that time, highly active antiretroviral therapy (HAART) regimens were beginning to be used to treat HIV, replacing non-HAART regimens. Thus, a key question was whether HAART would lead to better neurocognitive performance overall (compared with non-HAART) and whether any effect would be modified by history of alcohol use.

To address the question, we use data from the HIV Epidemiology Research (HER) Study, a multisite study of the natural history of HIV in U.S. women [191]. For women with CD4 cell count < 100 cells/μl, neurocognitive exams were administered every 6 months beginning with a baseline exam at 3 months after this threshold was reached. In total, 126 women completed at least two neurological exams in 1993–1996 and made up our study population.

Notation

At each semiannual visit, neurocognitive exams were administered. We focus on one test, Color Trail Making 1 total time (CTM). Our focus is on the change in scores over time. Our outcome of interest Y_i is the difference between CTM score on the last exam and the CTM score at baseline. A HAART drug regimen was defined using the guidelines of the US Public Health Service as a combination of either protease inhibitor plus two nucleoside analog reverse transcriptase inhibitors or a protease inhibitor plus a nucleoside analog reverse transcriptase inhibitor plus a non-nucleoside reverse transcriptase inhibitor [191]. Non-HAART consisted of monotherapy or dual nucleoside therapy without the above combinations. We define $A_i = 1$ if subject i received the HAART drug regime and $A_i = 0$ otherwise; the potential outcomes if treated with or without a HAART regimen are $Y_i(1)$ and $Y_i(0)$, respectively. Improvement is indicated by a negative change in time to complete the task. Possible confounders (L_i) that were identified a priori and collected at enrollment were age (continuous); intravenous drug use in the past six months (yes/no); any previous use of opiates, cocaine, amphetamines, barbiturates, and/or hallucinogens (yes/no); depression severity measured using The Center for Epidemiology Scale of Depression (CESD) (continuous); and CD4 cell count (continuous). A potential effect modifier (and possible confounder) of interest is alcohol use in the past 6 months (yes/no). Let V_i represent the alcohol use variable, where $V_i = 1$ represents the subject used alcohol in the last 6 months.

Causal model

We apply a marginal structural model (MSM) (cf. Chapter 1, Section 1.4) that consists of main effects for recent alcohol use and HAART drug regimen and an interaction term. Specifically, we propose

$$E\{Y_i(a) \mid V_i = v\} = \psi_0 + \psi_1 a + \psi_2 v + \psi_3 v \times a, \tag{10.1}$$

for $a = 0,1$ and $v = 0,1$. The interpretation of ψ_1 is the difference in mean of the potential outcome comparing HAART vs. non-HAART among women with no recent alcohol use ($v = 0$). Furthermore, ψ_3 indicates modification of this difference by recent alcohol use. Our goal is obtaining the posterior distribution for these parameters, while making minimal modeling assumptions.

Early analyses of the data relied on parametric assumptions about the mean function of covariates [192] or the outcome distribution [189]. Later, Roy et al. [35] analyzed the data using a tailored DDP-GP, which is the approach we will describe here.

Causal inference here relies on the validity of the assumptions described in Chapter 1. Specifically, we assume SUTVA, consistency, ignorability, and positivity. In this study we do not expect one person's treatment or outcome to affect other individuals, and therefore SUTVA is a reasonable assumption. Ignorability is probably the strongest assumption here. It requires that the treatment decision regarding HAART is independent of the potential outcomes given the confounders L listed above. Because we are most concerned about the validity of this assumption, we carry out a sensitivity analysis in what follows. We also discuss the positivity assumption in more detail below.

10.2 Methods

Models

We first describe a model for the conditional density $f(y \mid a, \ell, v, \theta)$ of $[Y_i \mid A_i = a, L_i = \ell, V_i = v, \theta]$, where L_i is the vector of confounders not including V_i. We specify a DDP+GP prior for this conditional density as described in Section 7.3. A key difference here is that we specify it in a way to accommodate the MSM specification in (10.1). Our DDP model, in stick-breaking form, is

$$f(y \mid v, a, \ell, \theta) = \sum_{k=1}^{\infty} w_k N\{y \mid \delta(a,v) + g_k(\ell), \sigma^2\} \tag{10.2}$$

where $\delta(a,v) + g_k(\ell)$ and σ are the mean and standard deviation of the kth component of the mixture. The function $\delta(a,v)$, given below, is specified to ensure the MSM in (10.1) holds. Recall that the prior for $(w_1, w_2, \ldots)$ is of the form

$$w_k = w'_k \prod_{j<k} (1 - w'_j) \qquad \text{where} \qquad w'_k \sim \text{Beta}(1, \alpha).$$

Because the mean function $g_k(\ell)$ has an unknown form, we specify a prior distribution for it as described in Chapter 7. We assume $g_k(\ell)$ is a Gaussian process $g_k(\cdot) \sim \text{GP}(m_k, \kappa)$, with $m_k(\ell) = \ell^\top \beta_k$ and the β_k's are unknown regression coefficients. We use the squared-exponential covariance function which takes the form $\kappa(\ell, \ell') = \eta \exp\left(-\rho \|\ell - \ell'\|^2\right) + b\,\delta_{\ell\ell'})$ where $\|x\|^2$ is the squared Euclidean norm of x, η and ρ are unknown positive parameters, and b is a small positive value (we use 0.01).

We next derive the form of the correction factor $\delta(a,v)$ given the following two constraints. First, the MSM given in equation (10.1) must hold. Second, the DDP+GP model implies

$$\begin{aligned} E\{Y_i(a) \mid V_i = v\} &= \int E\{Y_i(a) \mid V_i = v, L_i = \ell\}\, dF(\ell \mid v) \\ &= \int E(Y_i \mid A_i = a, V_i = v, L_i = \ell)\, dF(\ell \mid v) \\ &= \sum_k w_k \int \{\delta(a,v) + g_k(\ell)\}\, dF(\ell \mid v) \end{aligned}$$

Given these two constraints, the correction factor $\delta(a,v)$ is

$$\delta(a,v) = \psi_0 + \psi_1 a + \psi_2 v + \psi_3 v \times a - \int \sum_k w_k g_k(\ell) f(L_i = \ell \mid V_i = v)\, d\ell. \quad (10.3)$$

The empirical distribution (or the Bayesian bootstrap) for $[L \mid V = v]$ ($v \in \{0,1\}$) can be used to compute the integral within the MCMC algorithm, described below.

Priors

We specify the following normal priors for the prior mean parameters in the GP model: $\beta_k \sim N(\beta_0, \Sigma_0^\beta)$, where β_0 and Σ_0^β are known following the recommendations in [162]. We assume diffuse normal priors for the MSM parameters ψ. For the DP precision parameter α, we assume $\alpha \sim \text{Gam}(1,1)$. For the prior correlation-like parameter in the GP ρ, we assume $\rho \sim \text{Gam}(1,1)$. For the variance parameters, we assume the following: $\sigma^2 \sim \text{InvGam}(\lambda_1 = 1, \lambda_2 = 1)$ and $\eta \sim \text{Gam}(1,1)$.

Computations

In Chapter 7 we described a Gibbs sampler for a DDP+GP (Algorithm 7.3). However, here our DDP+GP model directly parameterizes the marginal causal effects, which leads to a modified sampling algorithm. The primary modifications are (i) replacing $g_k(\ell_i)$ with $\delta(a_i,v_i) + g_k(\ell_i)$ and (ii) updating $\delta(a_i,v_i)$ at each step of the Gibbs sampler. Here we provide the detailed steps of the algorithm. As in Chapter 7, we use a blocked Gibbs sampler to approximate the likelihood, a finite mixture with K components, $f(y \mid x, \theta) = \sum_{k=1}^K w_k N\{y \mid \delta(a,v) + g_k(\ell), \sigma^2\}$ (cf. Section 6.3.1). Denote by Z_i the random variable that denotes cluster membership for subject i. Given the current values of the parameters, we update as follows.

1. Generate Z_i for each subject ($i = 1, \ldots, n$) from

$$\Pr(Z_i = k \mid \text{rest}) = \frac{w_k N\{y_i \mid \delta(a_i,v_i) + g_k(\ell_i), \sigma^2\}}{\sum_{j=1}^K w_j N\{y_i \mid \delta(a_i,v_i) + g_j(\ell_i), \sigma^2\}} \qquad (k = 1, \ldots, K).$$

2. Update w'_k for $k = 1, \ldots, K-1$ from

$$\pi(w'_k \mid \text{rest}) \propto (w'_k)^{n_k} (1 - w'_k)^{n_{>k}} \times (1 - w'_k)^{\alpha - 1} \times \exp\left\{ -\frac{1}{2\sigma^2} \sum_i (y_i - \delta(a_i,v_i) - g_{Z_i}(\ell_i))^2 \right\}$$

where $n_k = \sum_i I(Z_i = k)$ and $n_{>k} = \sum_i I(Z_i > k)$ using (for example) slice sampling.

3. Update σ^2 from

$$\pi(\sigma^2 \mid \text{rest}) \propto \text{InvGam}\left\{\lambda_1 + n/2, \lambda_2 + \frac{1}{2}\sum_{i=1}^{n}(y_i - \delta(a_i, v_i) - g_{Z_i}(\ell_i))^2\right\}.$$

4. Update $\mathbf{g}_k$ for $k = 1, \ldots, K$ from

$$\pi(\mathbf{g}_k \mid \text{rest}) = N(\mu_k^{g,\text{post}}, \Sigma_k^{g,\text{post}}),$$

where

$$\Sigma_k^{g,\text{post}} = \left(\kappa(\boldsymbol{\ell},\boldsymbol{\ell})^{-1} + U_k/\sigma^2\right)^{-1},$$
$$\mu_k^{g,\text{post}} = (\Sigma^{g,\text{post}})\left(\kappa(\boldsymbol{\ell},\boldsymbol{\ell})^{-1}\boldsymbol{\ell}^\top\beta_k + U_k(\mathbf{y} - \boldsymbol{\delta})/\sigma^2\right),$$

U_k is a $n \times n$ matrix of 0's except for diagonal elements (i,i) that are set to 1 if $Z_i = k$, and $\boldsymbol{\delta} = (\delta(a_1, v_1), \ldots, \delta(a_n, v_n))$.

5. Update β_k for $k = 1, \ldots, K$ from $\pi(\beta_k \mid \text{rest}) = N(\beta_k \mid \mu_k^{\beta,\text{post}}, \Sigma_k^{\beta,\text{post}})$ where

$$\Sigma_k^{\beta,post} = \left\{\boldsymbol{\ell}^\top\kappa(\boldsymbol{\ell},\boldsymbol{\ell})^{-1}\boldsymbol{\ell} + (\Sigma_0^\beta)^{-1} + \boldsymbol{\ell}^\top\boldsymbol{\ell}w_k^2/\sigma^2\right\}^{-1},$$
$$\mu_k^{\beta,\text{post}} = (\Sigma_k^{\beta,post})\left\{\boldsymbol{\ell}^\top\kappa(\boldsymbol{\ell},\boldsymbol{\ell})^{-1}\mathbf{g}_k + (\Sigma_0^\beta)^{-1}\beta_0 + w_k\boldsymbol{\ell}^\top\left(\mathbf{y} - A^*\psi - \mathbf{g}_k + \sum_{j\neq k} w_j\boldsymbol{\ell}^\top\beta_j\right)/\sigma^2\right\}$$

and A^* is the design matrix from model (10.1).

6. Update ψ from a $N(\mu^{\psi,\text{post}}, \Sigma^{\psi,\text{post}})$ where

$$\Sigma^{\psi,\text{post}} = \left\{(A^*)^\top A^*/\sigma^2 + (\Sigma_0^\psi)^{-1}\right\}^{-1},$$
$$\mu^{\psi,\text{post}} = \Sigma^{\psi,\text{post}}\{(A^*)^\top/\sigma^2(-\mathbf{g}_z + y + \sum_{j=1}^{K} w_j\mathbf{g}_j),$$

where $\mathbf{g}_z = (g_{Z_1}(\ell_1), \ldots, g_{Z_n}(\ell_n))^\top$.

7. Update η and ρ from $\pi(\eta, \rho \mid \text{rest}) \propto \prod_{k=1}^K N\{\mathbf{g}_k \mid m_k, \kappa(\boldsymbol{\ell},\boldsymbol{\ell})\}\pi(\rho)\pi(\eta)$.

Note that Steps 1–5 each add more terms to the response residuals (cf. Algorithm 7.3) and Step 6 is new. After burn-in, draws of the causal parameters ψ are stored and used to approximate the posterior distribution.

Sensitivity analysis

To assess sensitivity to the ignorability assumption, we can modify within the mean of the DDP in (10.2) to condition on A: $f\{Y_i(a) \mid A_i = a, L_i = \ell\} = \sum_{k=1}^{\infty}\gamma_k N(y \mid \delta_{SA}(a,v) + A\phi + g_k(\ell), \sigma^2)$, where ϕ is a sensitivity parameter and the "SA" in δ_{SA}

is meant to distinguish the δ function in the sensitivity analysis from the one previously defined in (10.3). The correction factor to ensure parameters have marginal interpretations from (10.1) is now

$$\delta_{SA}(a,v) = \psi_0 + \psi_1 a + \psi_2 v + \psi_3 v \times a - \int\int \left(a\phi + \sum_{k=1}^{\infty} w_k \ell^\top \beta_k \right) f(A_i = a, L_i = \ell \mid V_i = v) \, da \, d\ell.$$

In the Gibbs sampler, we replace $\delta_{SA}(a,v)$ with $\delta(a,v)$ and otherwise proceed with the algorithm as before, inserting the term, $A_i\phi$, where necessary. If $\phi = 0$, then the ignorability assumption holds.

We calibrate ϕ using observed data summaries similar to causal mediation sensitivity as described in Chapter 4, Section 4.2.2. That is, we first calculate the total variance in Y_i explained by L_i (but not A_i). Denote this by R^2. We then assume that $|\phi| = |E\{Y_i(a) \mid A_i = 1, L_i\} - E\{Y_i(a) \mid A_i = 0, L_i\}|$ is less than $\sqrt{\text{Var}(Y_i)(1-R^2)k}$ (i.e., unmeasured confounding would account for less than 100k% of the remaining variance). We then specify a prior distribution for k or vary k along a grid.

We consider a Uniform$(0, 0.1)$ prior for k, which represents the prior belief that up to 10% of the remaining unexplained variance is explained by unmeasured confounders and that any value between 0% and 10% is equally likely. In addition, because we require that $|\phi| < \sqrt{\text{Var}(Y)(1-R^2)k}$, we consider both positive and negative values of ϕ. A positive value, for example, would indicate that subjects in the HAART group had a higher value of each potential outcome than did subjects in the non-HAART group.

10.3 Analysis

We first compare the confounders by treatment group. Figure 10.1 displays the standardized differences, sorted from largest to smallest. Standardized differences for CESD, any previous use of opiates, cocaine, amphetamines, barbiturates, and/or hallucinogens (other_drugs), and CD4 count were all well above 0.1. For the other confounders, the marginal differences between treatment groups were smaller. Based on these differences, it does indicate concerns about confounding as discussed in Chapter 1, Section 1.3.3.

As further exploratory analysis to assess the positivity assumption and overlap, we estimated the propensity score using BART using the confounders listed above. The minimum and maximum values of the propensity score were 0.12 and 0.87. We plotted the propensity score, stratified by treatment group in Figure 10.2. Overall there does not appear to be serious concerns about positivity violations, as there is a lot of overlap on the two plots and there are no extreme values of the propensity score.

We then fit the DDP-GP model to the observed data. Under each of the three specifications for the sensitivity parameter distribution (see Table 10.1), we obtained posterior distributions for the parameters from the MSM. The results are presented in Table 10.1. The rows with $\phi = 0$ are the results when ignorability (1.1) is assumed to

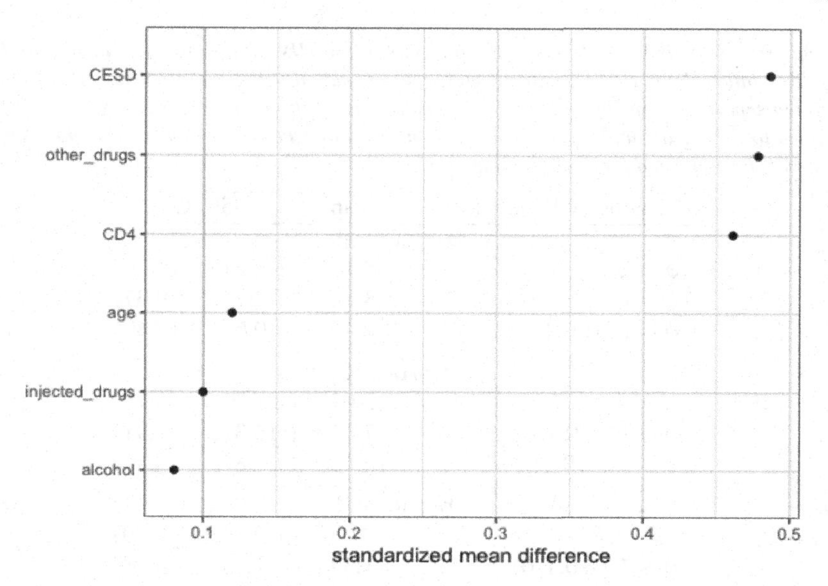

Figure 10.1 *Standardized mean differences for the covariates in the HIV study.*

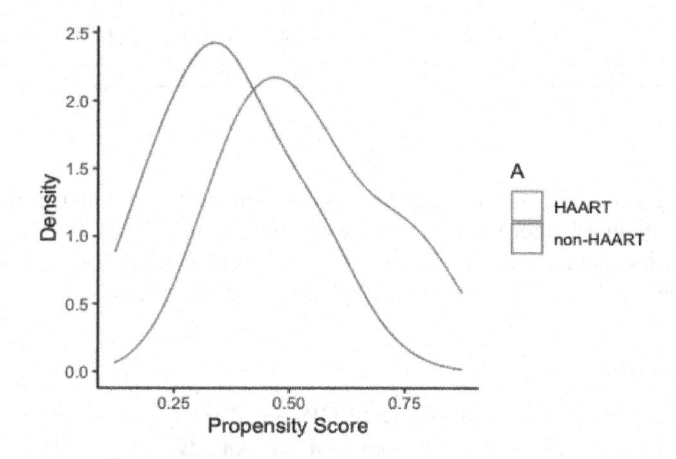

Figure 10.2 *Estimated propensity score stratified by treatment group in the HIV study.*

hold. There we see evidence of those treated with HAART having improved performance on Color Trail Making test, relative to non-HAART (negative coefficient of ψ_1 with 95% credible interval excluding 0). Although the point estimate of ϕ_3 was in the hypothesized direction (less of a benefit of HAART if alcohol user), the credible interval is wide and includes substantial area on both sides of 0. In the sensitivity

Table 10.1 *Sensitivity analysis for the analysis of the HIV data. Posterior median and 95% credible intervals of $\psi = (\psi_0, \psi_1, \psi_2, \psi_3)$ were estimated using the proposed BNP approach applied separately for different combinations of sensitivity parameters. The row $\phi = 0$ corresponds to the results under the ignorability assumption. Posterior median and 95% credible intervals are reported for each parameter.*

Sensitivity Parameters	Median	95% CI	
	ψ_0: intercept		
$\phi = 0$	5.10	(-1.21,	11.30)
$\phi \sim U(-0.1, 0)$	3.32	(-2.70,	9.37)
$\phi \sim U(0, 0.1)$	7.40	(0.67,	13.89)
	ψ_1: HAART		
$\phi = 0$	-12.01	(-21.30,	-3.15)
$\phi \sim U(-0.1, 0)$	-7.72	(-16.32,	0.91)
$\phi \sim U(0, 0.1)$	-16.22	(-25.88,	-6.47)
	ψ_2: alcohol		
$\phi = 0$	-0.03	(-9.71,	9.99)
$\phi \sim U(-0.1, 0)$	-1.37	(-11.12,	8.00)
$\phi \sim U(0, 0.1)$	-2.96	(-12.48,	6.62)
	ψ_3: HAART $\times$ alcohol		
$\phi = 0$	6.51	(-7.25,	21.67)
$\phi \sim U(-0.1, 0)$	7.14	(-6.04,	21.25)
$\phi \sim U(0, 0.1)$	9.44	(-4.54,	23.86)

analyses, we can see that the credible intervals tended to be wider, reflecting our uncertainty about whether the ignorability assumption is actually true. If ψ is positive, our main conclusions about HAART would not change. If HAART is negative, however, we see the credible interval for ψ_1 does now include 0.

10.4 Conclusions

We have demonstrated the use of a DDP-GP model for the point treatment setting and used it to estimate parameters from a MSM. An advantage of this approach is we avoid making parametric assumptions about either the regression mean function or the outcome distribution. In addition, informative priors can be used to capture uncertainty about non-identifiable sensitivity parameters. In further work, we explored the performance of the methodology in a variety of simulation studies and included comparisons with other (non-Bayesian) causal methods [35]. We found that the DDP+GP approach performed well, resulting in low bias and good frequentist coverage of credible intervals, including the scenario described by [193] where there were near violations of the positivity assumption.

Chapter 11

DPMs for dropout in longitudinal studies

DOI: 10.1201/9780429324222-11

11.1 Schizophrenia Clinical Trial

The *schizophrenia clinical trial* was a longitudinal clinical trial designed to assess the efficacy of a new drug on symptoms of acute schizophrenia. Subjects were assigned to one of three treatment arms: the test drug, an existing drug (called the "active control"), and a placebo. There were a total of 204 patients enrolled in the trial, with 45 assigned to the active control, 78 assigned to the placebo, and 81 assigned to the test drug. The primary clinical response of interest was the Positive and Negative Syndrome Scale (PANSS) score, which measures the severity of schizophrenia symptoms [194]. The PANSS score was scheduled to be measured at 0, 4, 7, 14, 21, and 28 days from baseline.

During the trial, many of the PANSS scores intended to be collected were missing. The majority of this missingness was due to dropout, i.e., patients were removed entirely from the study. Individuals dropped out of the trial for a variety of reasons, including pregnancy, withdrawal of patient consent, and progression of disease symptoms.

Notation We let Y_{ij} (for $i = 1, \dots N$ and $j = 1, \dots, J = 6$) denote the response of subject i at t_j days from baseline. We let $A_i = 1, 2, 3$ correspond to subject i being randomized to the active control, test treatment, and placebo, respectively. The causal effect of interest is the *intention to treat* effect on the change from baseline relative to the placebo,

$$\Delta_a = \eta_a - \eta_3, \tag{11.1}$$

where

$$\eta_a = E(Y_{iJ} - Y_{i1} \mid A_i = a).$$

As usual, we let $R_i = (R_{i1}, \dots, R_{iJ})$ denote the *missingness indicators* for each response such that Y_{ij} is observed if $R_{ij} = 1$. Because almost all missingness was due to dropout, we let $S_i = \max\{j : R_{ij} = 1\}$ denote the *follow-up time* for individual i. For any $j = 1, \dots, J$, we let $\overline{Y}_{ij} = (Y_{i1}, \dots, Y_{ij})$ denote the history of Y_i up-to-and-including time j. We let ω denote the collection of all parameters of the distribution of (Y_i, R_i).

11.2 Methods

To obtain a flexible model for the observed data, we use an infinite mixture model

$$f(\overline{y}_s, s) = \sum_{k=1}^{\infty} w_k N(\overline{y}_s \mid \mu_{1:s}^{(k)}, \Sigma_{1:s}^{(k)}) g(s \mid \overline{y}_s, \psi^{(k)}), \tag{11.2}$$

where $\mu_{1:s}^{(k)} = (\mu_1^{(k)}, \dots, \mu_s^{(k)})$ is a subvector of $\mu^{(k)} = (\mu_1^{(k)}, \dots, \mu_J^{(k)})$ and, similarly, $\Sigma_{1:s}^{(k)}$ is a submatrix of a $J \times J$, covariance matrix $\Sigma^{(k)}$. Here, $N(\overline{y}_s \mid \mu_{1:s}^{(k)}, \Sigma_{1:s}^{(k)})$ denotes a multivariate normal model for $\overline{Y}_s$ in cluster k. The model $g(s \mid \overline{y}_s, \psi) = \Pr(S_i = s \mid \overline{Y}_{is} = \overline{y}_s, \psi)$ is a parametric model for the follow-up time conditional on the history up to that time $\overline{y}_s$. Together, within each cluster k, $N(\cdot)$ and $g(\cdot)$ define a joint model for the observed data $(\overline{Y}_{S_i,i}, S_i)$. We choose the infinite mixture to correspond to a Dirichlet process mixture (cf. Chapter 6) and use the truncation approximation. That is, $w_k = w'_k \prod_{j<k}(1-w'_j)$ with $w'_1, \ldots, w'_{K-1} \overset{\text{iid}}{\sim} \text{Beta}(1,\alpha)$ and $w'_K \equiv 1$ for some moderate value of K. For the schizophrenia clinical trial, we set $K = 20$ and $\alpha = 1$ (note we could have also put a prior on α). The parameters $(\mu^{(k)}, \Sigma^{(k)}, \psi^{(k)})$ are iid from some base distribution $H = H_{\mu,\Sigma} \times H_\psi$. We describe our choice of H in the following sections. See Chapter 6, Section 6.3, for more details on Dirichlet process mixture models.

Prior specification: the response model

Rather than directly specifying priors for $\mu^{(k)}$ and $\Sigma^{(k)}$, the longitudinal structure of the data suggests that it may be more appropriate to work in terms of sequential linear regression models. Consider the one-to-one mapping $(\mu, \Sigma) \mapsto (\sigma, \phi, \mu)$ defined by the models

$$
\begin{aligned}
&Y_{i1} \sim N(\mu_1, \sigma_1^2),\\
&Y_{ij} \sim N\left\{\mu_j + \sum_{\ell=1}^{j-1} \phi_{j\ell}(y_{i\ell} - \mu_\ell), \sigma_\ell^2\right\} \qquad \text{given} \qquad \overline{Y}_{i(j-1)} = \overline{y}_{i(j-1)}.
\end{aligned}
$$

This reparameterization is based on the *modified Cholesky decomposition* of the precision matrix Σ^{-1} [195]. An advantage of this reparameterization is that we can shrink the model to certain *structured* covariance matrices [152]. For example, shrinking $\phi_{j\ell}$ toward 0 for all $\ell \neq j-1$ induces shrinkage toward a lag-1 autoregressive model for Y_{ij}.

Similar to the suggestions in Chapter 6, we specify $H_{\mu,\Sigma}$ by setting

$$
\begin{aligned}
\sigma_j^{-2(k)} &\sim \text{Gam}(\xi, \xi/\zeta_j),\\
\mu_j^{(k)} &\sim N(m_j^\mu, \tau_j^2),\\
\phi_{j\ell}^{(k)} &\sim N(m_\ell^\phi, \nu_\ell^2).
\end{aligned}
$$

By taking $\xi \sim \text{Gam}(5,1)$, we encourage homogeneity of $\sigma_j^{(k)}$'s across clusters, while our choice of $\zeta_j \sim \text{Gam}(4, \kappa_j)$ allows us to learn separate variances across times. The hyperparameter κ_j is chosen empirically so that $\sigma_k^{(j)}$'s prior is centered on the MLE of $\text{Var}(Y_j \mid \overline{Y}_{j-1})$ obtained from a multivariate normal fit to the data under MAR.

We specify our prior for the μ's and ϕ's after a pre-processing step in which we center and scale the observed Y_{ij}'s (pooling across all times). We then set $m_j^\mu \sim N(0, \sqrt{10}^2)$, $m_\ell^\phi = 0$ for $\ell < j-1$, and $m_{j-1}^\phi = m^\phi \sim \text{Cauchy}(0, 4/25)$. This centers the model for each cluster on a lag-1 autoregressive model. We use a further hyperprior on the τ_j's and ν_ℓ's described by [99], but the simpler option of setting $\nu, \tau \sim \text{Gam}(0.5, 0.5)$ with $\nu_\ell \equiv \nu$ and $\tau_j \equiv \tau$ is also adequate.

Prior Specification: $g(s \mid \bar{y}_s, \psi)$

As with the response model, we parameterize $g(\cdot)$ sequentially. Abusing notation slightly, we let $g(s \geq j \mid \bar{y}_{j-1}, \psi)$ denote the modeled probability of $S_i \geq j$ given $\bar{Y}_{i,j-1}$ and let $g(s = j \mid s \geq j, \bar{y}_j, \psi)$ denote the probability of $S_i = j$ given $S_i \geq j$ and $\bar{y}_j$. We use a lag-2 selection model

$$\text{logit}\{g(s = j \mid s \geq j, \bar{y}_j, \psi)\} = \zeta_j + \gamma_{j1} y_j + \gamma_{j2} y_{j-1}$$

where $\psi = \{\zeta_j, \gamma_{j1}, \gamma_{j2} : j = 1, \ldots, J-1\}$. All ζ and γ terms are given independent $N(\mu_\zeta, \sigma_\zeta^2)$ and $N(\mu_\gamma, \sigma_\gamma^2)$ distributions. The hyperparameters μ_ζ and μ_γ are given independent Cauchy priors with scales 5 and 2.5, respectively, while σ_ζ^{-2} and σ_γ^{-2} are given $\text{Gam}(1,1)$ priors. At time $j = 1$, we modify this setup by removing the γ_{12} term.

Modeling the missing data with non-future dependence

In the schizophrenia clinical trial there is a substantial amount of missingness in the PANSS scores, and moreover the missing PANSS scores are most likely MNAR; because the reason for dropout was recorded, we know that individuals who dropped out of the trial were more likely to be removed if they had a regression in symptoms. Hence, individuals who dropped out are likely to have higher PANSS scores than individuals who did not, even after accounting for their respective response histories. To deal with MNAR missingness, we will anchor analysis to an MAR modeled that is embedded in an interpretable class of *non-future dependence* (NFD) missingness assumptions.

As a starting point, we invoke the *non-future dependence* assumption [110], which we recall from Section 4.3.2 states that

$$\Pr(S_i = s \mid Y_i = y, \omega) = \Pr(S_i = s \mid \bar{Y}_{i(s+1)} = \bar{y}_{s+1}, \omega), \tag{11.3}$$

for all possible values of the complete data (y_i, s); in words, this states that the probability of dropout at time $S_i = s$ can depend on the *past* values of the outcome $\bar{y}_s$ and the *present* value y_{s+1}, but not on the future values. We complete the NFD specification by taking

$$[Y_{ij} \mid \bar{Y}_{i,j-1}, S_i = j-1, \omega] \stackrel{d}{=} [Y_{ij} + \xi_j \mid \bar{Y}_{i,j-1}, S_i \geq j, \omega].$$

As stated in Section 4.3.2, this leads to the following clinical interpretation of the parameter ξ_j:

> Suppose that individual A and individual B are observed to be identical up to time $j-1$ in terms of their response history and covariates, but that A drops out ($S_i = j-1$) while B does not ($S_i \geq j$). Then, on average, A will have PANSS score Y_{ij} which is ξ_j units higher at time j than B; other than this difference, the distribution of the responses is the same.

For the schizophrenia clinical trial, this amounts to a total of 15 sensitivity parameters: a ξ_j for $j = 2, \ldots, 5$ for each of the three treatment arms. For simplicity, we reduce the number of sensitivity parameters to 3 by taking $\xi_j \equiv \xi$ for each treatment (cf. Chapter 4).

Accommodating different reasons for dropout

In addition to observing the follow-up time S_i, we have auxiliary information in the form of the *reason* for the dropout. A simple way to account for this information is to regard S_i as being censored if an individual drops out for a "non-informative reason" (such as an accidental protocol violation). That is, S_i is redefined to be the time at which an individual would have been removed *for informative reasons*; S_i is then partially observed for some subjects, and partially observed S_i's can be sampled from their full conditional when using a Gibbs sampler to fit the model.

11.3 Posterior Computation

Markov chain Monte Carlo

To sample from the posterior of ω, we use the truncation approximation of the Dirichlet process described in Section 6.3 with $K = 20$. We use JAGS to carry out the sampling. We ran the chain for 55000 iterations, discarding the first 5000 to burn-in, and retained every 10th sample thereafter.

Some diagnostic plots for the MCMC are given in Figure 11.1. Based on these plots, the chains seem to mix well; the trace plots suggest that the treatment effect is close to its stationary distribution, and the autocorrelation plots show that samples from the chain are nearly independent (cf. Section 3.3).

Monte Carlo integration details

The causal effect Δ_a in (11.1) is not expressible as a closed-form function of the parameters ω. Hence, we have to address the computational problem of computing Δ_a from ω *for each sample of* ω *from the posterior*. We will resolve this problem by computing Δ_a using a Monte Carlo implementation of the g-formula. Details on how to implement the Monte Carlo integration for the truncated infinite mixture model given in (11.2) using the location-shift NFD restriction described in Section 4.3.2 can be found in Section 4.3.5.

Removing Monte Carlo Error We use the approach described in Section 3.6 for more efficient g-computation. Figure 11.2 gives a comparison of the posterior distribution of η with a reference normal distribution under the MAR assumption with $N_L = 100N$. Because the posterior distribution of η is virtually indistinguishable from the reference normal distribution, we can have confidence that the accelerated g-computation algorithm will introduce very little approximation error.

The gain from incurring this slight approximation error is a dramatic increase in speed. Table 11.1 shows the computational cost of fitting the observed data model, running the g-computation with $N_{\text{mc}} = 100N$, and running the accelerated

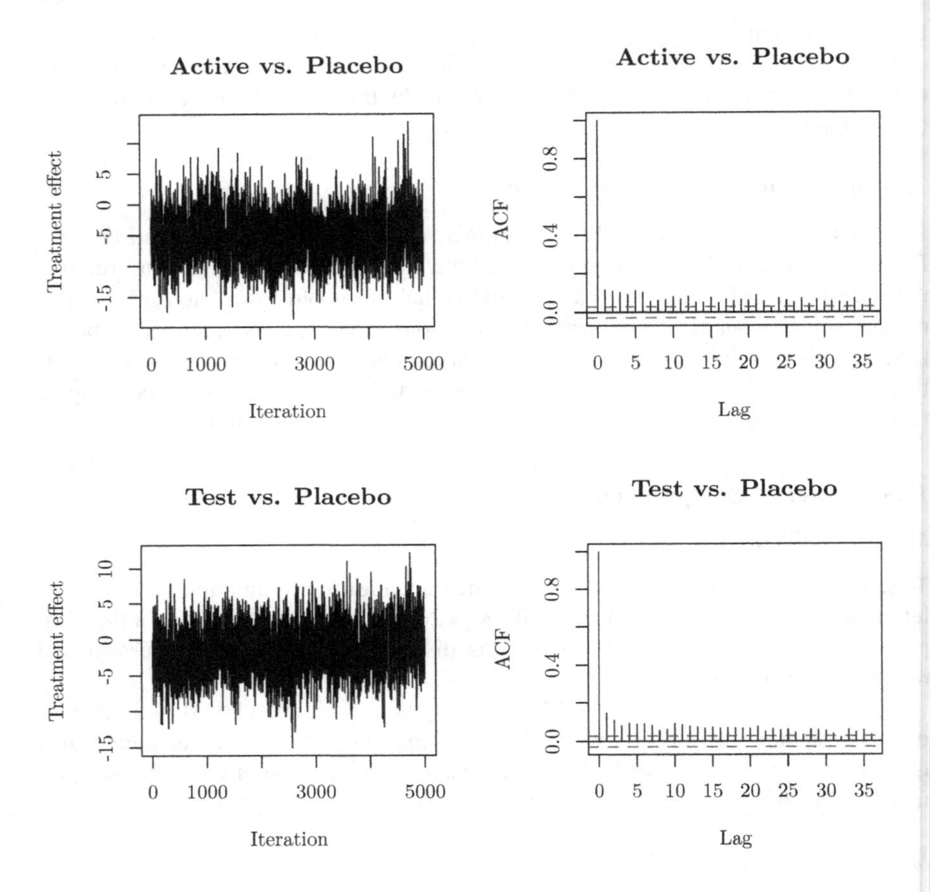

Figure 11.1 *Trace and autocorrelation plots for the treatment effect (relative to the placebo) under MAR.*

Table 11.1 *Computation time for running the MCMC, running the g-computation algorithm, and running the accelerated g-computation algorithm. Here, M denotes the number of values of the sensitivity parameter we wish to consider in a sensitivity analysis.*

Algorithm	MCMC	g-computation	Accelerated
Computational Time (seconds)	142	$68 \times M$	$1 \times M$

g-computation algorithm. To give a proper account of how long each step takes, it is important to remember that the g-computation algorithm must be run *for every value of ξ under consideration*. Hence, while g-computation is feasible for a single ξ (or using a prior), if we wish to compute ξ on a grid of size M it quickly becomes the dominant computational consideration.

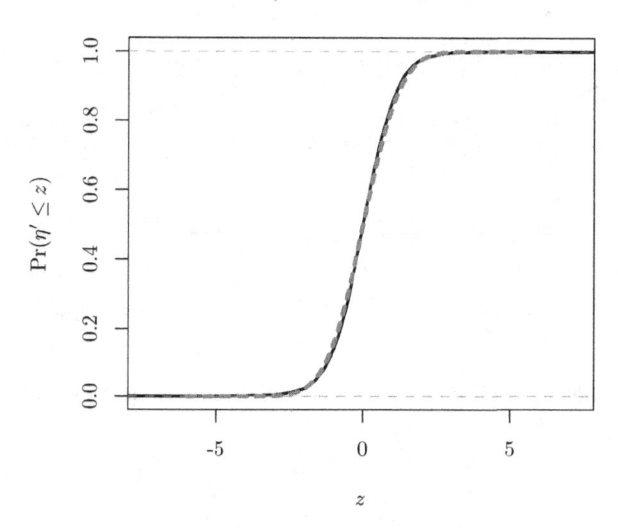

Figure 11.2 *Solid: estimate of the posterior distribution function of η', the parameter η centered and scaled by its posterior mean and standard deviation. Dashed: the associated $N(0,1)$ distribution.*

11.4 Analysis

Assessing model fit

The decision to use a Dirichlet process mixture is motivated by the fact that the observed data is not consistent with a multivariate normal distribution. Figure 11.3 shows several subpopulations of the data extracted from a Dirichlet process mixture model fit to the observed data, which suggests that a single multivariate normal distribution does not give an adequate description of the data. Because of this clustering structure, a formal test easily rejects the multivariate normality assumption across all three treatments. As such, we expect that the failure of the multivariate normality assumption will introduce bias into the estimation of the intention-to-treat effect.

To see how well our BNP model fits, we can verify that inferences obtained for *identified* components of the model are in agreement with model-free inferences. For example, we might expect that the posterior distribution of $E(Y_{i6} \mid S_i = 6, \omega)$ will be in rough agreement with the model-free estimate $\sum_{i:S_i=6} Y_{i6} / \sum_{i:S_i=6} 1$.

Figure 11.4 provides an assessment of how inferences from the model compare with model-free inferences for (i) the mean of Y_{ij} conditional on $S_i \geq j$ and (ii) the probability of missingness after time t_j for each arm of the study. Unbiased estimates and confidence intervals for these quantities can easily be computed without requiring any modeling assumptions. We see from Figure 11.4 that inferences from the model are largely in agreement with model-free inferences.

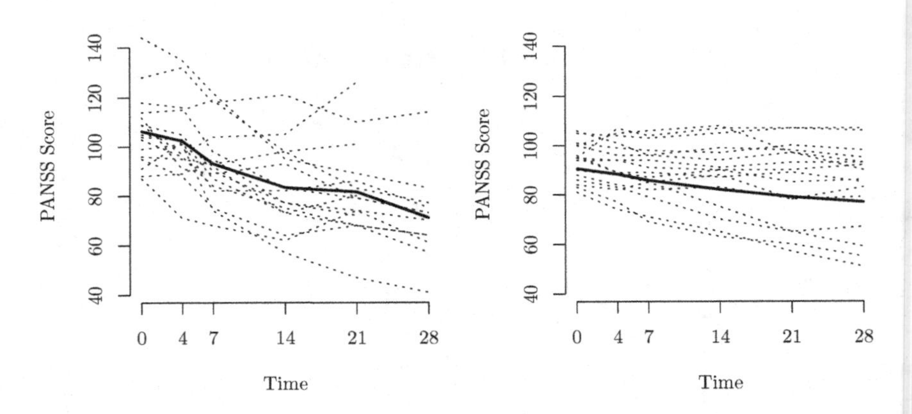

Figure 11.3 *Two subgroups identified by a Gaussian mixture model fit to the data from the schizophrenia clinical trial.*

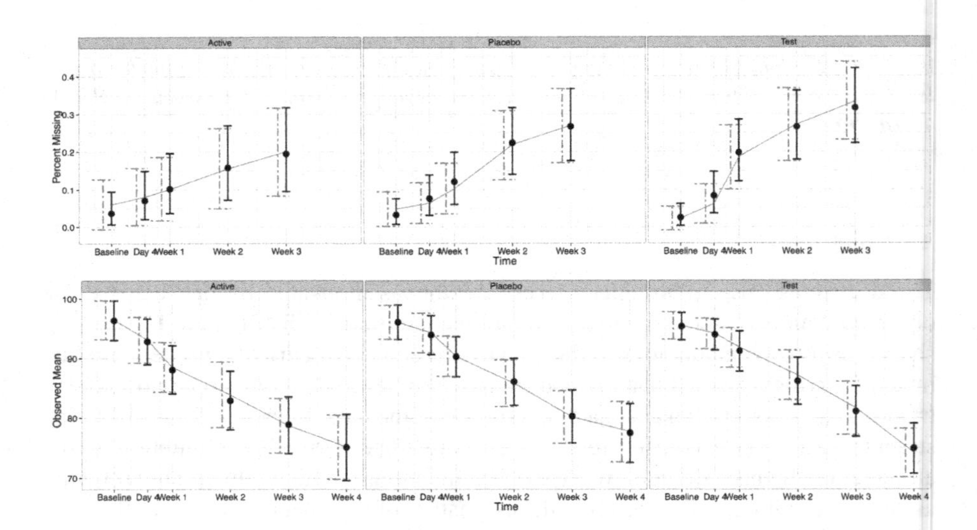

Figure 11.4 *Top: modeled dropout versus observed dropout over time. Bottom: modeled observed means versus empirical observed means. The solid line represents the model-free estimates, while solid dots represent model-based estimates. Dashed error bars represent frequentist 95% model-free confidence intervals, while solid error bars represent 95% posterior credible intervals obtained from the model.*

Primary analysis

We consider three specifications for the sensitivity parameter ξ. The first takes $\xi \equiv 0$ across all treatments, which reduces to the MAR model. The second sets

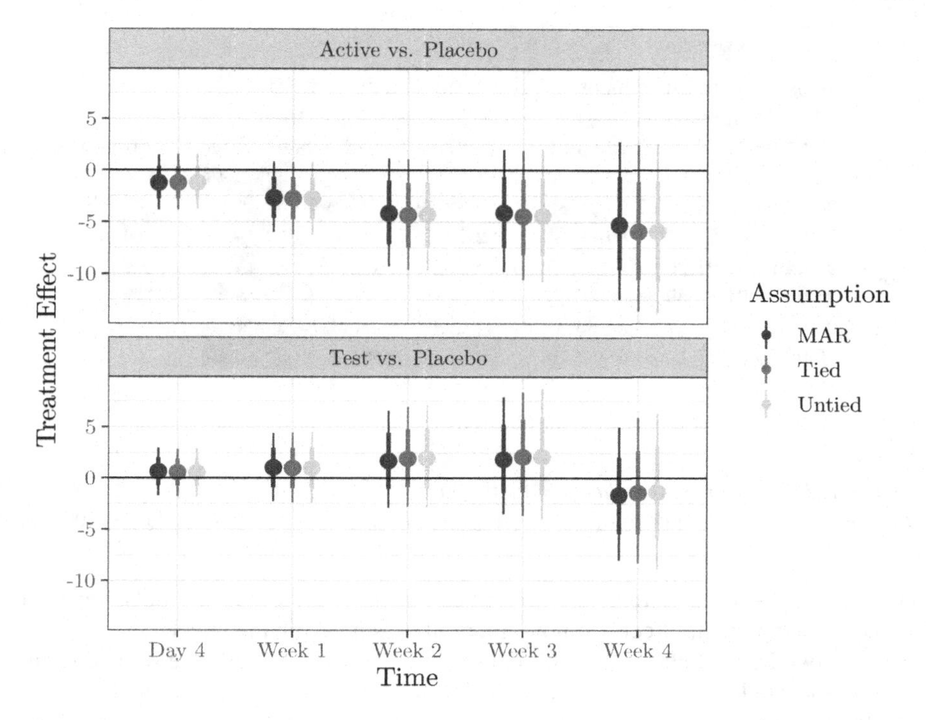

Figure 11.5 *Treatment effects* $\Delta_a = \eta_a - \eta_3$ *with associated credible bands across treatments and times. Points give the posterior mean, short whiskers give the posterior quartiles, and long whiskers give a posterior 95% interval. MAR, Tied, and Untied refer to inferences obtained by assuming MAR, setting* $\xi \sim \text{Uniform}(0,8)$ *with* ξ *shared across treatments, and* $\xi \sim \text{Uniform}(0,8)$ *with different* ξ*'s for each treatment, respectively.*

$\xi \sim \text{Uniform}(0,8)$ independently across treatments, and the third sets $\xi \sim \text{Uniform}(0,8)$ where ξ is *shared* across treatments. The value 8 was chosen by calibrating the effect of non-ignorability to the scale of the data: the estimated conditional standard deviation of Y_{ij} given $\bar{Y}_{i,j-1}$ under the normal model $Y_i \sim N(\mu, \Sigma)$ was roughly 8 across each treatment and time. We believe that, whatever the effect of non-ignorability is, it is unlikely to exceed the noise of the data, and therefore upper-bound $\xi \leq 8$. The lower bound $0 \leq \xi$ is based on the fact that we believe individuals who drop out will have higher PANSS scores than those who do not, and therefore ξ should be positive. This calibration based on an observed data summary follows "Strategy 2" from Chapter 4.

Results for the analysis are summarized in Figure 11.5. We first see that under the baseline assumption of MAR there is little evidence in favor of a treatment effect for either the active or test treatments relative to placebo at any of the times. If we regard MAR as plausible, we might stop at this point and conclude that there is no convincing evidence of a treatment effect. If our baseline belief is that

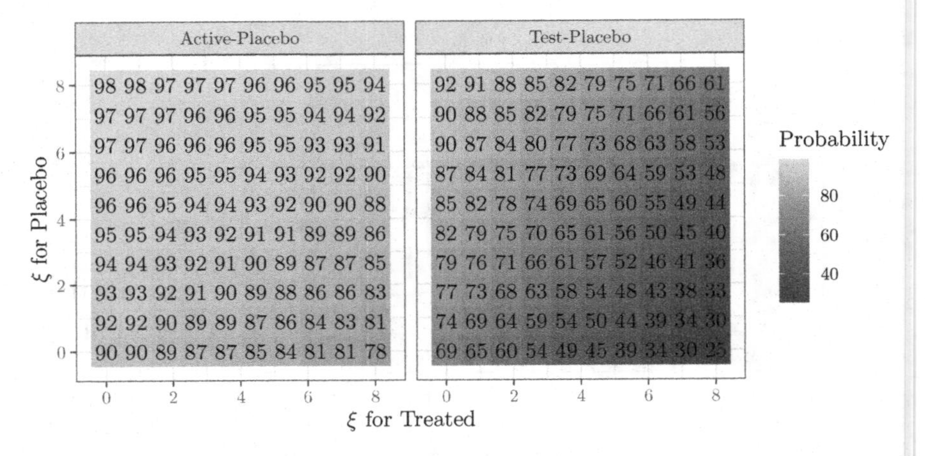

Figure 11.6 *Posterior probability* $p = \Pr(\eta \leq 0 \mid \mathscr{D})$ *using a different value of* ξ *for each treatment. Cells are colored by the size of the probability, and the text of the cell gives* $100 \times p$.

missingness is non-ignorable (as we believe in this case), then the only hope for detecting a treatment effect is if non-ignorability is masking the effect.

The two MNAR alternatives we considered lead to essentially the same conclusion as the MAR analysis, and, in particular, it does not matter if ξ is shared across treatments or not. We conclude that, relative to the two MNAR models we consider, the results of the MAR analysis are robust.

One feature of our MNAR priors for ξ is that the prior is the same regardless of treatment. If we believe that the effect of non-ignorability *varies* by treatment, then the priors we specified are not appropriate. We examine the effect of differences in ξ across treatments in the next section.

Sensitivity analysis

We are interested in two questions regarding the effect of ξ. First, how large must ξ be for our inferences to change, provided that ξ takes the same value for each treatment? Second, to what extent must ξ *differ* between treatments for us to find an effect? This is the *tipping point* strategy introduced in Chapter 4.

For our sensitivity analysis, we consider values of ξ in $[0, 8]$. We vary ξ independently for the active control, test, and placebo arms of the trial. The posterior probability $p_a = \Pr(\Delta_a \leq 0 \mid \mathscr{D})$ is given in Figure 11.6 for values of ξ along a grid. For the comparison between active control and a placebo, in order to detect even a marginally significant effect, we must have ξ quite large if it is not allowed to vary by treatment. We can have moderately significant evidence for an effect if ξ is large for the placebo and small for the active control as well. For the comparison of the test drug with the placebo, essentially no combination of ξ's for the two treatments will give strong evidence that the drug reduces PANSS scores, and in fact taking ξ large now *reduces* the evidence.

11.5 Conclusions

In this chapter, we combined a Dirichlet process mixture model with the non-future dependence assumption to provide a flexible model for the observed data while avoiding the identification of the extrapolation distribution. This necessitated overcoming several challenges: prior specification of the Dirichlet process mixture model, specification of clinically meaningful identifying restrictions, and posterior computation. An advantage of this approach is that it extends fairly easily to settings with non-monotone missingness; we will see this explored in the next case study (Chapter 12).

For the schizophrenia clinical trial, we ultimately conclude that there is little evidence for the effect of the test drug under a baseline assumption of MAR; in particular, the sign of the effect is uncertain. For the active control drug, there is more evidence that the effect is positive, but this is still not conclusive (posterior probability of an improvement in PANSS being $\approx 90\%$). These conclusions are robust to the choice of the sensitivity parameter ξ in the sense that there are no plausible values of ξ that lead to clear evidence in favor of the test drug.

Chapter 12

DPMs for non-monotone missingness

DOI: 10.1201/9780429324222-12

12.1 The Breast Cancer Prevention Trial (BCPT)

The Breast Cancer Prevention Trial (BCPT) was a large clinical trial that aimed to assess the efficacy of a 20 mg/day dose of tamoxifen at reducing the risk of developing breast cancer [196, 98]. The study enrolled 13,338 women who were either over the age of 60 or who had previously had breast cancer. A primary conclusion of this study was that tamoxifen substantially reduces the risk of breast cancer within the study population.

Prior to the BCPT, researchers were concerned that the use of tamoxifen may be associated with increased risk of clinical depression; because the women enrolled in the BCPT were otherwise healthy, if tamoxifen were associated with depression, then quality of life (QOL) considerations would require a more nuanced view of the benefits of tamoxifen as a prophylactic. Consequently, assessing QOL was a key secondary goal of this trial. Depression was measured using the Center for Epidemiologic Studies Depression Scale (CES-D) [197] and was scheduled to be measured at baseline, 3 months from baseline, 6 months from baseline, and subsequently measured every 6 months thereafter. The study was unblinded at the 36th month after an interim analysis found that individuals on the tamoxifen arm had substantially lower risk of developing breast cancer. We focus on the data collected up to this unblinding, with a total of $J = 8$ possible measurements for each individual.

Our goal in this chapter is to determine whether there is evidence of a difference in the probability of depression at the 36th month across the two arms of the study. The absence of a difference would have the important healthcare implication that tamoxifen can be used more freely as a prophylactic without the risk of causing depression in healthy women who may never develop breast cancer.

A large amount of the QOL data was not collected, including more than 30% of the CES-D scores at the 36-month follow-up. Missingness takes the form of both dropout, after which the subject is removed from the study entirely, and intermittent missingness. Roughly 15% of CES-D scores were intermittently missing. A priori, investigators worried that depression itself could be a cause of missingness – if a subject is depressed at the time of an assessment, then they will be less likely to attend scheduled meetings. If this type of behavior contributes to missingness, then the missing data will be MNAR because whether the individual is measured is tied to their current (potentially unobserved) depression status.

Notation and Estimand of interest For the purpose of this analysis, we say that subject i is clinically depressed at time t_j if their CES-D score is 16 or higher. We let $Y_i = (Y_{i1}, \dots, Y_{iJ})$ denote the observed depression status of the ith individual, with $Y_{ij} = 1$ if individual i is depressed at time t_j and $Y_{ij} = 0$ otherwise. As usual, R_{ij} denotes a missingness indicator. We let $A_i = 1$ if subject i is assigned to the treatment arm and $A_i = 0$ if subject i is assigned to the control arm. We focus on estimating the treatment effect on the change-from-baseline

$$\Delta = \eta_1 - \eta_0 \qquad \text{where} \qquad \eta_a = E(Y_{iJ} - Y_{i1} \mid A_i = a).$$

12.2 Methods

Observed data model

Due to MNAR missingness, we are required to define a joint model for R_i and Y_i. One the one hand, we have little reason a priori to expect that the relationship between Y_i and R_i can be adequately captured by a simple parametric model. On the other hand, we do not have sufficient data to estimate the required 2^{2J} contingency table to perform fully nonparametric inference; indeed, most possible combinations of (y_r, r) are not realized by any of the subjects in the study. As with our other examples, we proceed by shrinking a fully nonparametric model toward a simple baseline parametric model (using Bayesian nonparametrics) in order to stabilize the inference. We model the observed data (except for R_{i1}, which is always 1) using a Dirichlet process mixture of multinomial distributions as introduced in Example 6.3.3. Specifically,

$$f(y_r, r \mid \omega) = \sum_{k=1}^{\infty} w_k \left\{ \prod_{j=1}^{J} \gamma_{kj}^{r_j}(1-\gamma_{kj})^{1-r_j} \right\} \left\{ \prod_{j:r_j=1} \beta_{kj}^{y_j}(1-\beta_{kj})^{1-y_j} \right\}, \tag{12.1}$$

with $w \sim \mathrm{GEM}(\alpha)$ truncated to have $K = 50$ classes.

One might be tempted to interpret (12.1) as stating that the data arises from a latent class model, with $Z_i = k$ with probability w_k, $R_{ij} \sim \mathrm{Bernoulli}(\gamma_{kj})$, and $Y_{ij} \sim \mathrm{Bernoulli}(\beta_{kj})$ conditional on $Z_i = k$. There is some subtlety here, however, as we do not use (12.1) as a model for the missing data $Y_{i,-R_i}$. That is, we are not setting $f(y, r \mid \omega) = \sum_{k=1}^{\infty} w_k \left\{ \prod_{j=1}^{J} \gamma_{kj}^{r_j}(1-\gamma_{kj})^{1-r_j} \right\} \left\{ \prod_{j=1}^{J} \beta_{kj}^{y_j}(1-\beta_{kj})^{1-y_j} \right\}$. As we will show, the distribution of $Y_{i,-R_i}$ is determined in a complicated fashion by combining (12.1) and our chosen identifying restrictions.

Prior specification

The model parameter ω consists of the parameters $\{\beta_{kj}, \gamma_{kj} : 1 \le k \le K, 1 \le j \le J\}$ and w. We model $\gamma_{kj} \sim \mathrm{Beta}\{\rho_{\gamma_j} \lambda_{\gamma_j}, (1-\rho_{\gamma_j}), \lambda_{\gamma_j}\}$ and $\beta_{kj} \sim \mathrm{Beta}\{\rho_{\beta_j} \lambda_{\beta_j}, (1-\rho_{\beta_j})\lambda_{\beta_j}\}$. This shrinks the probability of $R_{ij} = 1$ and $Y_{ij} = 1$ toward a common mean ρ_{γ_j} across all latent classes. The weight parameter w is given a $\mathrm{GEM}(\alpha)$ prior with $\alpha = 1$.

The prior has hyperparameters $\{\rho_{\gamma_j}, \lambda_{\gamma_j}, \rho_{\beta_j}, \lambda_{\beta_j} : j = 1, \ldots, J\}$. To motivate our choice of hyperprior, consider the posterior mean of γ_{kj} given ρ_{γ_j} and λ_{γ_j}:

$$E(\gamma_{kj} \mid \mathrm{rest}) = \frac{\rho_{\gamma_j}\lambda_{\gamma_j} + n_k \widehat{\rho}_{jk}}{\lambda_{\gamma_j} + n_k} = \frac{\lambda_{\gamma_j}}{\lambda_{\gamma_j} + n_k}\rho_{\gamma_j} + \left(1 - \frac{\lambda_{\gamma_j}}{\lambda_{\gamma_j} + n_k}\right)\widehat{\rho}_{jk}$$

where $\widehat{\rho}_{jk}$ is the proportion of times $R_{ij} = 1$ in class k and n_k is the number of subjects assigned to class k. We see that the prior shrinks the mean of γ_{kj} toward an overall mean of ρ_{γ_j} with the total weight assigned to the overall mean given by the *shrinkage coefficient* $B_{\gamma_j} = \lambda_{\gamma_j} / (\lambda_{\gamma_j} + n_k)$.

The parameter ρ_{γ_j} is the prior mean of R_{ij} when all of the model parameters are integrated out. Accordingly, we express our prior ignorance about the overall mean of R_{ij} by setting $\rho_{\gamma_j} \stackrel{\text{iid}}{\sim} \text{Uniform}(0,1)$. Similarly, we set $\rho_{\beta_j} \stackrel{\text{iid}}{\sim} \text{Uniform}(0,1)$.

We express uncertainty in λ_{γ_j} in terms of the shrinkage coefficient B_{γ_j}. If we knew n_k, we could specify a *uniform shrinkage prior* [198] by setting $B_{\gamma_j} \sim \text{Uniform}(0,1)$, which induces the prior density $\pi(\lambda) = n_k/(\lambda + n_k)^2$. Unfortunately, we do not know n_k (the number of subjects assigned to cluster k), and moreover the prior itself depends on k. To resolve this, we note that under the GEM(1) prior, we expect there to be roughly ten unique values of Z_i under the Dirichlet process mixture prior [199], which suggests that n_k will be roughly 1000 on average over all non-empty latent classes. Because the number of observations in the smaller latent classes is expected to be substantially smaller than this, we consider independent priors for λ_{γ_j} and λ_{β_j} of the form $\pi_\sigma(\lambda) = \sigma/(\sigma + \lambda)^2$ with a scale $\sigma = 15$. Larger values of σ encourage heavier shrinkage of the β_{kj}'s and γ_{kj}'s toward their overall means. See [98] or [198] for more details on uniform shrinkage priors.

Identifying the treatment effect

We fit models under several different baseline assumptions as we discussed in Chapter 4 for non-monotone missing data. This constitutes a type of sensitivity analysis, where instead of introducing continuous sensitivity parameters we instead vary the baseline (or anchoring) assumption. Let $\bar{O}_{ij}$ denote the observed data (including the R_{ij}'s) up-to-and-including time t_j, and let $\widetilde{O}_{ij}$ denote the data we observe *strictly after* time t_j. We consider the following assumptions, which are also considered by [114] and discussed in Section 4.4.

- *Missing at Random (MAR):* We assume $f(y_{-r} \mid y_r, r, \omega) = f(y_{-r} \mid y_r, \omega)$.
- *Pairwise Missing at Random (PMAR):* We assume $f(y_{-r} \mid R_i = r, y_r, \omega) = f(y_{-r} \mid R_i = \mathbf{1}, y_r, \omega)$ for all r.
- *Sequential Explainability (SE):* We assume $f(y_j \mid \bar{o}_{j-1}, R_{ij} = 0, \omega) = f(y_j \mid \bar{o}_{j-1}, R_{ij} = 1, \omega)$.
- *Nearest Identified Pattern (NIP):* We assume $f(y_j \mid R_i = r, y_r, \omega) = f(y_j \mid R_i = r_j^\star, y_r, \omega)$ where $r_j^\star$ is equal to r but with the jth component fixed at 1.

See Section 4.4 for a motivation of these different assumptions; in the case of inference on Y_{iJ}, it turns out that NIP and SE give identical results, so we will only look at SE during our analysis.

Introducing sensitivity parameters

A benefit of the identifying restrictions above is that it is relatively simple to introduce sensitivity parameters as discussed in Chapter 4. For illustrative purposes, we will consider the *tilted last occasion* (TLO) identifying restriction

$$\text{logit}\{\Pr(Y_{iJ} = 1 \mid R_i = r, y_r, \omega)\} = \xi_r + \text{logit}\{\Pr(Y_{iJ} = 1 \mid R_i = \mathbf{1}, y_r, \omega)\} \quad (12.2)$$

whenever $r_J = 0$. This is quite similar to PMAR with two differences. First, the assumption concerns only the last measurement Y_{iJ} rather than all the missing data

$Y_{i,-r}$. Second, we have added a sensitivity parameter ξ that controls how much more likely depression is for individuals who are missing at t_J relative to those who are observed. It is straight forward to modify (12.2) so that either the SE assumption (4.22) or NIP assumption (4.21) is used for anchoring if this is desired. Rather than eliciting prior beliefs about ξ through (12.2), it may be more natural to elicit beliefs about the missing data mechanism as discussed in Chapter 4. Applying Bayes rule, (12.2) can be phrased in terms of the selection model as

$$\begin{aligned}&\text{Odds}(R_i = r \mid Y_{ir}, Y_{iJ} = 1, R_i \in \{r, \mathbf{1}\}) \\ &\quad = \text{Odds}(R_i = r \mid Y_{ir}, Y_{iJ} = 0, R_i \in \{r, \mathbf{1}\})e^{\xi}.\end{aligned} \tag{12.3}$$

That is, e^{ξ} can be thought of as the multiplicative effect of $Y_{iJ} = 1$ on the odds of being fully observed, given that subject i either has its actual missingness pattern ($R_i = r$) or is fully observed ($R_i = \mathbf{1}$). The equivalence of (12.3) and (12.2) follows from the symmetry properties of the odds ratio [200, and Section 2.2.4].

In our analysis we consider separate values ($\xi_a : a = 0, 1$) of ξ for each of the two treatment arms. We consider values of ξ_a in the interval $(0, 0.8)$, corresponding to the belief that the multiplicative effect of missingness on the odds of depression does not exceed $e^{0.8} = 2.2$. A more precise elicitation of prior beliefs from subject-matter expertise is performed by [98] and follows the third strategy from Chapter 4; our upper bound of 0.8 comes from the fact that none of the experts posited a value of ξ larger than 0.8.

12.3 Posterior Computation

The combination of our flexible modeling framework and our use of interpretable identifying restrictions introduces computational challenges. In total, we need to compute $E(Y_J \mid A = a)$ by averaging over all possible pattern values of (y, r) as $\sum_{y,r} y_J f(y, r \mid \omega)$. There are $2^{2J} = 65536$ terms in this summation. Given the size of the dataset, computing η_a for a fixed ω is not overly burdensome relative to fitting the model. If either (i) J were much larger, (ii) y were continuous, or (iii) we incorporated covariates, however, we would run into serious computational issues. For this reason, we will proceed as through computation of η_a exactly is computationally infeasible.

Fitting the observed data model

We use a blocked Gibbs sampler (see Section 6.3.1) to sample from the posterior distribution of $\omega = (\gamma, \beta, \rho, \lambda)$ and a latent mixture indicator $Z_i \sim \text{Categorical}(w)$. Using the conjugacy of the beta distribution, the full conditional of (γ, β) is given by

$$\begin{aligned}\gamma_{kj} &\stackrel{\text{ind}}{\sim} \text{Beta}\left\{\rho_{\gamma_j}\lambda_{\gamma_j} + \sum_{i:Z_i=k} R_{ij}, \rho_{\gamma_j}(1-\lambda_{\gamma_j}) + \sum_{i:Z_i=k}(1-R_{ij})\right\} \\ \beta_{kj} &\stackrel{\text{ind}}{\sim} \text{Beta}\left\{\rho_{\beta_j}\lambda_{\beta_j} + \sum_{i:Z_i=k} R_{ij}Y_{ij}, \rho_{\beta_j}(1-\lambda_{\beta_j}) + \sum_{i:Z_i=k} R_{ij}(1-Y_{ij})\right\},\end{aligned} \tag{12.4}$$

while full conditional of Z_i is given by

$$\Pr(Z_i = k \mid \text{rest}) = \frac{w_k \prod_j \gamma_{kj}^{R_{ij}} (1-\gamma_{kj})^{1-R_{ij}} \beta_{kj}^{R_{ij}Y_{ij}} (1-\beta_{kj})^{R_{ij}(1-Y_{ij})}}{\sum_{k'} w_{k'} \prod_j \gamma_{k'j}^{R_{ij}} (1-\gamma_{k'j})^{1-R_{ij}} \beta_{k'j}^{R_{ij}Y_{ij}} (1-\beta_{k'j})^{R_{ij}(1-Y_{ij})}}. \tag{12.5}$$

This leaves updating the hyperparameters (ρ, λ), which do not have full conditionals which are easy to sample from. The full conditionals of ρ and λ are (up to a normalizing constant)

$$\begin{aligned}
\lambda_{\beta_j} &\stackrel{\text{ind}}{\sim} \frac{\sigma}{(\sigma+\lambda_{\beta_j})^2} \prod_{k=1}^{K} \text{Beta}\{\beta_{kj} \mid \lambda_{\beta_j}\rho_{\beta_j}, \lambda_{\beta_j}(1-\rho_{\beta_j})\}, \\
\lambda_{\gamma_j} &\stackrel{\text{ind}}{\sim} \frac{\sigma}{(\sigma+\lambda_{\gamma_j})^2} \prod_{k=1}^{K} \text{Beta}\{\gamma_{kj} \mid \lambda_{\gamma_j}\rho_{\gamma_j}, \lambda_{\gamma_j}(1-\rho_{\gamma_j})\}, \\
\rho_{\beta_j} &\stackrel{\text{ind}}{\sim} \prod_{k=1}^{K} \text{Beta}\{\beta_{kj} \mid \lambda_{\beta_j}\rho_{\beta_j}, \lambda_{\beta_j}(1-\rho_{\beta_j})\}, \\
\rho_{\gamma_j} &\stackrel{\text{ind}}{\sim} \prod_{k=1}^{K} \text{Beta}\{\gamma_{kj} \mid \lambda_{\gamma_j}\rho_{\gamma_j}, \lambda_{\gamma_j}(1-\rho_{\gamma_j})\}.
\end{aligned} \tag{12.6}$$

We use slice sampling to update these parameters (see Section 3.3).

Effect computation via Monte Carlo integration

We compute η_a using the accelerated g-computation algorithm (AGC, see Example 3.6.4). We use the Monte Carlo estimator $\widehat{\eta}_a = (n')^{-1} \sum_i (Y_{iJ}^\star - Y_{i1}^\star)$ where $(Y_{i1}^\star, Y_{iJ}^\star)$ are samples of Y_{i1} and Y_{iJ} from the model.

How we simulate $Y_{iJ}^\star$ depends on the particular identifying restriction we apply. In Algorithm 12.1 we give a Monte Carlo integration algorithm for approximating η_a under the SE restriction. The accelerated g-computation approach described in Example 3.6.4 can also be used if it is computationally infeasible to set n' high enough to make the Monte Carlo error negligible; to accommodate this, Algorithm 12.1 also returns an estimate of the "within variance" of $\widehat{\eta}_a$.

Algorithm 12.1 is equally applicable to the other identifying restrictions we consider. The only modification required is to change step (d); for example, if the PMAR restriction (4.23) is used, we instead set

$$p_i = f(Y_{iJ}^\star = 1 \mid Y_{iR_i^\star}, R_i = \mathbf{1}, \omega) = \frac{\sum_k w_k \beta_{kJ} \gamma_{kJ} \prod_{j=1}^{J-1} \gamma_{kj} \beta_{kj}^{R_{ij}^\star Y_{ij}^\star} (1-\beta_{kj})^{R_{ij}^\star (1-Y_{ij}^\star)}}{\sum_k w_k \gamma_{kJ} \prod_{j=1}^{J-1} \gamma_{kj} \beta_{kj}^{R_{ij}^\star Y_{ij}^\star} (1-\beta_{kj})^{R_{ij}^\star (1-Y_{ij}^\star)}}.$$

12.4 Analysis

Assessing goodness of fit

We begin by verifying that the model is producing reasonable results by comparing the output of the model to the inferences one would obtain from a

Algorithm 12.1 Approximating η_a via Monte Carlo Integration Under Sequential Ignorability

1. Update ω by sampling from the full conditionals given in (12.4), (12.5), and (12.6).
2. For $i = 1, \ldots, n'$:
   (a) Sample a latent class $Z_i^\star \sim \text{Categorical}(w)$.
   (b) Sample $R_{ij}^\star \overset{\text{ind}}{\sim} \text{Bernoulli}(\gamma_{Z_i^\star j})$
   (c) For all j with $R_{ij}^\star = 1$, sample $Y_{ij}^\star \overset{\text{ind}}{\sim} \text{Bernoulli}(\beta_{Z_i^\star j})$.
   (d) If $R_{iJ}^\star = 0$, sample $Y_{iJ}^\star \sim \text{Bernoulli}(p_i)$ where $p_i = f(Y_{iJ}^\star = 1 \mid Y_{iR_i^\star}, R_{i,-J}^\star, R_{iJ}^\star = 1, \omega)$ is given by

$$\begin{aligned} p_i &= \frac{f(Y_{iJ}^\star = 1, Y_{iR_i^\star}, R_{i,-J}^\star = 1, R_{i,-J}^\star \mid \omega)}{f(Y_{iR_i^\star}, R_{i,-J}^\star = 1, R_{i,-J}^\star \mid \omega)} \\ &= \frac{\sum_k w_k \beta_{kJ} \gamma_{kJ} \prod_{j=1}^{J-1} \gamma_{kj}^{R_{ij}^\star} (1-\gamma_{kj})^{1-R_{ij}^\star} \beta_{kj}^{R_{ij}^\star Y_{ij}^\star} (1-\beta_{kj})^{R_{ij}^\star(1-Y_{ij}^\star)}}{\sum_k w_k \gamma_{kJ} \prod_{j=1}^{J-1} \gamma_{kj}^{R_{ij}^\star} (1-\gamma_{kj})^{1-R_{ij}^\star} \beta_{kj}^{R_{ij}^\star Y_{ij}^\star} (1-\beta_{kj})^{R_{ij}^\star(1-Y_{ij}^\star)}}. \end{aligned}$$

3. Set $\widehat{\eta}_a = (n')^{-1} \sum_i (Y_{iJ}^\star - Y_{i1}^\star)$ and $s_{\widehat{\eta}_a}^2 = \{n'(n'-1)\}^{-1} \sum_i (Y_{iJ}^\star - Y_{i1}^\star - \widehat{\eta}_a)^2$.
4. Return $(\widehat{\eta}_a, s_{\widehat{\eta}_a}^2)$.

---

fully-nonparametric model for the observed data. We compare samples of

$$E(Y_{ij} \mid R_{ij} = 1, \omega) = \frac{\sum_k w_k \beta_{kj} \gamma_{kj}}{\sum_k w_k \gamma_{kj}} \quad \text{and} \quad E(R_{ij} \mid \omega) = \sum_k w_k \gamma_{kj}$$

to their nonparametrically estimated counterparts

$$\widehat{E}(Y_{ij} \mid R_{ij} = 1) = \frac{\sum_i R_{ij} Y_{ij}}{\sum_i R_{ij}} \quad \text{and} \quad \widehat{E}(R_{ij}) = \frac{\sum_i R_{ij}}{N}.$$

Results for the mean of $Y_{ij} | A_i = a$ are given in Figure 12.1. We see that the infinite product multinomial mixture model faithfully reproduces the nonparametric inferences for the marginal distributions.

Sensitivity analysis

In Figure 12.2 we give inferences under the SE, MAR, PMAR, and TLO assumptions. For the TLO assumption, we specify the prior $\xi \sim \text{Uniform}(0, 0.8)$ with the upper bound specified as noted at end of Section 12.2 and allow ξ to vary independently across treatments.

We see that the PMAR, MAR, and TLO assumptions lead to essentially the same substantive conclusion: there is little evidence that tamoxifen is associated with an

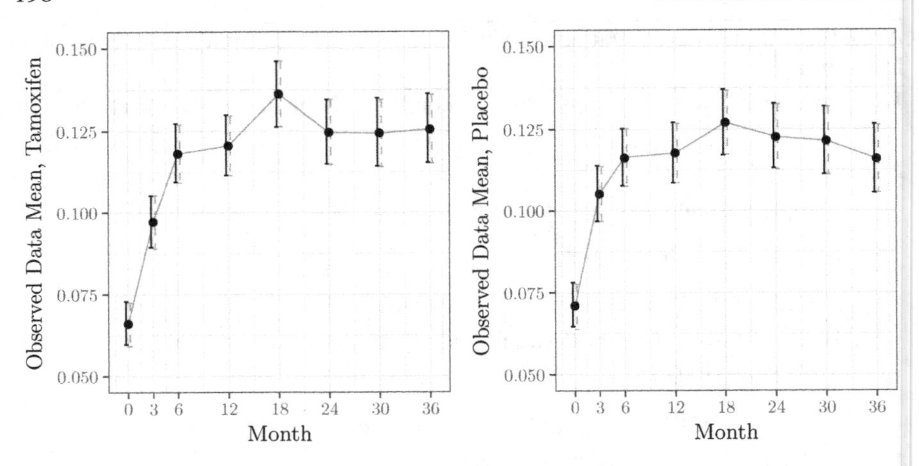

Figure 12.1 *Observed data means over time for the tamoxifen and placebo arms. Dots correspond to the posterior mean using the prior described in Section 12.2. Lines correspond to the empirical mean of the observed data for each time. Solid error bars give 95% credible intervals for the observed data means based on the model; dashed error bars give the usual distribution-free 95% confidence intervals based on asymptotic normality.*

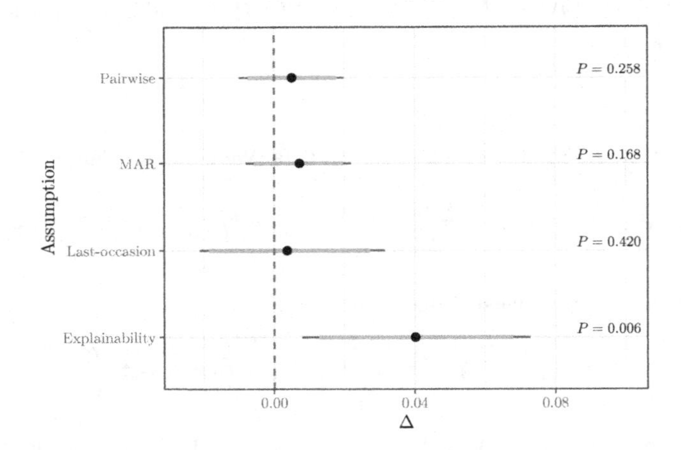

Figure 12.2 *Posterior credible intervals for Δ under different identifying restrictions. Dots give the posterior mean, green bars give two-sided 90% credible intervals, blue bars give two-sided 95% credible intervals. The posterior probability $P = \Pr(\Delta < 0 \mid \mathcal{O})$ (where $\mathcal{O}$ is the observed data) is given for each assumption on the right of the figure.*

increase in depression. On the other hand, SE leads to the opposite conclusion, with the evidence being strong that tamoxifen is associated with a higher risk of depression.

After showing that our conclusions are sensitive to our choice of identifying restriction, we have an obligation to determine which features of the data lead to this difference. Thinking in terms of the distributions we would use to impute missing data, SE differs significantly from the other identifying restrictions in which subjects should be used as part of the *donor pool* for constructing imputations. We can stratify the observed values of Y_{iJ} into three groups:

$$\begin{aligned} G_1 &= \{Y_{iJ} : R_{iJ} = 1 \text{ and no intermittent missingness}\}, \\ G_2 &= \{Y_{iJ} : R_{iJ} = 1 \text{ and some intermittent missingness}\}, \qquad \text{and} \\ G_3 &= \{Y_{iJ} : R_{iJ} = 0\}. \end{aligned}$$

When restricted to G_1, there is no evidence of a relationship between tamoxifen and depression (Fisher's exact test gives a P-value exceeding 0.5). On the other hand, within G_2, there is substantial evidence of a relationship (P-value ≈ 0.002).

Given that G_1 and G_2 lead to substantially different conclusions, we can now explain why SE gives different conclusions: under SE, subjects with Y_{iJ} missing and at least one other missing value will be imputed using information from observations in G_2, whereas under PMAR/TLO *all* missing values of Y_{iJ} will be imputed from G_1. Because there is no evidence for an effect in G_1, this leads to a lack of significance for PMAR/TLO. On the other hand, the strong evidence of an effect in G_2 leads to overall significance when SE is used.

Sensitivity analysis based on stratification

Having determined why the different identifying restrictions lead to different substantive conclusions, we can now perform a simplified sensitivity analysis that makes no modeling assumptions. Define π_g and μ_g to be the probability of being a member of strata G_g and the mean of Y_{iJ} given that i is in G_g, respectively. Then we can write

$$E(Y_J \mid A = a) = \sum_g \mu_g \pi_g. \tag{12.7}$$

The only parameter that is not identified in (12.7) is $\mu_3 = E(Y_{iJ} \mid R_{iJ} = 0, \omega)$. Given that we expect individuals who drop out to be more like individuals with missingness than individuals without missingness, we consider the assumption

$$\mu_3 = \lambda \mu_1 + (1 - \lambda)\mu_2 \quad \text{for some } \lambda < 1/2. \tag{12.8}$$

In reality, we expect that $\lambda < 0$ because we expect individuals who dropped out to have a higher risk of depression than those with intermittent missingness.

Results of a fully nonparametric analysis based on (12.7) and (12.8) are given in Figure 12.3. We see that all reasonable values of λ point to a significant effect of tamoxifen on depression

12.5 Conclusions

By appealing in the final analysis to the identifying restriction defined by (12.8), it may seem as though our final conclusions do not depend on the Bayesian nonparametric approach we have espoused. On the contrary, we were only led to this analysis

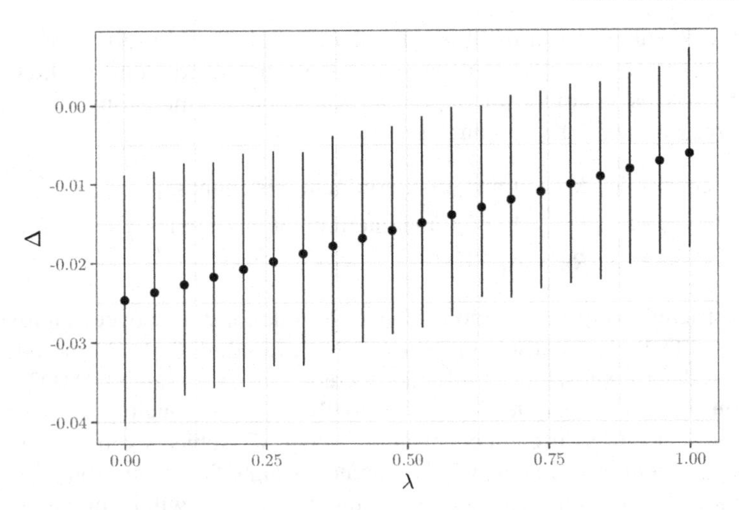

Figure 12.3 *Results of the second sensitivity analysis based on stratifying according to whether there is intermittent missingness or not. Dots correspond to point estimates of* Δ *with error bars giving a 95% confidence interval for* Δ.

by going through a detailed sensitivity analysis, which required both Bayesian nonparametric modeling and the use of a variety of identifying restrictions.

The final analysis is highly appealing for a variety of reasons. First, by reducing to an essentially univariate analysis, it is fully nonparametric, with no shrinkage-induced bias. Second, the restriction (12.8) is more easily justified on a subject-matter basis, stating only that individuals missing at the end of the study are more similar to individuals with intermittent missingness than fully observed individuals.

Chapter 13

Causal mediation using DPMs

DOI: 10.1201/9780429324222-13

13.1 STRIDE Project

STRIDE [201] was a randomized clinical trial to evaluate the effectiveness of three interventions designed to increase physical activity among sedentary adults. Participants ($n = 239$) were randomized to one of the following three arms: (i) telephone-based intervention; (ii) print-based intervention; or (iii) contact control. Given the telephone- and print-based interventions delivered the same theory-based, individually tailored physical activity intervention (but delivered via different channels), we combine telephone-based and print-based interventions into a single intervention group (161 participants) versus the contact control group (78 participants). Secondary outcomes were collected using physical activity questionnaires. These outcomes included behavioral processes (e.g., enlisting social support), cognitive processes (e.g., seeking out information on physical activity), decisional balance (i.e., weighing the pros and cons of adopting physical activity), and self-efficacy and were collected at baseline, 6, and 12 months.

In the original analyses, the secondary outcomes measured at 6 months were the proposed mediators. The primary outcome (Y_i) was the minutes per week of moderate to vigorous intensity physical activity at 12 months. We examine "behavioral processes" at 6 months as a potential mediator (M_i).

Behavioral processes were measured as follows. Subjects were instructed to think back over the past month and to rate the frequency of occurrence of each item in the questionnaire on a 5 point scale ranging from never (1) to always (5). The behavioral processes portion of the questionnaire consisted of 20 items. We use the average score on these questions as the mediator. We include baseline physical activity (measured in minutes per week), baseline behavioral processes and self-efficacy, age, and BMI as potential confounders L_i.

Estimands of interest The estimands of interest are the natural direct and indirect effects (cf. Section 1.6), as well as the total effect,

$$
\begin{aligned}
\text{NDE} &= E[Y_i\{1, M_i(0)\} - Y_i\{0, M_i(0)\}] \\
\text{NIE} &= E[Y_i\{1, M_i(1)\} - Y_i\{1, M_i(0)\}], \qquad \text{and} \\
\text{TE} &= \text{NDE} + \text{NIE},
\end{aligned}
$$

where we recall that $M_i(a)$ is the potential outcome for the mediator under treatment a and $Y_i(a,m)$ is the potential outcome of the response when exposed to treatment a and the mediator value m.

13.2 Methods

Observed data models The observed data models are specified as Dirichlet process mixtures (DPMs) of multivariate normals (cf. Example 6.3.2) for the q-dimensional joint distribution of $O_i = (Y_i, M_i, L_i)$ for each intervention $A_i = 0, 1$,

$$
\begin{aligned}
[Y_i, M_i, L_i \mid A_i = a, \mu_i, \Sigma_i, \theta] &\sim N(\mu_i, \Sigma_i), \\
[\mu_i, \Sigma_i \mid A_i = a, \theta] &\sim F_a, \\
F_a &\sim \text{DP}(\alpha_a, H_a).
\end{aligned} \tag{13.1}
$$

The base distribution H_a is taken to be the conjugate normal-inverse Wishart distribution (NIW), $N(\mu \mid m_a, S_a)\mathscr{W}^{-1}(\Sigma \mid \nu_a, \Psi_a)$. The inverse Wishart distribution is parameterized such that $E(\Sigma) = \Psi_a/(\nu_a - q - 1)$.

Priors We specify a Gam$(1,1)$ prior for α_a. For the hyper priors, we follow the specification suggested in [162] as described in Section 6.3.4.

Identifying assumptions We infer causal effects of mediation based on the sequential ignorability assumption as given in (1.5)-(1.6). Since it is a randomized study, (1.5) will hold and we only need to adjust for confounding between the behavioral processes (mediator) and physical activity minutes per week (outcome) in equation (1.6) under an assumption of no exposure-induced confounding.

Sensitivity analysis We use the strategy described in Chapter 4 (Section 4.2.2).

Computational details To compute NIE and NDE, we sample U sets of (w_a, μ_a, Σ_a) from the posterior distribution of $\pi(w_a, \mu_a, \Sigma_a \mid O)$, where $w_a = (w_{a1}, \ldots, w_{aK})$ is the collection of stick-breaking weights associated with F_a, $\mu_a = (\mu_{a1}, \ldots, \mu_{aK})$, and $\Sigma = (\Sigma_{a1}, \ldots, \Sigma_{aK})$ for $a = 0, 1$. The posterior sampling is done in `rstan` [82] (cf. Section 3.3) using the truncation approximation to the DPM (cf. Section 6.3.1); as `Stan` does not allow for discrete variables, however, we do not use the blocked Gibbs sampler but instead work directly with the density $f(y, m, \ell \mid a) = \sum_{k=1}^{K} w_k N\{(y, m, \ell) \mid \mu_k^*, \Sigma_k^*\}$.

Post-processing steps: g-computation Here we discuss the specifics of the post-MCMC (g-computation) steps (introduced in Example 3.6.2) for computing the posterior distribution of the causal effects when using the truncation approximation to the DPM. For each posterior sample of (w_a, μ_a, Σ_a), we use Algorithm 13.1 to compute the potential outcomes that define the causal effects, $E[Y_i\{a, M_i(a')\}]$ for $(a, a') \in \{(1,1), (0,0), (1,0)\}$.

13.3 Analysis

The outcome was log transformed due to skewness. Fifty-one participants had incomplete information on the primary outcome and/or mediator. For these, we assume ignorable missingness.

Computations

We approximate the infinite mixture of normals using a finite mixture of $K = 20$ normals for both $a = 0, 1$ which induced an error less than 0.01 (cf. Section 6.3).

Results

For the observed data posterior, we collected 10,000 samples from the posterior using `rstan` and used the last 5,000 iterations. We used a thinning interval of 10 to obtain $U = 500$ posterior samples.

We provide summaries of the causal effects in Table 13.1. The NDE was almost twice as large as the NIE, with the former having a credible interval excluding zero

Algorithm 13.1 G-computation for causal mediation with sensitivity parameters

For each posterior sample from the MCMC algorithm:

1. Sample n' sets of L_i from its marginal distribution:
   (a) Sample $A_i^\star \sim \text{Bernoulli}(\pi_a)$ where π_a is the sample proportion of $A_i = 1$.
   (b) Sample $L_i^\star$ from $f(\ell \mid A_i = A_i^\star, \theta) = \sum_{k=1}^K w_{ak} N(\ell \mid \mu_{ak,\ell}, \Sigma_{ak,\ell})$ where $\mu_{ak,\ell}$ consists of the components of μ_{ak} associated with L_i (and similarly for $\Sigma_{ak,\ell}$).
2. For each $L_i^\star = \ell$ sample

$$M_i^\star \sim \sum_{k=1}^K w_{a'k}(\ell) N(m \mid \ell; \gamma_{a'k,\ell}^\top \ell, \sigma_{a'k}^2)$$

   where

$$w_{a'k}(\ell) = \frac{w_{a'k} N(\ell \mid \mu_{a'k,\ell}, \Sigma_{a'k,\ell})}{\sum_{j=1}^K w_{a'j} N(\ell \mid \mu_{a'j,\ell}, \Sigma_{a'j,\ell})},$$

   where the mean and variance parameters from the normal distributions here are the corresponding marginals and conditionals from the multivariate normals in (13.1).
3. Sample (k_0, k_1) from the prior and compute g_0 and g_1 (see Section 4.2.2 for the definition of these quantities and sensitivity parameters).
4. Given the n' sets of $(M_i^\star, L_i^\star)$, compute

$$E[Y_i\{a, M(a')\}] = (n')^{-1} \sum_{j=1}^{n'} E(Y \mid A = a, M^\star = m, L^\star = \ell) + g_a(m, \ell) - g_a'(m, \ell)$$

   where

$$E(Y \mid A = a, M^\star = m, L^\star = \ell) = \sum_{k=1}^K w_{a,k}(m, \ell) \gamma_{a,(m,\ell),k}^\top (m, \ell),$$

$$w_{a,k}(m, \ell) = \frac{w_{a,k} N(m \mid \ell; \gamma_{a,\ell,k}, \sigma_{a,m|\ell,k}^2) N(\ell; \mu_{a,\ell,k}, \sigma_{a,\ell,k}^2)}{\sum_{j'=1}^K w_{a,j'} N(m \mid \ell; \gamma_{a,\ell,j'}, \sigma_{a,m|\ell,j'}^2) N(\ell; \mu_{a,j'\ell}, \sigma_{a,j'\ell}^2)}, \quad (13.2)$$

   and the mean and variance parameters from the normal distributions here are again the corresponding marginals and conditionals from the multivariate normals in (13.1).
5. Compute the posterior sample of the direct and indirect effects using $E[Y_i(a, M(a'))]$ the following formulas:

$$\begin{aligned}
\text{NDE} &= E[Y_i\{1, M_i(0)\} - Y_i\{0, M_i(0)\}], \\
\text{NIE} &= E[Y\{1, M(1)\} - Y\{1, M(0)\}], \\
\text{TE} &= \text{NIE} + \text{NDE}.
\end{aligned}$$

Table 13.1 *Posterior means and credible intervals for NDE, NIE, and TE under the parametric model (Param) and the DPM.*

	NIE	NDE	TE
Param	0.51 (0.14, 0.89)	0.44 (-0.27, 1.10)	0.95 (0.35, 1.54)
DPM	0.42 (-0.02, 0.97)	0.78 (0.19, 1.26)	1.20 (0.95, 1.45)

Table 13.2 *DICs from Parametric and DPM models for the STRIDE data.*

	$A=0$	$A=1$
DIC (DPM)	1424.6	2776.0
DIC (Parametric)	3194.2	6129.9

and the latter just including zero. These results suggest that a little more than a third of the total effect is through the mediator, behavioral processes.

Model comparisons

For comparison, a parametric multivariate normal model was also fit; Specifically, we fit the model

$$[Y_i, M_i, X_i \mid A_i = a, \theta] \sim N(\mu_a, \Sigma_a), \tag{13.3}$$

$$(\mu_a, \Sigma_a) \sim N(\mu_a \mid m_a, S_a)\, \mathscr{W}^{-1}(\Sigma_a \mid \nu_a, \Psi_a). \tag{13.4}$$

where hyperpriors are specified as for the base measure of the DPM.

We compare the fit of the parametric model with the DPM using the DIC from Section 3.4, where the observed data for subject i is $O_i = (Y_i, M_i, A_i, L_i)$. The DPM has lower DICs, indicating the need for the DPM for this data (Table 13.2). The results are quite differenta as well, with an attenuated TE and a much larger NIE (relative to the NDE), with the NIE having a credible interval that excludes zero (Table 13.1).

Assessing model fit

We use graphical posterior predictive checks (cf. Section 3.4) to examine the model's fit. Figures 13.1 and 13.2 display the observed outcome y along with eight replicated datasets y^{pred} from the posterior predictive distributions. Clearly the parametric model fits poorly, while the DPM model fits the data better. In particular, the DPM is better able to pick up the spike at zero than the parametric alternative.

Sensitivity analysis

For the sensitivity analysis to the sequential ignorability assumptions, we varied k_0 from 0 to 10 and k_1 from k_0 to 20. Recall that k_j is the percent of the total variability

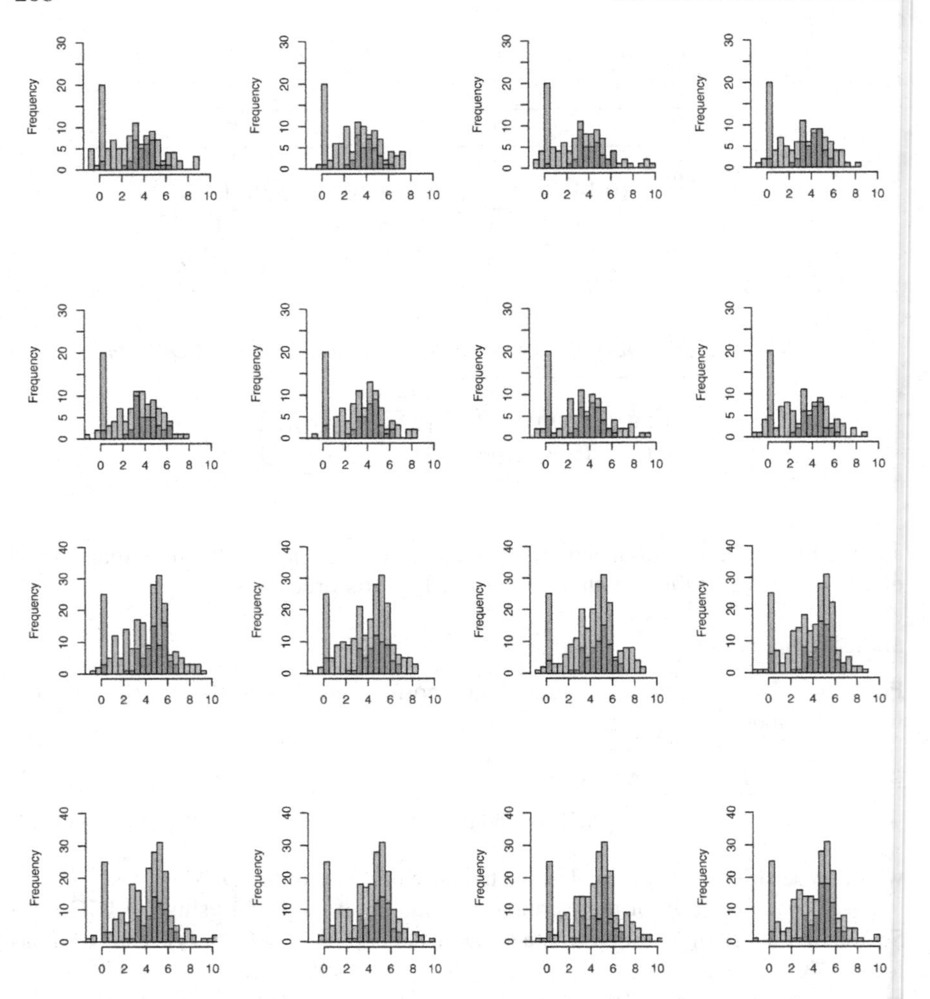

Figure 13.1 *Lighter shading (data) vs. Darker shading (predictive distribution). The first two rows : eight replications of the outcome under* $A_i = 0$ *from the posterior predictive distribution of the parametric model. The last two rows : eight replications of the outcome under* $A_i = 1$ *from the posterior predictive distribution of the parametric model.*

not explained by L; see Section 4.2.2 for details. The above specification is chosen so that $M(1)$ explains no more than 20% of the remaining variability in Y (after accounting for L) and $M(0)$ no more than 10% (after accounting for L). Within these ranges, we considered the following pairs of values for k_0 and k_1:

$$(k_0, k_1) \in \{(5,5),(5,10),(5,15),(10,10),(10,15),(15,20)\}.$$

We also considered the pair of non-degenerate priors, $k_0 \sim \text{Uniform}(0,10)$ and $[k_1 \mid k_0] \sim \text{Uniform}(k_0, 20)$. The estimands are again computed using Algorithm 13.1.

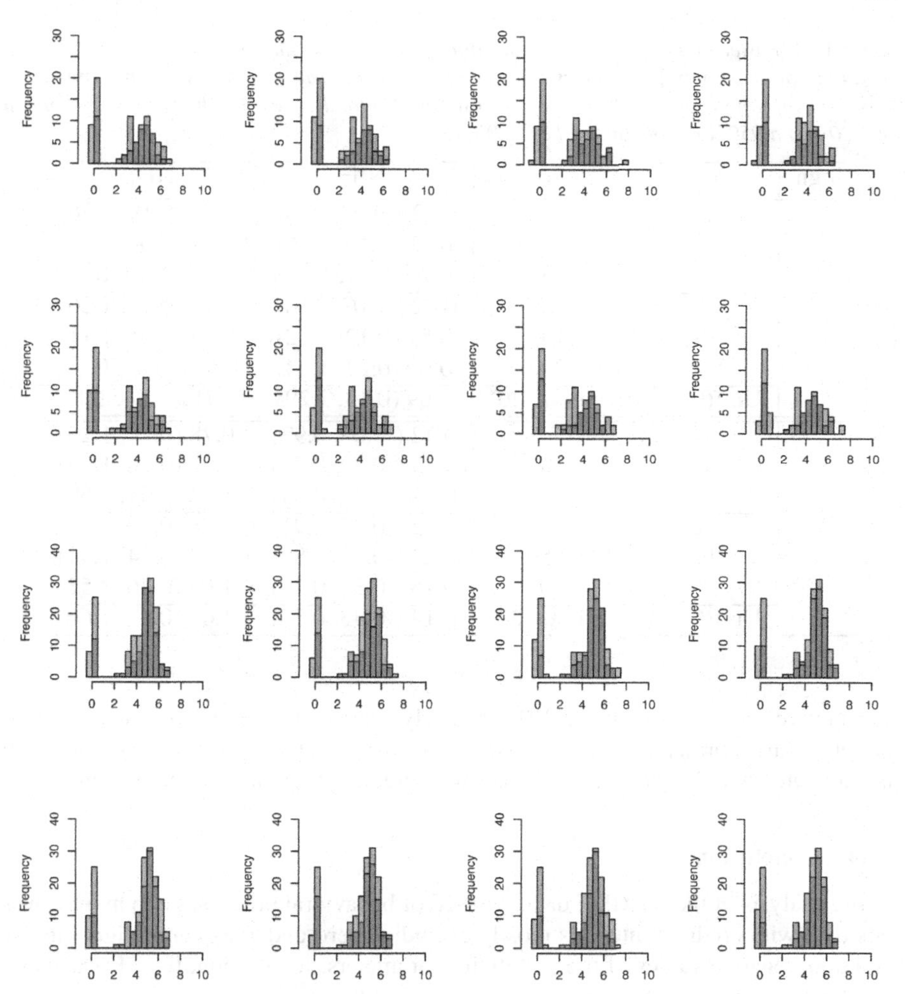

Figure 13.2 *Lighter shading (data) vs. Darker shading (predictive distribution). The first two rows : eight replications of the outcome under* $A_i = 0$ *from the posterior predictive distribution of the DPM model. The last two rows : eight replications of the outcome under* $A_i = 1$ *from the posterior predictive distribution of the DPM model.*

Table 13.3 contains the results under the different scenarios. The results indicate that as k_0 and k_1 get closer (i.e., $M_i(0)$ explains as much of the total variance of the outcome as $M_i(1)$) the NIE and the NDE approach their values under the sequential ignorability assumptions (cf. Table 13.1). On the contrary, if $M_i(1)$ explains much more of the total variance than $M_i(0)$, then the effects differ greatly from their values under the sequential ignorability assumption. Recall that sequential ignorability implies $k_0 = k_1 = 0$ and that the TEs are constant as functions of k_0 and k_1.

Table 13.3 also contains the results under the uniform distributions, Unif$(0,10)$ and Unif$(0,20)$, on k_0 and k_1, respectively, with the constraint that $k_0 \leq k_1$. The

Table 13.3 *Posterior means and credible intervals for NDE and NIE under different combinations of sensitivity parameters* (k_0, k_1) *and with uniform priors:* $k_0 \sim Uniform(0,10)$, $k_1|k_0 \sim Uniform(k_0, 20)$. *Note that for each setting the total effect (TE) had a posterior mean of 1.20 and a 95% credible interval of* $(0.95, 1.45)$.

Sign	k_0	k_1	NIE	NDE
+	5	5	0.42 (-0.03, 0.97)	0.78 (0.19, 1.26)
		10	0.61 (0.16, 1.16)	0.59 (0.01, 1.07)
		15	0.76 (0.31, 1.31)	0.45 (-0.13, 0.92)
	10	10	0.42 (-0.03, 0.97)	0.78 (0.19, 1.26)
		15	0.56 (0.12, 1.12)	0.64 (0.05, 1.11)
		20	0.69 (0.24, 1.24)	0.64 (-0.07, 0.99)
	Unif(0,10)	Unif(k_0, 20)	0.69 (0.13, 1.29)	0.51 (-0.24, 1.12)
-	5	5	0.42 (-0.03, 0.97)	0.78 (0.19, 1.26)
		10	0.23 (-0.22, 0.78)	0.97 (0.38, 1.45)
		15	0.08 (-0.37, 0.64)	1.12 (0.52, 1.60)
	10	10	0.42 (-0.03, 0.97)	0.78 (0.19, 1.26)
		15	0.27 (-0.18, 0.82)	0.93 (0.34, 1.41)
		20	0.15 (-0.30, 0.70)	1.05 (0.46, 1.53)
	Unif(0,10)	Unif(k_0, 20)	0.14 (-0.43, 0.77)	1.06 (0.43, 1.66)

substantive conclusions on the NIE are fairly robust to the positive sensitivity parameters (and corresponding prior), but for negative sensitivity parameters the NIE is attenuated and the NDE grows larger with credible intervals excluding zero.

13.4 Conclusions

In the analysis of the STRIDE data, the NIE of behavioral processes at 6 months was positive, with credible intervals mostly excluding zero under sequential ignorability and non-negative values of the sensitivity parameters. In addition, the TE was positive (indicating an increase in the minutes per week of moderate to vigorous physical activity for the individually tailored intervention(s) versus the control arm) with a credible interval excluding zero; by construction, the TE is invariant to the sensitivity parameters. We also demonstrated that the DPM fits the data much better than a parametric equivalent by computing the DIC and examining posterior predictive checks.

Note that the sequential ignorability assumption must hold without exposure-induced outcome and mediator confounders [105]. If this is not the case, within this framework, we can incorporate an additional mediator variable (the mediator-outcome confounder) into the observed data models [202] and use the assumptions introduced in [44].

An alternative to sequential ignorability is the mediator induction equivalence assumption. Details (and corresponding sensitivity analysis) can be found in [45].

The spike at zero in the observed outcomes for the minutes per week of moderate to vigorous intensity exercise was captured well by the DPM (see Figures 13.1 and 13.2); however, the DPM of normals for the outcome could have been replaced by a zero-inflated version similar to the specification in [203].

The approach here can be extended to mixed-type confounders using a Dirichlet process mixture of generalized linear models as described in Section 6.3.3, or to an EDPM [166, 202].

Chapter 14

Causal mediation using BART

DOI: 10.1201/9780429324222-14

14.1 Motivation

In this chapter we revisit the Medical Expenditure Panel Survey (MEPS) dataset described in Section 5.1.2, where we concluded that there was limited evidence about the direction of the effect of smoking on total medical expenditures in a cohort of adult women. This is somewhat counterintuitive, given that it is well known at this point that smoking is causally linked with many adverse health outcomes, and logically one expects that individuals with poor health will have higher medical expenditures.

Note, however, that the above intuition regarding *why* smoking should increase medical expenditures is actually an argument from mediation – if not for the negative effect of smoking on health, we might not expect there to be any effect of smoking on healthcare expenditures. Accordingly, we will find that when we perform a mediation analysis the effect of smoking on medical expenditures comes into much clearer focus. Specifically, a mediation analysis will reveal that the majority of the uncertainty in the average treatment effect in Figure 5.9 is attributable to uncertainty in the direct effect of smoking on medical expenditures, i.e., the effect *not attributable to the effect of smoking on health.* By contrast, the indirect effect, i.e., the effect *attributable to the effect of smoking on health*, is well identified from the data, with smoking causing bad health outcomes that in turn cause higher medical expenditures, as expected.

14.2 Methods

Notation and causal estimands

Following Section 1.6, we let A_i denote the treatment received by individual i ($A_i = 1$ if individual i is a smoker, $A_i = 0$ otherwise) and we let M_i denote the observed value of the mediator, which in this case we take to be the ith individual's self-perceived health status measured on a scale from 1 to 5 ("Poor" to "Excellent"). For simplicity, we ignore possible measurement error in this instrument and assume self-perceived health status to be a reliable omnibus measure of overall health for an individual. We let Y_i denote the logarithm of the total medical expenditure of individual i within the population of individuals who incurred at least some medical expenditure over the year. The potential outcome $M_i(a)$ represents the health status of individual i if their smoking status was fixed at a, while the potential outcome $Y_i(a,m)$ represents the logarithm of the total medical expenditure individual i would have incurred if subject i had their smoking status fixed at a and their health status fixed at m. Our interest will be in the population average natural direct and indirect effects,

$$NDE_a = E[Y_i\{1,M_i(a)\} - Y_i\{0,M_i(a)\}] \qquad \text{and}$$
$$NIE_a = E[Y_i\{a,M_i(1)\} - Y_i\{a,M_i(0)\}].$$

Assumptions

As a starting point, we will make use of the *sequential ignorability* assumption described in Section 1.6. To recap, sequential ignorability states the following:

1. *Ignorability of the treatment assignment mechanism:* $[\{Y_i(a',m), M_i(a)\} \perp\!\!\!\perp A_i \mid L_i = \ell]$ for all $a, a' \in \{0,1\}$, m, and ℓ.
2. *No unmeasured confounders of the mediator and outcome:* $[Y_i(a',m) \perp\!\!\!\perp M_i(a) \mid A_i = a, L_i = \ell]$ for all $a, a' \in \{0,1\}$, m, and ℓ.
3. *Overlap:* $f_\theta(M_i = m \mid A_i = a, L_i = \ell) > 0$ and $f_\theta(A_i = a \mid L_i = \ell) > 0$ for all $a \in \{0,1\}$, m, and ℓ.

As shown in Section 1.6, under these assumptions the natural direct and indirect mediation effects are identified via the formula

$$E_\theta[Y_i\{a', M_i(a)\}] = \iiint y\, f_\theta(Y_i = y \mid M_i = m, A_i = a', L_i = \ell) \\ \times f_\theta(M_i = m \mid A_i = a, L_i = \ell)\, f_\theta(L_i = \ell)\, dy\, dm\, d\ell,$$

which depends on the potential outcomes only through the distribution of the observed data (Y_i, M_i, A_i, L_i).

To proceed, we need to specify models for the observed data. Before doing so, in the following section we discuss some precautions we need to take when specifying BART models in causal mediation analysis.

14.3 Regularization-Induced Confounding and the Prior on Selection Bias

The same issues of regularization-induced confounding that arose in the setting of causal inference for the average treatment effect in an observational study in Section 5.4.2 also arise in the setting of mediation analysis. To make this point, it is useful to consider what occurs in the setting of a high-dimensional linear regression model of the form

$$\begin{aligned} Y_i(a,m) &= \beta_L^\top L_i + \beta_a a + \beta_m m + \varepsilon_i(a,m), \\ M_i(a) &= \alpha_L^\top L_i + \alpha_a a + \nu_i(a), \\ L_i &\sim N(0, \Sigma), \\ A_i &\sim \text{Bernoulli}(q). \end{aligned} \tag{14.1}$$

The assumptions that $L_i \sim N(0,\Sigma)$ and that A_i is randomized independently of L_i are made to allow for the derivation of closed-form expressions and are not essential to the point we make below. Note that BART models have a similar structure to (14.1), the only difference being that we replace the confounders L_i with random tree-based basis functions that are learned from the data. Under the model (14.1), the *NIE* and *NDE* have simple closed-form expressions as products of the regression coefficients:

$$NDE = \beta_a \qquad \text{and} \qquad NIE = \alpha_a \beta_m.$$

A natural approach to performing Bayesian inference with the model (14.1) when L_i is high dimensional is to specify the priors $\beta_L \sim N(0, \sigma_\beta^2 \mathbf{I}/p)$ and $\alpha_L \sim N(0, \sigma_\alpha^2 \mathbf{I}/p)$, where p is the number of predictors; normalization by p ensures that the total signal in the predictors is, on average, $\|\beta_L\|^2 \approx \sigma_\beta^2$ regardless of the dimension. This takes care of the high-dimensional part of the model, and we complete the model by specifying flat priors on the parameters of interest $(\beta_a, \beta_m, \alpha_a)$.

When L_i is high dimensional, we begin to run into problems and, in particular, the priors we have described above are highly informative about the amount of confounding bias that can occur. One issue is that, when p is large, the above prior encodes the belief that controlling for L_i is unnecessary to get approximately valid inference. To see this, suppose that (14.1) holds and we estimate NDE_a and NIE_a using the following misspecified models:

$$\begin{aligned} Y_i(a,m) &= \beta_a\, a + \beta_m\, m + \varepsilon_i(a,m), \\ M_i(a) &= \alpha_a\, a + \nu_i(a). \end{aligned} \tag{14.2}$$

The following proposition shows that basing inference on (14.2) when in fact (14.1) holds leads to biased estimates of the mediation effects.

Proposition 14.3.1. *Under the model* (14.1), *we have*

$$\begin{aligned} E(Y_i \mid M_i, A_i) &= (\beta_a - \rho\, \alpha_a) A_i + (\beta_m + \rho) M_i \qquad \text{and} \\ E(M_i \mid A_i) &= \alpha_a A_i. \end{aligned}$$

Hence, the product-of-coefficients estimate based on fitting (14.2) *via least squares converges to* $\widetilde{\zeta} = \beta_a - \rho\, \alpha_a$ *and* $\widetilde{\delta} = \alpha_a \beta_m + \rho\, \alpha_a$, *where*

$$\rho = \frac{\beta_L^\top \Sigma \alpha_L}{\alpha_L^\top \Sigma \alpha_L + \sigma_m^2}.$$

Proof. First, the identity $E(M_i \mid A_i) = \alpha_a A_i$ follows from the fact that under (14.1) we have $A_i \perp\!\!\!\perp L_i$ and $E(L_i) = 0$. The identity for $E(Y_i \mid M_i, A_i)$ follows from the fact that

$$[(L_i, M_i)^\top \mid A_i] \sim N\left\{ \begin{pmatrix} 0 \\ \alpha_a A_i \end{pmatrix}, \begin{pmatrix} \Sigma & \Sigma\alpha_L \\ \alpha_L^\top \Sigma & \alpha_L^\top \Sigma \alpha_L + \sigma_m^2 \end{pmatrix} \right\},$$

which can be derived by observing that L_i and M_i are jointly normal under (14.1) and then computing the relevant moments. From this expression, standard properties of the multivariate normal distribution give $E(L_i \mid M_i, A_i) = \frac{\Sigma\alpha_L}{\alpha_L^\top \Sigma \alpha_L + \sigma_m^2}(M_i - \alpha_a A_i)$ so that $\beta_L^\top E(L_i \mid M_i, A_i) = \rho(M_i - \alpha_a A_i)$. Hence, under (14.1),

$$E(Y_i \mid M_i, A_i) = \beta_L^\top E(L_i \mid M_i, A_i) + \beta_a A_i + \beta_m M_i = (\beta_a - \rho\alpha_a)A_i + (\beta_m + \rho)M_i.$$

By the consistency of least squares, the least squares estimates of $(\alpha_a, \beta_a, \beta_m)$ based on (14.2) will converge to the regression coefficients defined by the linear models for $E(Y_i \mid M_i, A_i)$ and $E(M_i \mid A_i)$, i.e., to $(\alpha_a, \beta_a - \rho\alpha_a, \beta_m + \rho)$. Forming the product-of-coefficients from this gives $\beta_a - \rho\, \alpha_a$ and $\alpha_a(\beta_m + \rho)$, completing the proof. □

With the above setup, we can now ask the following question: *how biased does our prior that takes* $\beta_L \sim N(0, \sigma_\beta^2 \mathbf{I}/p)$ *and* $\alpha_L \sim N(0, \sigma_\alpha^2 \mathbf{I}/p)$ *expect the misspecified least squares estimator based on* (14.2) *to be?* Note that the misspecified least

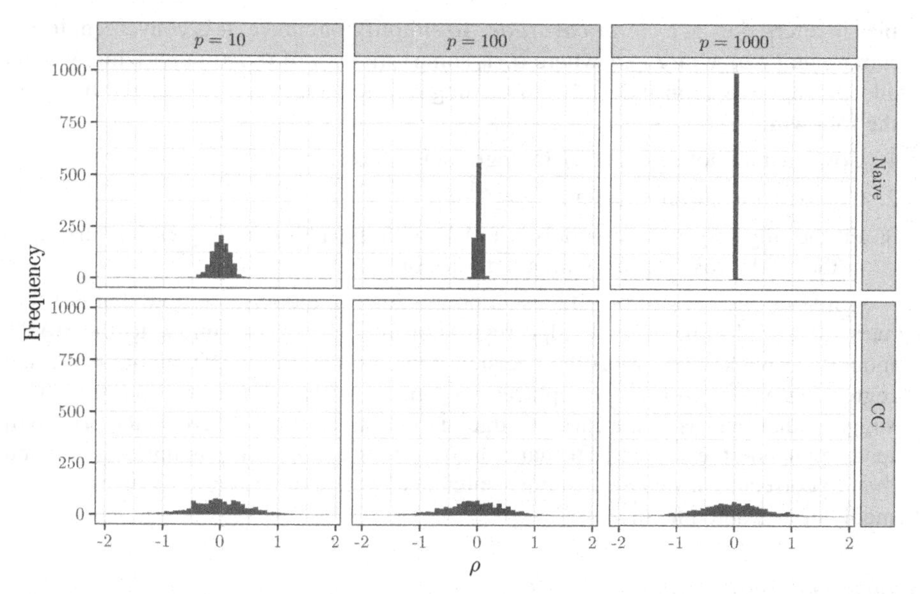

Figure 14.1 *Prior distribution of the selection bias parameter* ρ *under the "naive" prior (Naive), which places independent normal priors for the regression coefficients, and the model that includes the clever covariate (CC)* $\widehat{\alpha}_L^\top L_i$ *as a predictor.*

squares estimator is unbiased whenever $\rho = 0$. Figure 14.1 displays the prior distribution of ρ for various values of p when $\Sigma = \mathbf{I}$. We first see that the above prior specification, which is labeled "Naive," results in the prior distribution for ρ becoming increasingly concentrated near 0 for large p. Hence, the answer to the question is that our prior believes that controlling for confounding is unnecessary when p is large.

This places us in a paradoxical situation: by controlling for more and more confounders, our prior is encoding the belief that we might as well not control for any confounders at all! Fortunately, we can easily fix this problem by specifying the prior in such a way that ρ does not concentrate on 0. To do this, we simply shift the prior for β_L by taking $\beta_L \sim N(\eta\, \alpha_L, \sigma^2\mathbf{I}/p)$. The resulting prior on ρ when $\eta \sim N(0,1)$ is displayed in Figure 14.1 under the heading "CC." We see that, regardless of the dimension of L_i, the prior expects that controlling for confounding is necessary.

Another view of this modified prior is that we are adding a new covariate $L_i^\top \alpha_L$ into the model with regression coefficient η. To see this, note that if $\beta_L \sim N(\eta\, \alpha_L, \sigma_\beta^2\mathbf{I}/p)$ then we can write

$$\beta_L^\top L_i = L_i^\top \eta\, \alpha_L + L_i^\top b = \eta\,(L_i^\top \alpha_L) + L_i^\top b,$$

where $b \sim N(0, \sigma_\beta^2\mathbf{I}/p)$. In the parlance of the targeted maximum likelihood literature [4], a special covariate introduced for the sake of correcting for confounding bias is referred to as a *clever covariate*. To simplify inference, it is convenient to approximate $L_i^\top \alpha_L$ with $L_i^\top \widehat{\alpha}_L$ where $\widehat{\alpha}_L$ is some estimator of α_L. Summarizing, we can address regularization-induced confounding for the ridge regression model by doing the following:

1. Construct a pilot estimate $\widehat{\alpha}_L$ by regressing M_i on A_i and L_i.
2. Compute the clever covariate $C_i = L_i^\top \widehat{\alpha}_L$.
3. Fit the ridge regression models in (14.1) with the additional regression term $\eta\, C_i$ in the model for $Y_i(a,m)$, with a flat prior on η.

Linking this back to our Bayesian nonparametric models, we note that the scenario of a high-dimensional ridge regression model is very similar to the BART model – if we knew a priori the decision trees, in fact, the BART model *is* a ridge regression model with $L_i^\top \beta_L$ replaced by a random basis function expansion. This suggests that, just as in the case of ridge regression, an unmodified BART prior will tend to encode the prior belief that controlling for confounding is unnecessary and that, to correct this, we can introduce suitable clever covariates as covariates into the mediator and outcome models.

Bayesian Causal Mediation Forests

The *Bayesian causal mediation forest* (BCMF) model introduced by [204] starts by positing the following models for the observed data:

$$
\begin{aligned}
[Y_i \mid M_i, A_i, L_i] &\sim N\{r_y(M_i, A_i, L_i), \sigma_y^2\},\\
[M_i \mid A_i, L_i] &\sim N\{r_m(A_i, L_i), \sigma_m^2\},\\
\Pr_\theta(L_i = \ell_k) &= \varpi_k,
\end{aligned}
$$

where $\ell_1, \ldots, \ell_n$ ranges over the observed values of the covariates. We then specify a Bayesian bootstrap prior $\varpi \sim \text{Dirichlet}(0, \ldots, 0)$ and use BART priors for $r_y(\cdot)$ and $r_m(\cdot)$.

We make three modifications to the standard BART prior for $r_y(\cdot)$ and $r_m(\cdot)$ to control for the regularization-induced confounding described in Section 14.3.

1. To control for standard regularization-induced confounding that occurs in observational studies (see Section 5.4.2), we include an estimate of the propensity score $\widehat{e}_i \approx e(L_i)$ as a predictor in both $r_y(m,a,\ell)$ and $r_m(a,\ell)$.
2. To control for the regularization-induced confounding on the path from the mediator to the outcome, we include an estimate of the regression functions $\widehat{r}_{i0} \approx r_m(0, L_i)$ and $\widehat{r}_{i1} \approx r_m(1, L_i)$ as predictors in the outcome model.
3. Rather than treating the predictor A_i in the same fashion as L_i, we instead stratify by A_i, i.e., we specify separate BART models for $r_y(m,0,\ell)$, $r_y(m,1,\ell)$, $r_m(0,\ell)$, and $r_m(1,\ell)$. This ensures that the model makes use of A_i as a predictor.

We note that the clever covariates $(\widehat{e}_i, \widehat{r}_{i0}, \widehat{r}_{i1})$ are estimated *prior to fitting the models* and that it is not strictly required that they be estimated using the same models that we use for inference; rather, the clever covariates can be estimated using essentially any parametric or nonparametric estimators.

Other Notes on Prior Specification We use the default prior specification of [125] for each of the BART models under consideration here. We also use the variable selection prior described in Section 5.3.4, which used a sparsity-inducing Dirichlet prior on the probability of splitting on a given covariate.

Computation via MCMC

It is straight forward to fit a BCMF using existing softwares, such as the `BART`, `dbarts`, or `bartMachine` package. Under sequential ignorability, we can derive the following expression for the counterfactual mean:

$$\begin{aligned} & E[Y_i\{a', M_i(a)\}] \\ &= \int r_y(m, a', \ell)\, f_\theta(M_i = m \mid A_i = a, L_i = \ell)\, f_\theta(L_i = \ell)\; d\ell\; dm \\ &= \sum_i \varpi_i \int r_y(m, a', L_i)\, \phi\left\{m \mid r_m(a, L_i), \sigma_m^2\right\}\; dm. \end{aligned} \tag{14.3}$$

This expression can be computed either using Monte Carlo integration or, alternatively, by numeric integration. We use the `R` package `BartMediate`, which fits BCMFs using Monte Carlo integration with the AGC algorithm (see Example 3.6.4) to evaluate (14.3).

14.4 Results

We fit a Bayesian causal mediation forest to the MEPS data after controlling for the following possible confounders: age (male or female), body mass index, an ordinal measure of education level, income in dollars, geographical region (northeast, west, south, or midwest), marital status (married, divorced, single, separated, or widowed), race (White, Black, Pacific Islander, Indigenous, Black, or Asian), and frequency of wearing a seat belt (always, almost always, sometimes, never, seldom, or no car). The inclusion of seat belt usage as a confounder is intended to capture risk-taking behavior, with individuals who are less likely to wear a seat belt being perhaps more likely to smoke and less likely to seek medical care.

We first constructed the clever covariates $(\widehat{e}_i, \widehat{r}_{i0}, \widehat{r}_{i1})$ by fitting the models

$$A_i \sim \text{Bernoulli}[\Phi\{r_a(L_i)\}], \qquad \text{and} \qquad M_i \sim N\{r_m(A_i, L_i), \sigma_m^2\},$$

using independent BART models for $r_a(\ell), r_m(0,\ell)$, and $r_m(1,\ell)$; we then took $\widehat{e}_i$ to be the posterior mean of $\Phi\{r_a(L_i)\}$ and $\widehat{r}_{ia}$ to be the posterior mean of $r_m(a, L_i)$. We then fit a BCMF with these clever covariates and extracted estimates of the causal parameters of interest.

Results are summarized in Figure 14.2, which displays the posterior distribution of the *NDE*s, *NIE*s, and the total effect $TE = E[Y_i\{1, M_i(1)\} - Y_i\{0, M_i(0)\}]$. We see that the indirect effect of smoking on expenditures as mediated through health is positive, as expected. On the other hand, the direct effect of smoking is estimated to be negative, which results in the posterior distribution of the total effect (i.e., the average causal effect estimated in Section 5.4.1) having an uncertain sign.

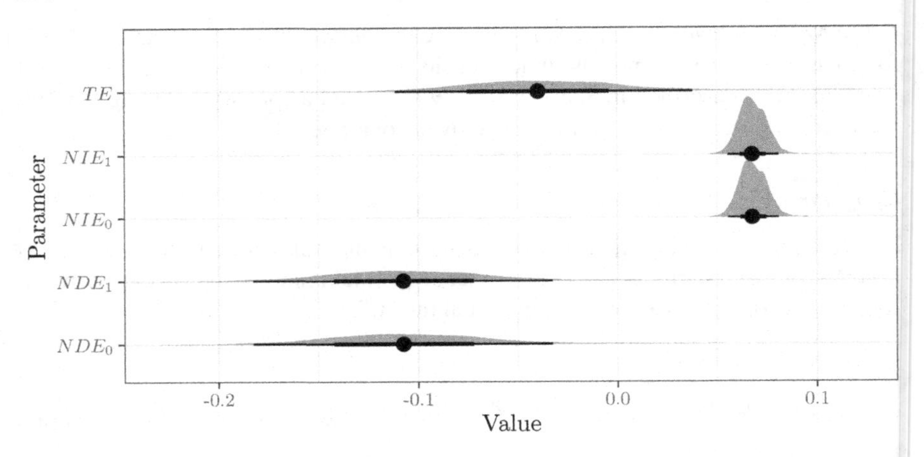

Figure 14.2 *Posterior distributions for the NDEs, NIEs, and the total effect. The thin black lines under each distribution cover 95% of the posterior mass, and the solid black dots correspond to the posterior mean.*

Intuitively, we might expect that the direct effect of smoking on healthcare expenditures should be zero, after accounting for the effect of smoking on health. This is contradicted to some extent by Figure 14.2, as the estimated direct effect is negative. This suggests to us that we may not have controlled for all necessary confounders; for example, our attempt to control for risk-taking behavior through the frequency of seat belt usage may be imperfect, or smokers as a group may be less inclined to seek medical care for other reasons. Alternatively, it is possible that there are post-treatment variables affected by smoking, such as stress, that also reduce medical expenditures. Next, we assess the robustness of these effects to the existence of unmeasured confounding.

Sensitivity to unmeasured confounding

In this section we perform a sensitivity analysis to understand the potential effect of unmeasured confounding on our conclusions. Of the four assumptions we have made, the most suspect are the assumptions of (i) no unmeasured common causes of A_i (the decision to smoke) and $\{M_i(a), Y_i(a', m)\}$ (the potential outcomes for health and total medical expenditures) (cf. equation 1.5) and (ii) no unmeasured common causes of health $M_i(a)$ and medical expenditures $Y_i(a', m)$ (cf. equation 1.6).

For simplicity, we will only consider the possibility of unmeasured confounding between the mediator and the outcome. One way to do this is to use the approach described in Section 4.2.2. We take a slightly different approach here by instead modeling the dependence between the potential outcome processes via

$$\begin{aligned} Y_i(a,m) &= r_y(m,a,L_i) + \lambda\{M_i(a) - m\} + \varepsilon_i(a,m), \\ M_i(a) &= r_m(a,L_i) + \nu_i(m). \end{aligned} \tag{14.4}$$

We note that λ is a sensitivity parameter in this setup, as this model still imposes the same observed data model $[Y_i \mid M_i, A_i, L_i] \sim N\{r_y(M_i, A_i, L_i), \sigma_y^2\}$ regardless of the value of λ.

One way to understand this model specification is to consider the case where $r_y(m,a,\ell)$ is linear, i.e., $Y_i(a,m) = \beta_L^\top L_i + \beta_m m + \beta_a a + \lambda\{M_i(a) - m\} + \varepsilon_i(a,m)$. Under this model, we can rearrange terms to get

$$Y_i(a,m) = \beta_L^\top L_i + (\beta_m - \lambda)m + \beta_a a + \lambda M_i(a) + \varepsilon_i(a,m). \tag{14.5}$$

Hence the effect of varying λ from 0 is to shift the causal effect of varying the mediator m (captured by β_m) by λ units into dependence between $Y_i(a,m)$ and $M_i(a)$.

Let $NIE_a(\lambda)$ and $NDE_a(\lambda)$ denote the natural indirect/direct effects at exposure level a with a fixed value of λ in (14.4). Then, conveniently, we can derive closed-form expression for $NIE_a(\lambda)$ and $NDE_a(\lambda)$ as functions of $NIE_a(0)$ and $NDE_a(0)$. For example, $NDE_a(\lambda)$ is given by

$$\begin{aligned}
NDE_a(\lambda) &= E[Y_i\{1, M_i(a)\} - Y_i\{0, M_i(a)\}] \\
&= E\big[r_y\{M_i(a), 1, L_i\} + \lambda\{M_i(1) - M_i(a)\} \\
&\qquad - r_y\{M_i(a), 0, L_i\} - \lambda\{M_i(0) - M_i(a)\}\big] \\
&= NDE_a(0) + \lambda E\{M_i(1) - M_i(0)\} \\
&= NDE_a(0) + \lambda\,\tau_M,
\end{aligned}$$

where $\tau_M = E\{M_i(1) - M_i(0)\}$ denotes the average causal effect of the treatment on the mediator. For BCMFs, τ_M is given by

$$\tau_M = \sum_i \varpi_i\,\{r_m(1, L_i) - r_m(0, L_i)\}.$$

It can similarly be shown that $NIE_a(\lambda) = NIE_a(0) - \lambda\,\tau_M$. Note that this implies that the total effect decomposition $TE = NDE_a(\lambda) + NIE_{1-a}(\lambda)$ does not depend on λ; this is intuitive, as TE only requires ignorability of the assignment mechanism to be identified.

Given the expressions for $NIE_a(\lambda)$ and $NDE_a(\lambda)$ above, it is straight forward to post-process the MCMC output of the BCMF model fit under sequential ignorability to produce valid samples for non-zero values of λ: we simply collect samples of $(NIE_a(0), NDE_a(0), \tau_M)$ as we normally would, and then collect samples of $NIE_a(\lambda)$ and $NDE_a(\lambda)$ by applying the formulas $NIE_a(\lambda) = NIE_a(0) + \lambda\,\tau_M$ and $NDE_a(\lambda) = NDE_a(0) - \lambda\,\tau_M$ along a grid of λ.

Calibrating the Sensitivity Parameter Using the observed data summary calibration strategy from Chapter 4, we choose a range of plausible values for λ by relating it to the regression coefficient β_m in the model (14.5). To do this, we fit the model (14.5) to the data and extract the least squares estimate $\widehat{\beta}_m$. We then vary λ from $[-\widehat{\beta}_m, \widehat{\beta}_m]$, the intuition being that we do not expect that the effect of unmeasured confounding is substantially larger than the total effect attributable to *both* unmeasured confounding and the mediator when taken together.

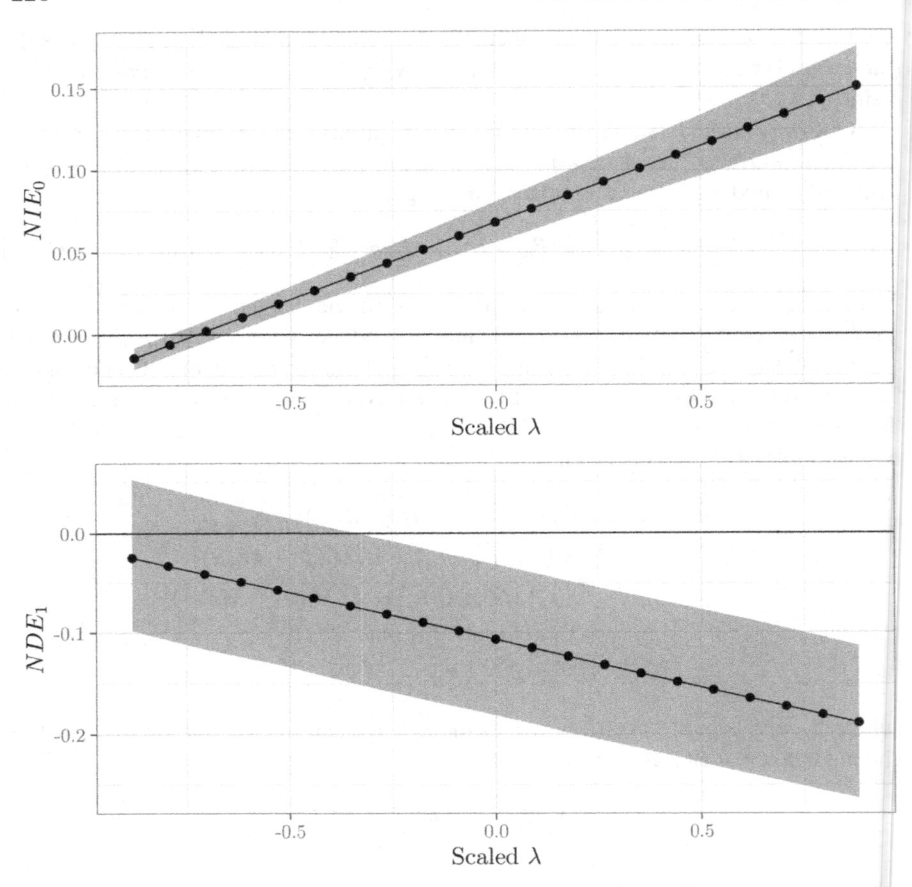

Figure 14.3 *Posterior mean and 95% credible interval for NIE_0 for different values of λ. Scaled λ denotes $\lambda/\widehat{\beta}_m$ where $\widehat{\beta}_m$ is the least squares estimate of β_m in* (14.5).

Sensitivity Analysis Results The impact of varying λ on NIE_0 and NDE_1 is displayed in Figure 14.3. We first observe that λ must be large and negative for the effect of smoking on medical expenditures to not be clearly positive; we need roughly $\lambda \leq -0.6\,\widehat{\beta}_m$ for this to occur, making the indirect effect quite robust to unmeasured confounding. The direct effect is less robust, requiring only that $\lambda \leq -0.35\,\widehat{\beta}_m$ before the sign of NDE_1 is indeterminate. Hence there are some values of λ that are consistent with both a null direct effect and a positive indirect effect of smoking.

14.5 Conclusions

A full description of this methodology is given in [204], which also gives an in-depth simulation experiment that justifies the inclusion of the clever covariates as necessary for accurate inference about the mediation effects.

There are several improvements to the analysis proposed here that we did not pursue for the sake of simplicity. For example, we chose to model M_i as a continuous variable, despite the fact that self-perceived health was measured on a Likert scale from 1 to 5 ("Poor" to "Excellent"). A more faithful model would take M_i to be either ordinal or categorical. For example, if M_i is ordinal with K categories, we could have used an *ordinal probit* model of the form

$$\Pr_\theta(M_i = k \mid A_i = a, L_i = x) = \Phi\{\alpha_k + r_m(a,x)\} - \Phi\{\alpha_{k-1} + r_m(a,x)\},$$

where $-\infty = \alpha_0 < \alpha_1 < \cdots < \alpha_{K-1} < \alpha_K = \infty$. This corresponds to a *latent variable* model, where Z_i is an unmeasured continuous measure of health such that we observe $M_i = k$ if $\alpha_{k-1} < Z_i \leq \alpha_k$. Along the lines of the ordinal probit model, one might regard the latent health score Z_i as being the actual mediator, with M_i being an imprecise measure of Z_i; this extension is not particularly easy to implement with current softwares, however.

A second possible improvement is to reparameterize the model so that it more closely resembles the Bayesian causal forest (BCF) model of [143] (see Section 5.4.2). The BCF can be understood as providing a middle ground between viewing the treatment of interest as a "regular covariate" on the one hand versus stratifying on the treatment on the other. One possible reparameterization, if we are willing to tolerate the mediator entering linearly and not interacting with the treatment, is to take

$$Y_i(a,m) \sim N\{\mu_y(x) + a\,\zeta(x) + m\,\beta(x), \sigma_y^2\} \qquad \text{and}$$
$$M_i(a) \sim N\{\mu_m(x) + a\,\tau_m(x), \sigma_m^2\}.$$

This parameterization makes it straight forward to separately regularize the direct effect $\zeta(x)$ and the indirect effect $\delta(x) = \beta(x)\,\tau_m(x)$ where $\tau_m(x)$ is the conditional average causal effect of the treatment on the mediator. This allows us to control both the magnitude and degree of heterogeneity of the treatment effects by using differing amounts of regularization for the components $\zeta(\cdot), \beta(\cdot), \tau_m(\cdot), \mu_y(\cdot)$, and $\mu_m(\cdot)$. We could similarly specify extensions of this model that allow for mediator and treatment interaction in the outcome model.

Another simplification we made was to consider only those individuals who incurred some medical expenditure, i.e., we excluded from the analysis all individuals who incurred no medical expense over the year. This concession allowed us to model the logarithm of the total medical expenditure, which [205] show to have a distribution which is very well approximated by a normal distribution. But this strategy targets a different population than we started with (those who incurred some expenditure), and one also worries that the selection into the population may be affected by the decision to smoke (i.e., smokers may be more or less likely to incur zero medical expenditures). A final possible extension is to explicitly model the indicator W_i that there is some medical expenditure by specifying a model for $\Pr_\theta(W_i = 1 \mid M_i, A_i, X_i)$ (say, a probit BART regression) and then model $[Y_i \mid W_i = 1, M_i, A_i, X_i]$ using a BART model. Letting $T_i = \exp(Y_i)$, we can then define the usual causal mediation effects directly on T_i, although given the skewness of T_i it may make more sense to define quantile causal mediation effects instead (see, e.g., [206] for a definition of the quantile mediation effects).

Chapter 15

Causal analysis of semicompeting risks using a principal stratification estimand and DDP+GP

DOI: 10.1201/9780429324222-15

15.1 Brain Cancer Clinical Trial

This case study involves data from a randomized (placebo-controlled) phase II trial for patients with recurrent gliomas ($n = 222$) who were scheduled for tumor resection [207]. Patients enrolled in the trial had a single focus of tumor in the cerebrum, had a Karnofsky score greater than 60, had completed radiation therapy, had not taken nitrosoureas within six weeks of enrollment, and had not had systematic chemotherapy within 4 weeks of enrolment. Baseline prognostic variables and a baseline evaluation of cerebellar function were measured. The former includes age, race, Karnofsky performance score, local vs. whole brain radiation, percent of tumor resection, previous use of nitrosoureas, and tumor histology (glioblastoma, anapestic astrocytoma, oligodendroglioma, or other) at implantation. Patients were randomized to receive surgically implanted biodegradable polymer discs with or without 3.85% of carmustine and followed for (up to) one year.

Semi competing risks

Semi competing risks [208] occur in studies where observation of a non-terminal event (e.g., progression) may be pre-empted by a terminal event (e.g., death), but not vice versa. In randomized clinical trials to evaluate treatments of life-threatening diseases, patients are often observed for specific types of disease progression and survival. Often, the primary outcome is patient survival, resulting in data analyses focusing on the terminal event using standard survival analysis tools. However, there may also be interest in understanding the effect of treatment on non-terminal outcomes such as progression. In the brain cancer trial here, an important progression endpoint is based on deterioration of the cerebellum. An important feature of this progression endpoint is that it is biologically plausible that a patient could die without cerebellar deterioration. Thus, analyzing the effect of treatment on progression needs to account for the fact that progression is not well defined after death. We introduce our causal estimand next.

Estimand of interest

To define the estimand of interest, we first need to introduce some notation. Let $Y_{iP}(a)$, $Y_{iD}(a)$, and $C(a)$ denote progression time, death time, and censoring time, under treatment a where $a = 0, 1$ represents the control and treatment groups, respectively. Our goal is to estimate the causal effect of treatment on time to cerebellar progression. A key feature of our setting is that $Y_{iP}(a) \not> Y_{iD}(a)$ (i.e., progression cannot happen after death) which will impact the specification of our causal estimand.

The causal estimand of interest is the function $\tau(u)$, given by

$$\tau(u) = \frac{\Pr\{Y_{iP}(1) < u \mid Y_{iD}(0) \geq u, Y_{iD}(1) \geq u\}}{\Pr\{Y_{iP}(0) < u \mid Y_{iD}(0) \geq u, Y_{iD}(1) \geq u\}}, \tag{15.1}$$

where $\tau(u)$ is a smooth function of u. This estimand contrasts the risk of progression prior to time u for treatment 1 relative to treatment 0 among patients who survived to time u under both treatments. This estimand is an example of a principal stratum causal effect (cf. Section 1.5).

15.2 Methods

Observed data models

We first introduce some notation for the observed data. Let A_i denote the treatment assignment and X_i a vector of baseline covariates. Let $Y_{iP} = Y_{iP}(A_i)$, $Y_{iD} = Y_{iD}(A_i)$, and $C_i = C_i(A_i)$. Let $T_{i1} = \min(Y_{iP}, Y_{iD}, C_i)$, $\delta = I\{Y_{iP} < \min\{Y_{iP}, C_i\}\}$, $T_{i2} = \min(Y_{iD}, C_i)$, and $\xi_i = I(Y_{iD} < C_i)$ denote the observed event times and event indicators. As such, the observed data for each patient is $O_i = (T_{i1}, T_{i2}, \delta_i, \xi_i, A_i, X_i)$.

For each treatment group, we specify a DDP+GP for $[Y_{iP}, Y_{iD}, \mid A_i = a, X_i, \theta]$ as in Section 7.3. It is composed of a DPM of bivariate normals (for each a) with the mean function given a GP prior. Specifically, we let

$$f(Y_{iP} = y_P, Y_{iD} = y_D \mid A_i = a, X_i = x, \theta) = \sum_{k=1}^{\infty} w_k N(y \mid g_{ak}(x), \Sigma_a),$$

with

$$g_{ak,j}(\cdot) \sim \text{GP}(m_{ak,j}, \kappa_a)$$

where $m_{ak,j}(x) = x^{\top}\beta_{ak,j}$ and

$$\kappa_a(x, x') = \eta \exp(-\rho \|x - x'\|^2) + 0.1 I(x = x').$$

Priors

For the priors for $\beta_{ak,j} : j \in \{P, D\}$ in the GP mean function, we specify $\beta_{ak,j} \sim N(\beta_{a0,j}, \Lambda_{a0,j})$ with $\Lambda_{a0,j} \sim \mathscr{W}^{-1}(\lambda_{a0}, \Psi_a)$, where $E(\Lambda_{a0,j}) = \frac{\Psi_a}{\lambda_{a0}-3}$. We estimate the hyperparameters by fitting a bivariate normal distribution for the outcomes of patients under treatment a, $[Y_{ij} \mid A_i = a] \sim N(x\beta_{aj}, \Sigma_{a\beta_0})$ and set $\beta_{a0,j} = \hat{\beta}_{aj}$. We choose λ_{a0} and Ψ_a in the inverse Wishart prior so that the prior mean of $\Lambda_{a0,j}$ matches the estimate inflated by a factor of 10, $\Psi_a = 10 \times \hat{\Sigma}_{a\beta_0}$, and the degrees of freedom are the smallest to maintain a proper prior, $\lambda_{a0} = 4$. For the precision parameter, we specify $\alpha \sim \text{Gam}(1, 1)$.

Identifying assumptions

We make the following assumptions that are sufficient for identifying the causal estimand in (15.1).

Assumption 1. *Non-informative censoring:* for all x and a,

$$C_i(a) \perp\!\!\!\perp \{Y_{iP}(a), Y_{iD}(a)\} \mid X_i = x, A_i = a$$

and $\Pr\{C_i(a) > Y_{iP}(a), C_i(a) > Y_{iD}(a) \mid X_i = x > 0\}$. This assumption is reasonable when most (all) of the censoring is administrative (e.g., due to the study ending).

Assumption 2: The distribution function of $(Y_{iD}(0), Y_{iD}(1))$ given $X_i = x$ and θ, G_x, follows a Gaussian copula model, i.e.,

$$G_x(v, w; \rho) = \Phi_{2,\rho}[\Phi^{-1}\{G_x^0(v)\}, \Phi^{-1}\{G_x^1(w)\}], \tag{15.2}$$

where Φ is a standard normal c.d.f. and $\Phi_{2,\rho}$ is a bivariate normal c.d.f. with mean 0, marginal variance 1, and correlation ρ.

The copula specification, which is uncheckable from the observed data, provides a parsimonious way to construct the joint distribution of $(Y_D(0), Y_D(1))$ given X while allowing the two marginals to be estimated flexibly using the DDP+GP; it contains a single sensitivity parameter $\rho \in [-1, 1]$, which simplifies prior specification as well. This joint is needed to identify the principal strata $\{Y_D(0) \geq u, Y_D(1) \geq u\}$ in (15.1).

Assumption 3: The progression time under treatment a is conditionally independent of death time under treatment $1-a$, given the death time under treatment a and covariates $X = x$, i.e., for $a \in \{0, 1\}$ we have

$$Y_P(a) \perp\!\!\!\perp Y_D(1-a) \mid Y_D(a), X = x.$$

Under these three assumptions, the causal estimand in (15.1) is identified; details can be found in [176].

Sensitivity Analysis

The assumptions above contain one explicit sensitivity parameter, ρ, from Assumption 2. With a lack of strong prior knowledge, default priors can be specified. If the correlation is expected to be positive, a uniform prior $\rho \sim \text{Uniform}(0, 1)$ can be specified; alternatively, if strong positive correlation is expected instead, a triangular prior on $[0, 1]$ with a mode at 1 could be specified instead. We discuss priors further in Section 15.4.

Computational Details

We use a truncation approximation for posterior inference as described in Section 7.3. To compute the causal quantity $\tau(u)$, which is conditional on covariates X, we integrate over the covariates using the empirical distribution; note we could also use the Bayesian bootstrap (see Section 3.2.3). Further details can be found in [176].

15.3 Analysis

There were 219 patients with complete baseline measures; of these, 304 died and 100 progressed prior to death. Of the 15 patients who did not die, 4 were observed to have cerebellar progression. For the modeling, the event times were log transformed.

Model Comparisons

We compared the observed data model of the proposed approach to a parametric approach (Param) assuming bivariate normality with linear regression, using LPML (described in Section 3.4). Overall, the LPML indicated better performance for the BNP approach (-442 vs. -446).

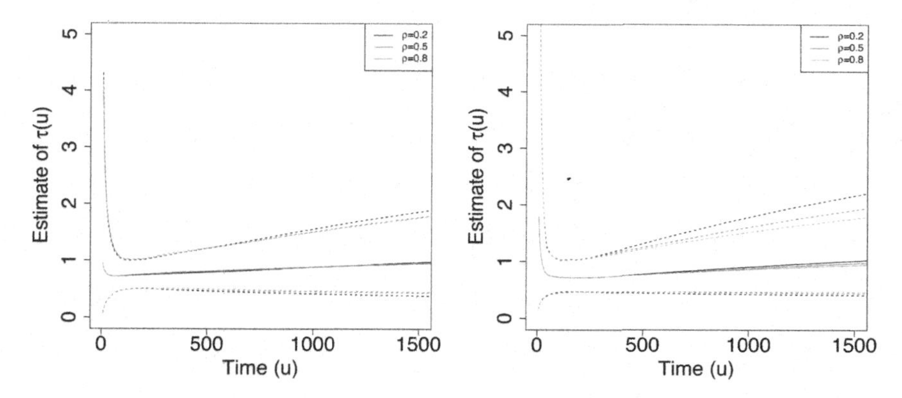

Figure 15.1 *Posterior estimates of $\tau(u)$ for different values of u and ρ under the proposed BNP approach (left) and the parametric approach (right). The solid lines represent the posterior estimate for $\tau(u)$, and the dashed lines represent 95% pointwise credible intervals.*

Results and sensitivity analysis

Posterior inference for the causal estimand $\tau(u)$ in (15.1) depends on the sensitivity parameter ρ. We first examine the posterior of $\tau(u)$ for different choices of ρ (recall (15.2)), in particular the values $\rho \in \{0.2, 0.5, 0.8\}$; these correspond to weak, moderate, and strong positive associations between death time under treatment and control, i.e., $(Y_D(0), Y_D(1))$. Figure 15.1 plots the causal estimand $\tau(u)$ as a function of u (in log(days)). The posterior distributions are quite similar for the different values of ρ, and for all scenarios there is no evidence of a significant treatment effect on progression based on examining 95% credible intervals. For this example, there turned out not to be large differences in inferences between the BNP and parametric approaches.

15.4 Conclusions

For the analysis of the brain cancer clinical trial, there was no evidence of a significant treatment effect on progression for any value of the sensitivity parameters considered. Determining the plausible values of ρ in an actual trial analysis depends on the nature of the application; since the sensitivity parameter is a correlation and ranges from $[-1, 1]$, at worst, a uniform prior could be specified over the range (or a subset of the range); to place more weight at the endpoints, triangular priors can also be considered.

We could have also used an EDPM for the observed data instead. This would alleviate concerns about ignorably missing baseline covariates. If the treatment was not randomized, and additional ignorability assumption would also be required (1.5), Assumption 3 could also be weakened to explore further sensitivity. This would add additional sensitivity parameters to construct the joint distribution of $[(Y_P(a), Y_D(1-a)) \mid Y_D(a), X = x]$ that is needed for the conditional distributions in (15.1).

Bibliography

[1] S. R. Seaman and I. R. White, "Review of inverse probability weighting for dealing with missing data," *Statistical Methods in Medical Research*, vol. 22, no. 3, pp. 278–295, 2013.

[2] P. C. Austin and E. A. Stuart, "Moving towards best practice when using inverse probability of treatment weighting (iptw) using the propensity score to estimate causal treatment effects in observational studies," *Statistics in Medicine*, vol. 34, no. 28, pp. 3661–3679, 2015.

[3] H. Bang and J. M. Robins, "Doubly robust estimation in missing data and causal inference models," *Biometrics*, vol. 61, no. 4, pp. 962–973, 2005.

[4] M. J. Van der Laan, S. Rose, *et al.*, *Targeted Learning: Causal Inference for Observational and Experimental Data*, vol. 10. Springer, 2011.

[5] A. Tsiatis, *Semiparametric Theory and Missing Data.* Springer New York, 2006.

[6] M. J. van der Laan and J. M. Robins, *Unified Methods for Censored Longitudinal Data and Causality.* Springer New York, 2003.

[7] M. Hernan and J. Robins, *Causal Inference: What If.* CRC Press, 2020.

[8] G. W. Imbens and D. B. Rubin, *Causal inference in statistics, social, and biomedical sciences.* Cambridge University Press, 2015.

[9] P. Rosenbaum, *Observational Studies.* Springer Series in Statistics, Springer, 2 ed., 2002.

[10] H. Shin and J. Antonelli, "Improved inference for doubly robust estimators of heterogeneous treatment effects," 2021.

[11] V. Chernozhukov, D. Chetverikov, M. Demirer, E. Duflo, C. Hansen, and W. Newey, "Double/Debiased/Neyman machine learning of treatment effects," *American Economic Review*, vol. 107, pp. 261–265, may 2017.

[12] R. Wyss, S. Schneeweiss, M. Van Der Laan, S. D. Lendle, C. Ju, and J. M. Franklin, "Using super learner prediction modeling to improve high-dimensional propensity score estimation," *Epidemiology*, vol. 29, no. 1, pp. 96–106, 2018.

[13] E. H. Kennedy, "Semiparametric doubly robust targeted double machine learning: a review," *arXiv preprint arXiv:2203.06469*, 2022.

[14] D. B. Rubin, "Estimating causal effects of treatments in randomized and nonrandomized studies.," *Journal of educational Psychology*, vol. 66, no. 5, p. 688, 1974.

[15] R. Neugebauer, B. Fireman, J. A. Roy, P. J. O'Connor, and J. V. Selby, "Dynamic marginal structural modeling to evaluate the comparative effectiveness of more or less aggressive treatment intensification strategies in adults with type 2 diabetes," *Pharmacoepidemiology and Drug Safety*, vol. 21, no. S2, pp. 99–113, 2012.

[16] M. G. Hudgens and M. E. Halloran, "Toward causal inference with interference," *Journal of the American Statistical Association*, vol. 103, pp. 832–842, jun 2008.

[17] S. R. Cole and C. E. Frangakis, "The consistency statement in causal inference: A definition or assumption?," *Epidemiology*, vol. 20, no. 1, pp. 3–5, 2009.

[18] B. C. Sauer, M. A. Brookhart, J. Roy, and T. VanderWeele, "A review of covariate selection for non-experimental comparative effectiveness research," *Pharmacoepidemiology and drug safety*, vol. 22, no. 11, pp. 1139–1145, 2013.

[19] B. Sauer, M. Brookhart, J. Roy, and T. VanderWeele, "A review of covariate selection for non-experimental comparative effectiveness research," *Pharmacoepidemiology & Drug Safety*, vol. 22, pp. 1139–1145, 2013.

[20] J. Robins, "A new approach to causal inference in mortality studies with a sustained exposure period—application to control of the healthy worker survivor effect," *Mathematical modelling*, vol. 7, no. 9-12, pp. 1393–1512, 1986.

[21] J. Robins and M. Hernán, "Estimation of the causal effects of time-varying exposures," in *Advances in Longitudinal Data Analysis* (G. Fitzmaurice, M. Davidian, G. Verbeke, and G. Molenberghs, eds.), pp. 553–599, Boca Raton, FL: Chapman & Hall, 2009.

[22] R. Daniel, S. Cousens, B. De Stavola, M. Kenward, and J. Sterne, "Methods for dealing with time-dependent confounding," *Statistics in Medicine*, vol. 32, pp. 1584–1618, 2013.

[23] J. Young, L. Cain, J. Robins, E. O'Reilly, and M. Hernán, "Comparative effectiveness of dynamic treatment regimes: an application of the parametric g-formula," *Statistics in the Biosciences*, vol. 3, pp. 119–143, 2011.

[24] P. Rosenbaum and D. B. Rubin, "The central role of the propensity score in observational studies for causal effects," *Biometrika*, vol. 70, no. 1, pp. 41–55, 1983.

[25] P. C. Austin, "An introduction to propensity score methods for reducing the effects of confounding in observational studies," *Multivariate Behavioral Research*, vol. 46, no. 3, pp. 399–424, 2011.

[26] G. Zhang and R. Little, "Extensions of the penalized spline of propensity prediction method of imputation," *Biometrics*, vol. 65, no. 3, pp. 911–918, 2009.

[27] D. Xu, M. J. Daniels, and A. G. Winterstein, "A Bayesian nonparametric approach to causal inference on quantiles," *Biometrics*, vol. 74, no. 3, pp. 986–996, 2018.

[28] L. McCandless, I. Douglas, S. Evans, and L. Smeeth, "Cutting feedback in Bayesian regression adjustment for the propensity score," *International Journal of Biostatistics*, vol. 6, p. article 16, 2010.

[29] C. M. Zigler, K. Watts, R. W. Yeh, Y. Wang, B. A. Coull, and F. Dominici, "Model feedback in Bayesian propensity score estimation," *Biometrics*, vol. 69, p. 263–273, 2013.

[30] Y. Zhu, R. Hubbard, J. Chubak, J. Roy, and N. Mitra, "Core concepts in pharmacoepidemiology: Violations of the positivity assumption in the causal analysis of observational data: Consequences and statistical approaches," *Pharmacoepidemiology and Drug Safety*, vol. 30, no. 11, pp. 1471–1485, 2021.

[31] R. K. Crump, V. J. Hotz, G. W. Imbens, and O. A. Mitnik, "Dealing with limited overlap in estimation of average treatment effects," *Biometrika*, vol. 96, pp. 187–199, 2009.

[32] M. Petersen, K. Porter, S. Gruber, Y. Wang, and M. van der Laan, "Diagnosing and responding to violations in the positivity assumption," *Statistical methods in medical research*, vol. 21, pp. 31–54, 2012.

[33] J. Robins, M. Hernan, and B. Brumback, "Marginal structural models and causal inference in epidemiology," *Epidemiology*, vol. 11, no. 5, pp. 550–560, 2000.

[34] D. O. Scharfstein, A. Rotnitzky, and J. M. Robins, "Adjusting for nonignorable drop-out using semiparametric nonresponse models," *Journal of the American Statistical Association*, vol. 94, no. 448, pp. 1096–1120, 1999.

[35] J. Roy, K. J. Lum, and M. J. Daniels, "A Bayesian nonparametric approach to marginal structural models for point treatments and a continuous or survival outcome," *Biostatistics*, vol. 18, no. 1, pp. 32–47, 2017.

[36] O. Saarela, D. A. Stephens, E. E. M. Moodie, and M. B. Klein, "On Bayesian estimation of marginal structural models," *Biometrics*, vol. 71, no. 2, pp. 279–288, 2015.

[37] C. Frangakis and D. Rubin, "Principal stratification in causal inference," *Biometrics*, vol. 58, pp. 21–29, 2002.

[38] R. Gallop, D. S. Small, J. Y. Lin, M. R. Elliott, M. Joffe, and T. R. Ten Have, "Mediation analysis with principal stratification," *Statistics in Medicine*, vol. 28, no. 7, pp. 1108–1130, 2009.

[39] J. M. Robins and S. Greenland, "Identifiability and exchangeability for direct and indirect effects," *Epidemiology*, vol. 3, no. 2, pp. 143–155, 1992-03.

[40] C. Kim, M. Daniels, J. Hogan, C. Choirat, and C. Zigler, "Bayesian methods for multiple mediators: Relating principal stratification and causal mediation in the analysis of power plant emission controls," *Annals of Applied Statistics*, vol. 13, pp. 1927–1956, 2019.

[41] K. Imai, L. Keele, and D. Tingley, "A general approach to causal mediation analysis.," *Psychological Methods*, vol. 15, no. 4, p. 309, 2010.

[42] C. Avin, I. Shpitser, and J. Pearl, "Identifiability of path-specific effects," *IJCAI-05, Proceedings of the Nineteenth International Joint Conference on Artificial Intelligence; Edinburgh, Scotland, UK.*, p. 357–363, 2005.

[43] E. Tchetgen Tchetgen and T. VanderWeele, "On identification of natural direct effects when a confounder of the mediator is directly affected by exposure," *Epidemiology*, vol. 25, pp. 282–291, 2014.

[44] G. Hong, F. Yang, and X. Qin, "Posttreatment confounding in causal mediation studies: A cutting-edge problem and a novel solution via sensitivity analysis," *Biometrics*, 00, 1–15, 2022. https://doi.org/10.1111/biom.13705.

[45] C. Kim, M. Daniels, B. Marcus, and J. Roy, "A framework for Bayesian nonparametric inference for causal effects of mediation," *Biometrics*, vol. 73, pp. 401–409, 2017.

[46] T. J. VanderWeele, "Simple relations between principal stratification and direct and indirect effects," *Statistics & Probability Letters*, vol. 78, no. 17, pp. 2957–2962, 2008.

[47] J. Pearl, "Direct and indirect effects," *Proceedings of the Seventeenth Conference on Uncertainty and Artificial Intelligence*, pp. 411–420, 2001.

[48] I. Díaz and N. S. Hejazi, "Causal mediation analysis for stochastic interventions," *Journal of the Royal Statistical Society: Series B (Statistical Methodology)*, vol. 82, pp. 661–683, feb 2020.

[49] S. R. Kay, A. Flszbein, and A. Opfer, "The positive and negative syndrome scale (PANSS) for schizophrenia," *Schizophrenia Bulletin*, vol. 13, no. 2, p. 261, 1987.

[50] D. B. Rubin, "Causal inference through potential outcomes and principal stratification: application to studies with "censoring" due to death," *Statistical Science*, vol. 21, no. 3, pp. 299–309, 2006.

[51] B. L. Egleston, D. O. Scharfstein, E. E. Freeman, and S. K. West, "Causal inference for non-mortality outcomes in the presence of death," *Biostatistics*, vol. 8, pp. 526–545, sep 2006.

[52] M. J. Daniels and J. W. Hogan, *Missing Data in Longitudinal Studies: Strategies for Bayesian Modeling and Sensitivity Analysis*. Chapman and Hall/CRC, 2008.

[53] B. M. Marlin and R. S. Zemel, "Collaborative prediction and ranking with non-random missing data," *Proceedings of the Third ACM Conference on Recommender Systems*, pp. 5–12, 2009.

[54] M. Sadinle and J. P. Reiter, "Itemwise conditionally independent nonresponse modelling for incomplete multivariate data," *Biometrika*, vol. 104, no. 1, pp. 207–220, 2017.

[55] A. R. Linero, "Bayesian nonparametric analysis of longitudinal studies in the presence of informative missingness," *Biometrika*, vol. 104, no. 2, pp. 327–341, 2017.

[56] I. Shpitser, "Consistent estimation of functions of data missing non-monotonically and not at random," in *Advances in Neural Information Processing Systems*, pp. 3144–3152, 2016.

[57] National Research Council, *The Prevention and Treatment of Missing Data in Clinical Trials*. The National Academies Press, 2010.

[58] The International Conference on Harmonization, "ICH E9 Statistical principles for clinical trials: Addendum: Estimands and sensitivity analysis in clinical trials," tech. rep., Food and Drug Administration, HHS, 2017.

[59] D. B. Rubin, "Inference and missing data," *Biometrika*, vol. 63, pp. 581–592, 1976.

[60] M. Daniels, C. Wang, and B. Marcus, "Fully bayesian inference under ignorable missingness in the presence of auxiliary covariates," *Biometrics*, vol. 70, no. 1, pp. 62–72, 2014.

[61] S. Seaman, J. Galati, D. Jackson, and J. Carlin, "What is meant by "missing at random"?," *Statistical Science*, vol. 28, no. 2, pp. 257–268, 2013.

[62] D. Horvitz and D. Thompson, "A generalization of sampling with replacment from a finite universe," *Journal of the American Statistical Association*, vol. 47, pp. 663–685, 1952.

[63] G. Molenberghs, C. Beunckens, C. Sotto, and M. G. Kenward, "Every missingness not at random model has a missingness at random counterpart with equal fit," *Journal of the Royal Statistical Society: Series B (Statistical Methodology)*, vol. 70, no. 2, pp. 371–388, 2008.

[64] R. J. A. Little, "A test of missing completely at random for multivariate data with missing values," *Journal of the American Statistical Association*, vol. 83, no. 404, pp. 1198–1202, 1988.

[65] J. M. Robins and Y. Ritov, "Toward a curse of dimensionality appropriate (CODA) asymptotic theory for semi-parametric models," *Statistics in Medicine*, vol. 16, pp. 285–319, 1997.

[66] J. J. Heckman, "Sample selection bias as a specification error.," *Econometrica*, vol. 47, pp. 153–161, 1979.

[67] A. Azzalini, *The Skew-Normal and Related Families*, vol. 3. Cambridge University Press, 2013.

[68] L. Tierney and J. Kadane, "Accurate approximations for posterior moments and marginal densities," *Journal of the American Statistical Association*, vol. 81, pp. 82–86, 1986.

[69] J. H. Albert and S. Chib, "Bayesian analysis of binary and polychotomous response data," *Journal of the American Statistical Association*, vol. 88, no. 422, pp. 669–679, 1993.

[70] L. Held and C. C. Holmes, "Bayesian auxiliary variable models for binary and multinomial regression," *Bayesian analysis*, vol. 1, no. 1, pp. 145–168, 2006.

[71] N. G. Polson, J. G. Scott, and J. Windle, "Bayesian inference for logistic models using Polya-Gamma latent variables," *Journal of the American Statistical Association*, vol. 108, no. 504, pp. 1339–1349, 2013.

[72] A. W. Van der Vaart, *Asymptotic statistics*, vol. 3. Cambridge University Press, 2000.

[73] R. Christensen, W. Johnson, A. Branscum, and T. E. Hanson, *Bayesian ideas and data analysis: an introduction for scientists and statisticians*. CRC Press, 2011.

[74] A. Gelman *et al.*, "Prior distributions for variance parameters in hierarchical models (comment on article by Browne and Draper)," *Bayesian analysis*, vol. 1, no. 3, pp. 515–534, 2006.

[75] A. E. Gelfand and A. F. M. Smith, "Sampling-based approaches to calculating marginal densities," *Journal of the American Statistical Association*, vol. 85, pp. 398–409, jun 1990.

[76] S. Brooks, A. Gelman, G. Jones, and X.-L. Meng, *Handbook of Markov chain Monte Carlo*. CRC press, 2011.

[77] R. M. Neal, "Slice sampling," *The Annals of Statistics*, vol. 31, no. 3, pp. 705–741, 2003.

[78] G. O. Roberts, A. Gelman, and W. R. Gilks, "Weak convergence and optimal scaling of random walk Metropolis algorithms," *Ann. Appl. Probab.*, vol. 7, pp. 110–120, 1997.

[79] R. M. Neal, "Markov chain sampling methods for dirichlet process mixture models," *Journal of computational and graphical statistics*, vol. 9, no. 2, pp. 249–265, 2000.

[80] P. Damlen, J. Wakefield, and S. Walker, "Gibbs sampling for Bayesian non-conjugate and hierarchical models by using auxiliary variables," *Journal of the Royal Statistical Society: Series B (Statistical Methodology)*, vol. 61, no. 2, pp. 331–344, 1999.

[81] M. Betancourt, "A conceptual introduction to Hamiltonian Monte Carlo," *arXiv preprint arXiv:1701.02434*, 2017.

[82] Stan Development Team, "RStan: the R interface to Stan, version 2.5.0," 2014.

[83] B. Leimkuhler and S. Reich, *Simulating hamiltonian dynamics*, vol. 14. Cambridge university press, 2004.

[84] A. Nishimura, D. B. Dunson, and J. Lu, "Discontinuous Hamiltonian Monte Carlo for discrete parameters and discontinuous likelihoods," *Biometrika*, vol. 107, no. 2, pp. 365–380, 2020.

[85] A. Gelman, D. B. Rubin, *et al.*, "Inference from iterative simulation using multiple sequences," *Statistical Science*, vol. 7, no. 4, pp. 457–472, 1992.

[86] G. L. Jones and J. P. Hobert, "Honest Exploration of Intractable Probability Distributions via Markov Chain Monte Carlo," *Statistical Science*, 16(4), 312–334, 2001.

[87] M. K. Cowles and B. P. Carlin, "Markov chain Monte Carlo convergence diagnostics: a comparative review," *Journal of the American Statistical Association*, vol. 91, no. 434, pp. 883–904, 1996.

[88] M.-H. Chen and Q.-M. Shao, "Monte Carlo estimation of Bayesian credible and HPD intervals," *Journal of Computational and Graphical Statistics*, vol. 8, no. 1, pp. 69–92, 1999.

[89] C. Han and B. P. Carlin, "Markov chain Monte Carlo methods for computing Bayes factors: A comparative review," *Journal of the American Statistical Association*, vol. 96, no. 455, pp. 1122–1132, 2001.

[90] S. Geisser and W. F. Eddy, "A predictive approach to model selection," *Journal of the American Statistical Association*, vol. 74, no. 365, pp. 153–160, 1979.

[91] A. E. Gelfand and D. K. Dey, "Bayesian model choice: asymptotics and exact calculations," *Journal of the Royal Statistical Society: Series B (Methodological)*, vol. 56, no. 3, pp. 501–514, 1994.

[92] A. Vehtari, J. Gabry, M. Magnusson, Y. Yao, P., Bürkner, T. Paananen, and A. Gelman, "loo: Efficient leave-one-out cross-validation and WAIC for Bayesian models." R package version 2.5.1, 2022. https://mc-stan.org/loo/.

[93] G. Celeux, F. Forbes, C. P. Robert, and D. M. Titterington, "Deviance information criteria for missing data models," *Bayesian Analysis*, vol. 1, no. 4, pp. 651–673, 2006.

[94] D. Xu, A. Chatterjee, and M. Daniels, "A note on posterior predictive checks to assess model fit for incomplete data," *Statistics in Medicine*, vol. 35, no. 27, pp. 5029–5039, 2016.

[95] M. J. Daniels, A. S. Chatterjee, and C. Wang, "Bayesian model selection for incomplete data using the posterior predictive distribution," *Biometrics*, vol. 68, no. 4, pp. 1055–1063, 2012.

[96] S. M. O'Brien and D. B. Dunson, "Bayesian multivariate logistic regression," *Biometrics*, vol. 60, no. 3, pp. 739–746, 2004.

[97] M. Josefsson and M. J. Daniels, "Bayesian semi-parametric g-computation for causal inference in a cohort study with MNAR dropout and death," *Journal of the Royal Statistical Society: Series C (Applied Statistics)*, vol. 70, no. 2, pp. 398–414, 2021.

[98] C. Wang, M. J. Daniels, D. O. Scharfstein, and S. Land, "A Bayesian shrinkage model for incomplete longitudinal binary data with application to the breast cancer prevention trial," *Journal of the American Statistical Association*, vol. 105, no. 492, pp. 1333–1346, 2010.

[99] A. R. Linero and M. J. Daniels, "A flexible Bayesian approach to monotone missing data in longitudinal studies with informative dropout with application to a schizophrenia clinical trial," *Journal of the American Statistical Association*, vol. 110, no. 1, pp. 45–55, 2015.

[100] A. R. Linero, "Simulation-based estimators of analytically intractable causal effects," *Biometrics*, vol. 78, pp. 1001–1017, 2022. In revision.

[101] D. Scharfstein, A. McDermott, W. Olson, and F. Wiegand, "Global sensitivity analysis for repeated measures studies with informative dropout: A fully parametric approach," *Statistics in Biopharmaceutical Research*, vol. 6, no. 4, pp. 338–348, 2014.

[102] M. J. Daniels, M. Lee, and W. Feng, "Dirichlet process mixture models for the analysis of repeated attempt designs," arXiv 2305.05099, 2023.

[103] B. A. Brumback, M. A. Hernán, S. J. Haneuse, and J. M. Robins, "Sensitivity analyses for unmeasured confounding assuming a marginal structural model for repeated measures," *Statistics in Medicine*, vol. 23, no. 5, pp. 749–767, 2004.

[104] J. M. Albert and W. Wang, "Sensitivity analyses for parametric causal mediation effect estimation," *Biostatistics*, vol. 16, no. 2, pp. 339–351, 2015.

[105] T. J. VanderWeele and Y. Chiba, "Sensitivity analysis for direct and indirect effects in the presence of exposure-induced mediator-outcome confounders," *Epidemiology, Biostatistics and Public Health*, vol. 11, no. 2, pp. 1–16, 2014.

[106] J. Roy, J. W. Hogan, and B. H. Marcus, "Principal stratification with predictors of compliance for randomized trials with 2 active treatments," *Biostatistics*, vol. 9, no. 2, pp. 277–289, 2007.

[107] G. Molenberghs, B. Michiels, M. G. Kenward, and P. J. Diggle, "Monotone missing data and pattern-mixture models," *Statistica Neerlandica*, vol. 52, no. 2, pp. 153–161, 1998.

[108] H. Thijs, G. Molenberghs, B. Michiels, G. Verbeke, and D. Curran, "Strategies to fit pattern-mixture models," *Biostatistics*, vol. 3, no. 2, pp. 245–265, 2002.

[109] R. J. Little, "Pattern-mixture models for multivariate incomplete data," *Journal of the American Statistical Association*, vol. 88, no. 421, pp. 125–134, 1993.

[110] M. G. Kenward, G. Molenberghs, and H. Thijs, "Pattern-mixture models with proper time dependence.," *Biometrika*, vol. 90, pp. 53–71, 2003.

[111] J. M. Robins and R. D. Gill, "Non-response models for the analysis of non-monotone ignograble missing data," *Statistics in Medicine*, vol. 16, pp. 39–56, 1997.

[112] S. Vansteelandt, A. Rotnitzky, and J. Robins, "Estimation of regression models for the mean of repeated outcomes under nonignorable nonmonotone nonresponse," *Biometrika*, vol. 94, no. 4, pp. 841–860, 2007.

[113] O. Harel and J. L. Schafer, "Partial and latent ignorability in missing-data problems," *Biometrika*, vol. 96, no. 1, pp. 37–50, 2009.

[114] A. R. Linero and M. J. Daniels, "Bayesian approaches for missing not at random outcome data: the role of identifying restrictions," *Statistical Science*, vol. 33, no. 2, pp. 198–213, 2018.

[115] J. Birmingham, A. Rotnitzky, and G. M. Fitzmaurice, "Pattern–mixture and selection models for analysing longitudinal data with monotone missing patterns," *Journal of the Royal Statistical Society: Series B (Statistical Methodology)*, vol. 65, no. 1, pp. 275–297, 2003.

[116] E. J. T. Tchetgen, L. Wang, and B. Sun, "Discrete choice models for nonmonotone nonignorable missing data: Identification and inference," *Statistica Sinica*, vol. 28, no. 4, pp. 2069–2088, 2018.

[117] M. J. Funk, D. Westreich, C. Wiesen, T. Stuürmer, A. M. Brookhart, and M. Davidian, "Doubly robust estimation of causal effects," *American Journal of Epidemiology*, vol. 173, no. 7, pp. 761–767, 2011.

[118] J. Antonelli, G. Papadogeorgou, and F. Dominici, "Causal inference in high dimensions: A marriage between bayesian modeling and good frequentist properties," *Biometrics*, vol. 78, no. 1, pp. 100–114, 2022.

[119] S. Athey and G. W. Imbens, "Machine learning methods that economists should know about," *Annual Review of Economics*, vol. 11, pp. 685–725, 2019.

[120] V. Dorie, J. Hill, U. Shalit, M. Scott, and D. Cervone, "Automated versus do-it-yourself methods for causal inference: Lessons learned from a data analysis competition," *Statistical Science*, vol. 34, no. 1, pp. 43–68, 2019.

[121] R. A. Sparapani, B. R. Logan, R. E. McCulloch, and P. W. Laud, "Nonparametric survival analysis using Bayesian additive regression trees (BART)," *Statistics in medicine*, vol. 35, no. 16, pp. 2741–2753, 2016.

[122] Y. Li, A. R. Linero, and J. Murray, "Adaptive conditional distribution estimation with Bayesian decision tree ensembles," *Journal of the American Statistical Association*, 2022. doi: 10.1080/01621459.2022.2037431.

[123] E. George, P. Laud, B. Logan, R. McCulloch, and R. Sparapani, "Fully nonparametric Bayesian additive regression trees," Topics in Identification, Limited Dependent Variables, Partial Observability, Experimentation, and Flexible Modeling: Part B (Advances in Econometrics, Vol. 40B), Emerald Publishing Limited, Bingley, pp. 89–110, 2019. https://doi.org/10.1108/S0731-90532019000040B006.

[124] M. T. Pratola, H. A. Chipman, E. I. George, and R. E. McCulloch, "Heteroscedastic BART via multiplicative regression trees," *Journal of Computational and Graphical Statistics*, vol. 29, pp. 405–417, nov 2019.

[125] H. A. Chipman, E. I. George, and R. E. McCulloch, "BART: Bayesian additive regression trees," *The Annals of Applied Statistics*, vol. 4, no. 1, pp. 266–298, 2010.

[126] G. Wahba, *Spline Models for Observational Data.* Society of Industrial and Applied Mathematics, 1990.

[127] H. A. Chipman, E. I. George, and R. E. McCulloch, "Bayesian CART model search," *Journal of the American Statistical Association*, vol. 93, no. 443, pp. 935–948, 1998.

[128] J. Friedman, T. Hastie, and R. Tibshirani, *The Elements of Statistical Learning*. Springer, 2 ed., 2008.

[129] Y. Freund and R. E. Schapire, "Experiments with a new boosting algorithm," *Thirteenth International Conference on Machine Learning*, vol. 96, pp. 148–156, 1996.

[130] A. R. Linero, "Bayesian regression trees for high-dimensional prediction and variable selection," *Journal of the American Statistical Association*, vol. 113, no. 522, pp. 626–636, 2018.

[131] M. T. Pratola, "Efficient Metropolis-Hastings proposal mechanisms for Bayesian regression tree models," *Bayesian Analysis*, vol. 11, no. 3, pp. 885–911, 2016.

[132] G. V. Kass, "An exploratory technique for investigating large quantities of categorical data," *Journal of the Royal Statistical Society: Series C (Applied Statistics)*, vol. 29, no. 2, pp. 119–127, 1980.

[133] L. Breiman, J. H. Friedman, R. A. Olshen, and C. J. Stone, *Classification and Regression Trees*. Wadsworth & Brooks/Cole Advanced Books & Software, 1984.

[134] P. Bühlmann and B. Yu, "Boosting with the L_2 loss: regression and classification," *Journal of the American Statistical Association*, vol. 98, no. 462, pp. 324–339, 2003.

[135] R. E. Schapire, "Explaining AdaBoost," in *Empirical Inference: Festschrift in Honor of Vladimir N. Vapnik* (B. Schölkopf, Z. Luo, and V. Vovk, eds.), pp. 37–52, Springer, 2013.

[136] J. Friedman, T. Hastie, R. Tibshirani, *et al.*, "Additive logistic regression: a statistical view of boosting," *The Annals of Statistics*, vol. 28, no. 2, pp. 337–407, 2000.

[137] J. H. Friedman, "Stochastic gradient boosting," *Computational statistics & data analysis*, vol. 38, no. 4, pp. 367–378, 2002.

[138] L. Breiman, "Random forests," *Machine Learning*, vol. 45, no. 1, pp. 5–32, 2001.

[139] L. Breiman, "Bagging predictors," *Machine Learning*, vol. 24, no. 2, pp. 123–140, 1996.

[140] S. Wager and S. Athey, "Estimation and inference of heterogeneous treatment effects using random forests," *Journal of the American Statistical Association*, vol. 113, no. 523, pp. 1228–1242, 2018.

[141] D. F. McCaffrey, G. Ridgeway, and A. R. Morral, "Propensity score estimation with boosted regression for evaluating causal effects in observational studies," *Psychological methods*, vol. 9, no. 4, p. 403, 2004.

[142] J. L. Hill, "Bayesian nonparametric modeling for causal inference," *Journal of Computational and Graphical Statistics*, vol. 20, no. 1, pp. 217–240, 2011.

[143] P. Hahn, J. S. Murray, and C. M. Carvalho, "Bayesian regression tree models for causal inference: regularization, confounding, and heterogeneous effects," *Bayesian Analysis*, vol. 15, pp. 965–1056, 2020.

[144] J. S. Murray, "Log-linear Bayesian additive regression trees for multinomial logistic and count regression models," *Journal of the American Statistical Association*, 116:534, 756–769, 2021. doi: 10.1080/01621459.2020.1813587.

[145] A. R. Linero, P. Basak, Y. Li, and D. Sinha, "Bayesian survival tree ensembles with submodel shrinkage," *Bayesian Analysis*, vol. 17, pp. 997–1020, 2022.

[146] V. Bonato, V. Baladandayuthapani, B. M. Broom, E. P. Sulman, K. D. Aldape, and K.-A. Do, "Bayesian ensemble methods for survival prediction in gene expression data," *Bioinformatics*, vol. 27, no. 3, pp. 359–367, 2011.

[147] N. C. Henderson, T. A. Louis, G. L. Rosner, and R. Varadhan, "Individualized treatment effects with censored data via fully nonparametric bayesian accelerated failure time models," *Biostatistics*, vol. 21, no. 1, pp. 50–68, 2020.

[148] D. Xu, M. J. Daniels, and A. G. Winterstein, "Sequential BART for imputation of missing covariates," *Biostatistics*, vol. 17, no. 3, pp. 589–602, 2016.

[149] J. Sethuraman, "A constructive definition of dirichlet priors," *Statistica Sinica*, vol. 4, no. 2, pp. 639–650, 1994.

[150] W. J. Ewens, "Population genetics theory-the past and the future," in *Mathematical and statistical developments of evolutionary theory*, pp. 177–227, Springer, 1990.

[151] H. Ishwaran and L. F. James, "Approximate dirichlet process computing in finite normal mixtures: smoothing and prior information," *Journal of Computational and Graphical statistics*, vol. 11, no. 3, pp. 508–532, 2002.

[152] M. J. Daniels and M. Pourahmadi, "Bayesian analysis of covariance matrices and dynamic models for longitudinal data," *Biometrika*, vol. 89, no. 3, pp. 553–566, 2002.

[153] P. Müller, A. Erkanli, and M. West, "Bayesian curve fitting using multivariate normal mixtures," *Biometrika*, vol. 83, no. 1, pp. 67–79, 1996.

[154] D. B. Dunson and C. Xing, "Nonparametric Bayes modeling of multivariate categorical data," *Journal of the American Statistical Association*, vol. 104, no. 487, pp. 1042–1051, 2009.

[155] Y. Si and J. P. Reiter, "Nonparametric Bayesian multiple imputation for incomplete categorical variables in large-scale assessment surveys," *Journal of Educational and Behavorial Statistics*, vol. 38, no. 5, pp. 499–521, 2013.

[156] H. Ishwaran and L. F. James, "Gibbs sampling methods for stick-breaking priors," *Journal of the American Statistical Association*, vol. 96, no. 453, pp. 161–173, 2001.

[157] S. G. Walker, "Sampling the dirichlet mixture model with slices," *Communications in Statistics - Simulation and Computation*, vol. 36, no. 1, pp. 45–54, 2007.

[158] H. Ishwaran and M. Zarepour, "Markov chain Monte Carlo in approximate dirichlet and beta two-parameter process hierarchical models," *Biometrika*, vol. 87, no. 2, pp. 371–390, 2000.

[159] M. D. Escobar and M. West, "Bayesian density estimation and inference using mixtures," *Journal of the american statistical association*, vol. 90, no. 430, pp. 577–588, 1995.

[160] B. Shahbaba and R. Neal, "Nonlinear models using dirichlet process mixtures," *Journal of Machine Learning Research*, vol. 10, no. Aug, pp. 1829–1850, 2009.

[161] J. S. Murray and J. P. Reiter, "Multiple imputation of missing categorical and continuous values via bayesian mixture models with local dependence," *Journal of the American Statistical Association*, vol. 111, no. 516, pp. 1466–1479, 2016.

[162] M. A. Taddy, *Bayesian nonparametric analysis of conditional distributions and inference for Poisson point processes*. PhD thesis, 2008.

[163] J. Roy, K. J. Lum, B. Zeldow, J. D. Dworkin, V. L. Re III, and M. J. Daniels, "Bayesian nonparametric generative models for causal inference with missing at random covariates," *Biometrics*, vol. 74, no. 4, pp. 1193–1202, 2018.

[164] S. Wade, D. B. Dunson, S. Petrone, and L. Trippa, "Improving prediction from dirichlet process mixtures via enrichment," *The Journal of Machine Learning Research*, vol. 15, no. 1, pp. 1041–1071, 2014.

[165] S. Wade, S. Mongelluzzo, S. Petrone, *et al.*, "An enriched conjugate prior for bayesian nonparametric inference," *Bayesian Analysis*, vol. 6, no. 3, pp. 359–385, 2011.

[166] S. Roy, M. J. Daniels, B. J. Kelly, and J. Roy, "A Bayesian nonparametric approach for causal inference with multiple mediators. 2022. *arXiv preprint arXiv:2208.13382*.

[167] N. Burns and M. J. Daniels, "Truncation Approximation for Enriched Dirichlet Process Mixture Models. 2023. *arXiv preprint arXiv:2305.01631*.

[168] C. Rasmussen and C. Williams, *Gaussian Processes for Machine Learning*. MIT Press, 2006.

[169] M. Ebden, "Gaussian processes: A quick introduction," 2015. *arXiv preprint arXiv:1505.02965*.

[170] R. B. Gramacy and H. K. H. Lee, "Bayesian treed gaussian process models with an application to computer modeling," *Journal of the American Statistical Association*, vol. 103, no. 483, pp. 1119–1130, 2008.

[171] S. Srivastava, V. Cevher, Q. Dinh, and D. Dunson, "WASP: Scalable Bayes via barycenters of subset posteriors," in *Proceedings of the Eighteenth International Conference on Artificial Intelligence and Statistics* (G. Lebanon and S. V. N. Vishwanathan, eds.), vol. 38 of *Proceedings of Machine Learning Research*, pp. 912–920, PMLR, 2015.

[172] M. M. Zhang and S. A. Williamson, "Embarrassingly parallel inference for gaussian processes," *Journal of Machine Learning Research*, vol. 20, no. 169, pp. 1–26, 2019.

[173] A. Y. Zhu, N. Mitra, and J. Roy, "Addressing positivity violations in causal effect estimation using Gaussian process priors," *Statistics in Medicine*, 2023; 42(1): 33–51. doi:10.1002/sim.9600.

[174] S. N. MacEachern, "Dependent nonparametric processes," *ASA Proceedings of the Section on Bayesian Statistical Science*, pp. 50–55, 1999.

[175] P. Mueller and A. Rodriguez, *Chapter 5: Dependent Dirichlet Processes and*

Other Extensions, vol. Volume 9 of *NSF-CBMS Regional Conference Series in Probability and Statistics*, pp. 53–75. Institute of Mathematical Statistics and American Statistical Association, 2013.

[176] Y. Xu, D. O. Scharfstein, P. Müller, and M. Daniels, "A Bayesian nonparametric approach for evaluating the causal effect of treatment in randomized trials with semi-competing risks.," *Biostatistics*, vol. 23, pp. 34–49, 2022.

[177] Y. Xu, P. Müller, A. S. Wahed, and P. F. Thall, "Bayesian nonparametric estimation for dynamic treatment regimes with sequential transition times," *Journal of the American Statistical Association*, vol. 111, no. 515, pp. 921–950, 2016. PMID: 28018015.

[178] C. Luo and M. J. Daniels, "BNPqte: A Bayesian nonparametric approach to causal inference on quantiles in R," *arXiv preprint arXiv:2106.14599*, 2021.

[179] Z. Zhang, Z. Chen, J. F. Troendle, and J. Zhang, "Causal inference on quantiles with an obstetric application," *Biometrics*, vol. 68, no. 3, pp. 697–706, 2012.

[180] I. Díaz, "Efficient estimation of quantiles in missing data models," *Journal of Statistical Planning and Inference*, vol. 190, pp. 39–51, 2017-11.

[181] C. M. Zigler, "The central role of bayes' theorem for joint estimation of causal effects and propensity scores," *The American Statistician*, vol. 70, no. 1, pp. 47–54, 2016.

[182] C. Luo, *Variable selection and causal inference using flexible Bayesian models*. PhD thesis, University of Florida 2021.

[183] S. L. Fultz, M. Skanderson, L. A. Mole, N. Gandhi, K. Bryant, S. Crystal, and A. Justice, "Development and verification of a "virtual" cohort using the national va health information system," *Medical Care*, vol. 44, pp. 25–30, 2006.

[184] M. J. Daniels and X. Luo, "A note on compatibility for inference with missing data in the presence of auxiliary covariates," *Statistics in Medicine*, vol. 38, no. 7, pp. 1190–1199, 2019.

[185] D. B. Rubin, "Multiple Imputation After 18+ Years," *Journal of the American Statistical Association*, 91(434), 473–489, 1996. https://doi.org/10.2307/2291635.

[186] S. van Buuren and K. Groothuis-Oudshoorn, "mice: Multivariate imputation by chained equations in r," *Journal of Statistical Software*, vol. 45, no. 3, pp. 1–67, 2011.

[187] J. Liu, A. Gelman, J. Hill, Y.-S. Su, and J. Kropko, "On the stationary distribution of iterative imputations," *Biometrika*, vol. 101, pp. 155–173, nov 2014.

[188] S. Gruber and M. van der Laan, "tmle: An R package for targeted maximum likelihood estimation," *Journal of Statistical Software*, vol. 51, pp. 1–35, 2012.

[189] R. Cohen, R. Boland, R. Paul, K. Tashima, E. Schoenbaum, D. Celentano, P. Schuman, D. Smith, and C. Carpenter, "Neurocognitive performance enhanced by highly active antiretroviral therapy in hiv-infected women," *AIDS*, vol. 16, pp. 341–5, 2001.

[190] R. S. Durvasula, H. F. Myers, K. Mason, and C. Hinkin, "Relationship between alcohol use/abuse, HIV infection and neuropsychological performance in African American men," *Journal of Clinical and Experimental Neuropsychology*, vol. 28, no. 3, pp. 383–404, 2006.

[191] D. K. Smith, D. L. Warren, D. Vlahov, P. Schuman, M. D. Stein, B. L. Greenberg, and S. D. Holmberg, "Design and baseline participant characteristics of the human immunodeficiency virus epidemiology research (her) study: a prospective cohort study of human immunodeficiency virus infection in us women," *American Journal of Epidemiology*, vol. 146, no. 6, pp. 459–469, 1997.

[192] J. Roy, X. Lin, and L. Ryan, "Scaled marginal models for multiple continuous outcomes," *Biostatistics*, vol. 4, pp. 371–383, 2003.

[193] J. D. Y. Kang and J. L. Schafer, "Demystifying double robustness: a comparison of alternative strategies for estimating a population mean from incomplete data," *Statistical Science*, vol. 22, no. 4, pp. 523–539, 2007.

[194] S. R. Kay, A. Fiszbein, and L. A. Opler, "The positive and negative syndrome scale (PANSS) for schizophrenia," *Schizophrenia Bulletin*, vol. 13, no. 2, pp. 261–276, 1987.

[195] M. Pourahmadi, "Joint mean-covariance models with applications to longitudinal data: Unconstrained parameterisation," *Biometrika*, vol. 86, no. 3, pp. 677–690, 1999.

[196] B. Fisher, J. P. Costantino, D. L. Wickerham, C. K. Redmond, M. Kavanah, W. M. Cronin, V. Vogel, A. Robidoux, N. Dimitrov, J. Atkins, and other National Surgical Adjuvant Breast and Bowel Project Investigators, "Tamoxifen for prevention of breast cancer: report of the National Surgical Adjuvant Breast and Bowel Project P-1 Study," *Journal of the National Cancer Institute*, vol. 90, no. 18, pp. 1371–1388, 1998.

[197] W. W. Eaton, C. Smith, M. Ybarra, C. Muntaner, and A. Tien, "Center for epidemiologic studies depression scale: review and revision (CESD and CESD-R).," in *The use of psychological testing for treatment planning and outcomes assessment: Instruments for adults* (M. E. Maruish, ed.), pp. 363–377, Lawrence Erlbaum Associates Publishers, 2004.

[198] M. J. Daniels, "A prior for the variance in hierarchical models," *Canadian Journal of Statistics*, vol. 27, no. 3, pp. 567–578, 1999.

[199] D. I. Ohlssen, L. D. Sharples, and D. J. Spiegelhalter, "Flexible random-effects models using Bayesian semi-parametric models: applications to institutional comparisons," *Statistics in medicine*, vol. 26, no. 9, pp. 2088–2112, 2007.

[200] A. Agresti, *Categorical Data Analysis*. John Wiley & Sons, 3rd ed., 2013.

[201] B. H. Marcus, M. A. Napolitano, A. C. King, B. A. Lewis, J. A. Whiteley, A. Albrecht, A. Parisi, B. Bock, B. Pinto, C. Sciamanna, J. Jakicic, and G. D. Papandonatos, "Telephone versus print delivery of an individualized motivationally tailored physical activity intervention: Project stride.," *Health Psychology*, vol. 26, no. 4, p. 401, 2007a.

[202] W. Bae and M. J. Daniels, "A Bayesian non-parametric approch for causal mediation with a post-treatment confounder," 2023, arXiv 2305.05017.

[203] A. Oganisian, N. Mitra, and J. A. Roy, "A Bayesian nonparametric model for zero-inflated outcomes: Prediction, clustering, and causal estimation," *Biometrics*, vol. 77, pp. 125–135, 2021.

[204] A. R. Linero and Q. Zhang, "Mediation analysis using Bayesian tree ensembles," *Psychological Methods*, 2022. Advance online publication. https://doi.org/10.1037/met0000504.

[205] A. R. Linero, D. Sinha, and S. R. Lipsitz, "Semiparametric Mixed-Scale Models Using Shared Bayesian Forests," *Biometrics*, vol. 77, no. 1, pp. 131–144, 2020.

[206] L. Rene, A. R. Linero, and E. Slate, "Causal mediation and sensitivity analysis for mixed-scale data," *arXiv e-prints*, 2021. https://doi.org/10.48550/arXiv.2111.03907.

[207] H. Brem, S. Piantadosi, P. Burger, M. Walker, R. Selker, N. Vick, K. Black, M. Sisti, S. Brem, G. Mohr, *et al.*, "Placebo-controlled trial of safety and efficacy of intraoperative controlled delivery by biodegradable polymers of chemotherapy for recurrent gliomas," *The Lancet*, vol. 345, no. 8956, pp. 1008–1012, 1995.

[208] J. P. Fine, H. Jiang, and R. Chappell, "On semi-competing risks data," *Biometrika*, vol. 88, pp. 907–919, dec 2001.

Index

Printed in the United States
by Baker & Taylor Publisher Services